Earthquakes and Volcanic Activity on Islands

This volume examines the impact of and responses to historic earthquakes and volcanic eruptions in the Azores. Study is placed in the contexts of: the history and geography of this fascinating archipelago; progress being made in predicting future events and policies of disaster risk reduction.

This is the only volume to consider the earthquake and volcanic histories of the Azores across the whole archipelago and is based, not only on contemporary published research, but also on the detailed study of archival source materials. The authors seek to show how extreme environmental events, as expressed through eruptions, earthquakes and related processes operating in the past may be considered using both complementary scientific and social scientific perspectives in order to reveal the ways in which Azorean society has been shaped by both an isolated location in the middle of the Atlantic Ocean and the ever present threat of environmental uncertainty. Chapter 2, which analyses in depth the geology and tectonics of the islands is of more specialist interest, but technical terms are fully explained so as to widen the accessibility of this material.

Figure 0.1 Furnas. Photo by Rúben Cabral.

The audience for this volume includes all those who are interested in the geology, geography, history and hazard responses in the Azores. It is written, not just for the educated general reader, but for the specialist earth scientist and hazard researcher.

David Chester is a graduate of the universities of Durham and Aberdeen and for many years he has carried out research on disasters, particularly those produced by earthquakes and volcanic eruptions. He is currently Professor Emeritus at Liverpool Hope University and a Senior Fellow at the University of Liverpool.

Angus Duncan has a degree in Geology from Durham University and a PhD in Volcanic Geology from University College London. He has worked on volcanoes in Southern Italy and then the Azores for over 45 years. He retired as Professor in Volcanology from the University of Bedfordshire in 2013 and subsequently was an Honorary Research Fellow at the University of Liverpool.

Rui Coutinho graduated in Geology from the University of Coimbra, obtained his MSc in Applied and Economic Geology from the University of Lisbon and holds a PhD in Geology specialising in volcanology from the University of the Azores. He is an Assistant Professor in the University of the Azores and a member of the Institute for Research in Volcanology and Risk Assessment (IVAR).

Nicolau Wallenstein has a degree in Geology from the University of Oporto and a PhD in Geology, specialising in volcanology, from the University of the Azores. He is currently an Assistant Professor in the University of the Azores, coordinates the MSc course in Volcanology and Geological Risks and is a member of the Institute for Research in Volcanology and Risk Assessment (IVAR).

Routledge Studies in Hazards, Disaster Risk and Climate Change

Series Editor: Ilan Kelman, *Reader in Risk, Resilience and Global Health at the Institute for Risk and Disaster Reduction (IRDR) and the Institute for Global Health (IGH), University College London (UCL)*

This series provides a forum for original and vibrant research. It offers contributions from each of these communities as well as innovative titles that examine the links between hazards, disasters and climate change, to bring these schools of thought closer together. This series promotes interdisciplinary scholarly work that is empirically and theoretically informed, with titles reflecting the wealth of research being undertaken in these diverse and exciting fields.

Crisis and Emergency Management in the Arctic
Navigating Complex Environments
Edited by Natalia Andreassen and Odd Jarl Borch

Disaster Deaths
Trends, Causes and Determinants
Bimal Kanti Paul

Disasters and Life in Anticipation of Slow Calamity
Perspectives from the Colombian Andes
Reidar Staupe-Delgado

The Invention of Disaster
Power and Knowledge in Discourses on Hazard and Vulnerability
JC Gaillard

Earthquakes and Volcanic Activity on Islands
History and Contemporary Perspectives from the Azores
David Chester, Angus Duncan, Rui Coutinho and Nicolau Wallenstein

For more information about this series, please visit: https://www.routledge.com/Routledge-Studies-in-Hazards-Disaster-Risk-and-Climate-Change/book-series/HDC

Earthquakes and Volcanic Activity on Islands

History and Contemporary Perspectives from the Azores

David Chester, Angus Duncan, Rui Coutinho and Nicolau Wallenstein

Routledge
Taylor & Francis Group

LONDON AND NEW YORK

First published 2022
by Routledge
2 Park Square, Milton Park, Abingdon, Oxon OX14 4RN

and by Routledge
605 Third Avenue, New York, NY 10158

Routledge is an imprint of the Taylor & Francis Group, an informa business

© 2022 David Chester, Angus Duncan, Rui Coutinho and Nicolau Wallenstein

British Library Cataloguing-in-Publication Data
A catalogue record for this book is available from the British Library

Library of Congress Cataloging-in-Publication Data
A catalog record has been requested for this book

ISBN: 978-0-367-13678-9 (hbk)
ISBN: 978-1-032-13964-7 (pbk)
ISBN: 978-0-429-02800-7 (ebk)

DOI: 10.4324/9780429028007

Typeset in Bembo
by codeMantra

Contents

Figures

Tables

Preface

Since the early 1990s, the authors have become increasingly aware that earthquakes and eruptions that have occurred in the Azores since the islands were first settled in the fifteenth century are important for a number of reasons. First, such events and peoples' reactions to them have profoundly influenced Azorean history, demography, economy and culture. Indeed, it is our contention that, without an understanding of these environmental extremes it is not possible fully to understand the cultural *milieu* of these singular islands and their distinctive landscapes, which have been shaped by a combination of natural processes and centuries of human endeavour. Second, it is upon knowledge of these often-catastrophic historical events that long-term forecasting of future earthquakes and eruptions is based. Accurate interrogation of the archival record, confirmed where possible by field work, is vital if such forecasts are to have any veracity and utility in disaster planning. A third reason for studying historic earthquakes and eruptions is because the ways in which people have coped with historic emergencies, the successes and failures of local leadership and later of the State, hold important lessons for future disaster management both in the Azores and further afield. A final reason is that some of the historic eruptions and earthquakes in the Azores are of global significance in understanding natural processes and their impacts, eruptions of Furnas in 1630 and Capelinhos in 1957–1958 and the Angra do Heroísmo earthquake of 1980, for example, being cases in point.

The research upon which this volume is based is the culmination of more than a decade of investigation. We are all earth scientists by training, but were inspired by the example of our friend and mentor, Professor John Guest (1938–2011), who not only addressed questions of pure research, but also ones falling within the more applied spheres of forecasting and policy development. John also possessed a keen awareness of local culture and was committed to human betterment. We have received both encouragement and friendship from colleagues in our respective universities in Liverpool and the Azores, particularly from members of **Instituto de Investigação em Vulcanologia e Avaliação de Riscos** (The Institute for Research in Volcanology and Risk Assessment – IVAR), based in Ponta Delgada on São Miguel Island which has been ably led for many years by our friend and colleague Professor João-Luís Gaspar. Any deficiencies in this volume are of course ours alone.

In a work of such wide scope, we are conscious of harvesting other peoples' flowers. In particular, we acknowledge our debt to the priest and scholar Gaspar Frutuoso (1522–1591†). Much of what is known about early emergencies was recorded by him, with later records being compiled and edited by the nineteenth-century polymath, Ernesto do Canto (1831–1900). David Chester and Angus Duncan also acknowledge the financial support of the *British Academy* (SG 120266) and financial and logistical support from IVAR. All of us are also aware that this volume has a hidden author, Sandra Mather, who has so expertly drafted all its maps and diagrams.

This volume is aimed at a readership interested in disaster risk management, the pure and applied geology of the Azores and the geography of the islands. To fulfil the needs of readers who are not earth science specialists, we have provided a brief glossary of terms (Table 2.1) which may assist those reading Chapter 2. In addition, sections of the book on such specialist topics as *Magma Genesis*, which provide contexts for earth science specialists, are not necessary for an appreciation of other parts of the volume which focus on historic eruptions and earthquakes, hazards, coping with disasters and development of disaster reduction policies. Finally, we trust that those reading this volume will come to appreciate the long-term struggle between people and an unstable environment which make the Azores at the same time friendly, fascinating and sometimes awesome.

April 2021
David Chester, Angus Duncan, Rui Coutinho
and Nicolau Wallenstein

1   The Azores and the Azoreans

Isolation is a recurrent theme in the history, geography and hazard responses of the inhabitants of the Azores to earthquakes and volcanic eruptions. The Azores, an archipelago of nine islands in the middle of the Atlantic Ocean (Figure 1.1), lie approximately 850 km NW of Madeira; 1,400 km west of the European mainland; 3,300 km east of North America and 4,500 km NNW of Brazil. The archipelago is about 600 km in width, being aligned WNW-ESE, and crosses the Mid-Atlantic Ridge in a zone of contact between the Eurasian, Nubian and North American lithological plates (see Chapter 2). From west to east, the islands are divided into three groups (Figure 1.1): the western – Flores (area 141 km^2) and Corvo (area 17 km^2); the central – Faial (area 173 km^2), Pico (area 445 km^2), São Jorge (area 244 km^2), Graciosa (area 61 km^2) and Terceira (area 400 km^2) and the eastern – São Miguel (area 745 km^2) and Santa Maria (area 97 km^2). All islands are of volcanic origin, but since the islands were first permanently settled in the fifteenth century and because of the evolution of tectonic processes over time (see Chapter 2), only Faial, Pico, São Jorge, Terceira and São Miguel have been impacted by onshore eruptions. Destructive earthquakes have affected the same sub-set of islands together with Graciosa (Gaspar et al. 2015).

Although the islands' morphology is fashioned by volcanic and seismic events, erosional and depositional processes have also moulded and continue to shape the Azores. In geological terms, the islands are young[1] and attain considerable heights[2] over short distances from their coastlines which, in combination with high precipitation totals and pronounced seasonality, make the islands subject to flashy river regimes, flooding and processes of gravity-induced mass-movement, including landsliding and the generation of mudflows. Volcanic islands remain dangerous between eruptions and seismic episodes (Wallenstein et al. 2007), and instability is exacerbated by the unconsolidated nature of many volcanic products and certain land uses, particularly the planting of exotic shallow-rooted trees such as *Cryptomeria japonica* (Japanese red-cedar). Major landslides have been triggered by volcanic and seismic events and have produced many large debris flows including coastal platforms (*fajãs*) formed from landslide deposits (see Chapter 2), but instability is also associated with erosional processes especially heavy rainfall, and,

DOI: 10.4324/9780429028007-1

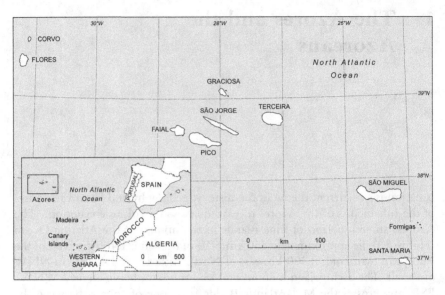

Figure 1.1 Azores: General location map.

for instance, on 31 October 1997 a major landslide in the Furnas area of São Miguel killed 29 people in the coastal settlement of Riberia Quente, left 114 homeless and cut the village off from the rest of the island for more than 12 hours (Malheiro 2006; Marques et al. 2008). Coastal platforms can also be formed by lava flowing down to the shore and building a bench of solidified lava out into the sea as occurred in the 1562–1564 eruption on Pico island (see Chapter 3). Coasts are also exposed to high intensity Atlantic storms as well as historic and pre-historic tsunamis, and erosion is enhanced by cliffs frequently being composed of volcanic materials with differing resistances to erosion. Today marine erosion is particularly problematic along the north coast of São Miguel and the south coast of Faial (Borges 2003; Malheiro 2006).

In this opening chapter, the environmental conditions which faced the first Portuguese settlers when they encountered the Azores will be outlined and discussion will focus on prevailing climatic, vegetational and pedological conditions. The development of Azorean society and economy will then be traced from the fifteenth century to the present-day, paying particular attention to the ways in which the distinctive culture of the archipelago – forged in isolation – has enabled the islanders to cope with frequent and often devastating disasters.

The environment of the Azores at the time of European contact

The Holocene was a momentous time in the Azores, not only have major eruptions been recognised using geological field investigations and radiometric dating (Chapter 2), but in a region so tectonically active the probabilistic nature of seismic activity means that the archipelago is also likely to have been affected by many large earthquakes. It was also a time of significant environmental change. Although sedimentary archives in the form of lake cores provide valuable palaeoenvironmental information for the later Holocene,[3] there is a dearth of detailed evidence for its earlier phases. Using data from other archipelagos in Macronesia,[4] it seems likely that sea-level rose rapidly in the early Holocene, that the rate was reduced from *c.*7,000 and by 5,500 BP it approximated to its present position (Meco et al. 2003). Establishing palaeo-sea levels on the Azores is complicated by the fact that, in addition to eustatic sea level fluctuations, there has been localised volcanic uplift (Azevedo & Portugal Ferreira 1999; Duncan et al. 2015). During the last hundred years, there has been a small increase above the late Holocene background rate of sea-level rise of 1.7 mm yr^{-1}, and it has been argued that this is due to global climatic change (Engelhart et al. 2011). It seems likely that during Pleistocene cold phases the Azores anti-cyclone, the principal factor affecting climate (see below), was displaced to the south and, though much milder than on the continents, colder conditions caused both species impoverishment within the primordial forest[5] constraining its areal distribution. The gradual re-establishment of warmer conditions during the early Holocene allowed forests to became more wide-spread and changed the areal and altitudinal distributions of their species (Fernándes-Palacios et al. 2011).[6]

Climate

By the fifteenth-century CE, the climate of the Azores was similar to that experienced today both in terms of temperature and precipitation (Cropper & Hanna 2014). The principal characteristics of the climate have been discussed by a number of scholars (Agostinho 1938, 1942, 1948; Admiralty 1945; Bettencourt 1979; Chazarra et al. 2012; Cropper 2013; Antunes & Carvalho 2018), and data on temperature and precipitation are summarised in Table 1.1. It is upon these sources that the following account is based. Being located in the middle of the North Atlantic Ocean between Latitude 37° and 39°N, the climate of the Azores is largely dictated by the sub-tropical high-pressure system known as the Azores anticyclone. For most of the year, the archipelago is on the northern side of the anticyclone and in winter the islands are affected by the passage of depressions associated with the polar front, bringing high precipitation totals between September and March which account for 75% of the total. Storms are common in winter and most frequent in the Western Azores, but winds are light in summer and on about one in every six days conditions are calm, although gales may occasionally occur between July and September produced by tropical cyclones that usually track between NE and SE. In summer, the Azores anticyclone is dominant and brings dryer and warmer weather. The Azores lie on the southern margin of the warm ocean current known as the North Atlantic Drift, and this contributes to equable temperatures throughout the year.

Table 1.1 Azores: summary temperature and precipitation data (based on Cropper (2013) and other sources)

Temperature (°C)

	Jan	Feb	Mar	April	May	June	July	Aug	Sept	Oct	Nov	Dec	Height (m)
Santa Maria	15	14	15	16	17	19	22	23	22	20	17	16	100
São Miguel	14	14	14	15	16	19	21	22	21	19	17	15	36
Terceira	14	14	14	15	17	19	21	22	21	19	17	15	55
Faial	15	15	15	16	17	19	22	23	22	20	17	16	41
Flores	15	14	15	16	17	19	22	23	22	19	17	15	29

Precipitation (mm)

	Jan	Feb	Mar	April	May	June	July	Aug	Sept	Oct	Nov	Dec	Total
Santa Maria	75	76	65	57	45	50	22	34	59	71	88	106	749
Ponta Delgada	79	76	82	69	53	32	23	42	79	87	99	109	829
Terceira	116	136	116	75	61	51	49	57	97	128	139	159	1184
Faial	87	91	82	58	64	53	30	50	78	96	100	102	891
Flores	170	146	134	99	97	88	49	72	118	159	168	211	1512

Figures are from: for Santa Maria – airport 1973–2012; São Miguel – Ponta Delgada 1973–2012; Terceira – airport 1947–2012; Faial – Horta 1976–2012 and Flores – airport 1979–2012.

Within the archipelago, there are spatial and altitudinal variations in climate. Lapse rates of temperature with height are steep, of the order 0.6°C for every 100 m of altitude and rainfall increases (Chester et al. 1995) and, for instance, on São Miguel, precipitation totals are 829 mm at Ponta Delgada (elevation 36 m), rising to 1,758 mm at Furnas village (elevation 290 m) and 3,197 mm on Monte Escuro (station elevation 688 mm; Chazarra et al. 2012, p. 64). Higher elevations on Flores and Pico have precipitation totals of up to 5,000 mm per annum, and there may be frosts above 1,000 m anywhere in an Azorean winter. Rainfall is highest in the west (Table 1.1), declines to the east and relative humidity is high throughout the year averaging over 80% at Ponta Delgada.

The climate of the Azores, though broadly stable since human contact has shown some variations. The instrumental record only dates from 1865, but shows that before 1960 there was no noticeable trend in temperature, from 1960 to 1975 mean temperatures fell by c.0.5°C and since then have risen by c.1.4°C. As far as precipitation is concerned, there is no evidence of a long-term trend since 1865, although the North Atlantic Oscillation has produced shorter-term variations including a positive increase between 1981 and 2011 (Cropper & Hanna 2013).

In terms of possible future trends, using the same global climate models (GCMs) as the United Nations' *Intergovernmental Panel on Climate Change*, Cropper (2013) has proposed that for Macronesia as a whole by 2100 temperature increases could be between 1.5°C and 2.8°C. In contrast to the other Macronesian archipelagos, precipitation in the Azores will show relatively little change according to the predictions of the GCMs, reflecting the effects of migration of the Azores anticyclone, which is forecast to produce a different result for the archipelago when compared to other parts of Macronesia (Cropper 2013, p. 306).

Vegetation and soils

Isolation and comparative geological 'youth' mean that the Azores Islands have only a small number of indigenous plant species (i.e. 197) of which 70 are endemic. It is estimated that wind (40%) and migrating birds (58%) brought seeds principally from Africa and Europe (Schaefer 2005; Connor et al. 2012). The priest and scholar Gaspar Frutuoso (1522–1591†) first described the appearance of the islands at the time of settlement in the fifteenth century[7]. As well as recording the colonisation of the islands and some of the early eruptions and earthquakes, Frutuoso also described in detail the vegetation of the archipelago based on documents, oral traditions and his travels around the islands. Later another priest, the Jesuit António Cordeiro (1641–1722), used Frutuoso's research and additional sources to produce a history of the Portuguese Islands (Cordeiro 1981), while in the nineteenth century a local man of letters, Ernesto do Canto (1831–1900) published historical material in *Archivo dos Açores*, a journal he founded, funded and edited. According to these writers, when the Portuguese first encountered the Azores they described the vegetation as being impenetrable and comprising forests of: *Laurus azorica* (Azores laurel); *Juniperus brevifolia* (Azores Juniper); *Prunus azorica* (Azores cherry) and *Morella faya* (fire tree, also known as *Myrica faia* – Azorean candleberry myrtle) (Costa 1950), which did not appear to be zoned by altitude, although Frutuoso recognised herbaceous vegetation at high altitudes on Flores (Schirone et al. 2010). As Connor et al. (2012, p. 1009) have argued, this may be accurate but could alternatively reflect the introduction of grazing animals during the time that elapsed between the settlement of the archipelago and the time when Gaspar Frutuoso was writing more than a hundred years later. When settled the only indigenous land mammals were bats.

Later, botanists visiting the Azores and using putative remnants of pre-settlement vegetation have argued strongly for altitudinal zones (e.g. Guppy 1917; Tutin 1953). More comprehensive accounts of vegetation were produced by Dansereau (1970) and Dias (1996) the latter, not only recognising altitudinal belts, but also classifying vegetation according to differences based on climate, geology, soil and topography. During the past decade, information from lake cores has become available (Connor et al. 2012; Connor et al. 2013; Rull et al. 2017a, 2017b), allowing a far more detailed picture of

Table 1.2 Reconstruction of vegetation at the time of European colonisation in the fifteenth century CE

Vegetation	Characteristics	Mean temperature (°C)	Mean precipitation (mm)	Altitude (m)
Erica-Morella coastal woodlands	Dominated by *Erica azorica* (Azores heather) and *Morella faya* (fire tree – also known as Myrica faia or Azorean Candleberry myrtle) – woodland 2–4 m in height. Occurs in the driest areas. Largest potential distribution on Pico, São Miguel and Terceira.	18	1074	0–100
Piconia-Morella lowland forest	Dominated by *Piconia azorica* (white wood) and *Morella faya*. At the upper altitudinal limit, *Laurus azorica* (Azorean laurel) becomes more common. Potentially present in all the islands and second largest area of occupation in the Azores.	17	1618	100–300
Laurus submontane forest	Height varies from 6–8 m and is dominated by *Laurus azorica* but may also contain most native tree types. Potentially the most common vegetation type (up to 42% of area of the Azores).	15	2203	300–600
Juniperus-Ilex montane forest	Height varies from 3 to 5 m and is dominated by *Juniperus brevifolia* (Azores juniper) and *Ilex perado* subsp. *azorica* (Madeira holly). About 14% of the area of the Azores could be covered by this type of vegetation, but would not have been found on Santa Maria and Graciosa because of height considerations.	12	3239	600–900
Juniperus montane woodland	Dominance of small *Juniperus brevifolia* trees (2–4 m) with *Myrsine africana* (Cape myrtle or African boxwood) and *Vaccinium cylindraceum* (*Azores blueberry*) are dominant shrubs. Height range depends on the island. Occur in mountainous areas with very high rainfall and most islands do not possess the required climatic conditions.	12	4440	700–900
Calluna-Juniperus altimontane scrublands	Scrublands are small sized communities usually less than 1 m in height. Dominated by *Calluna vulgaris* (heather) and *Juniperus brevifolia* and other herbaceous plants.	11	3373	900–1100
Calluna-Erica sub-alpine scrubland	Present only on Pico where average temperatures are below 9°C. Most frequently occurring shrub species are *Calluna vulgaris*; *Erica azorica* (Azores heather) and *Daboecia azorica* (St. Dabeoc's heath)	9	4350	1200–1700
Calluna alphine scrubland	*Calluna vulgaris* and *Thymus caespititius* (tyme) dominant, but with much bare rock surface. Only found on Pico above 1700 m. Ice cover present most of winter and average winter temperatures are 1°C–2°C.	6	2787	1700–2351

The mean temperature and precipitation figures are approximated means for sampled plots. Based on data in Elias et al. (2016).

vegetation at the time of European contact to be reconstructed, whilst Elias et al. (2016, p. 108) have studied 'established stands that persist without (or with little) direct or indirect human intervention, to estimate the potential distribution of natural vegetation.' Eight vegetation belts are recognised, and their characteristics are summarised in Table 1.2, while Figure 1.2 uses the examples of São Miguel, Terceira and Faial to show the distribution of potential natural vegetation (PNV).

One interesting finding from the palaeoenvironmental record is that there is now good evidence for human impact on the environment of the Azores before the generally accepted date of first human settlement. In a pollen core recovered from the caldera of Sete Cidades volcano (São Miguel, Rull et al. 2017a, p. 163), materials dating to between 1273 CE ± 40 and 1358 CE ± 40, evidence the cultivation of cereals, the burning of forests and possibly some pastoralism. This represents a time *c.*70 to *c.*150 years before the generally accepted date for the discovery São Miguel (i.e. 1431 CE – Cordeiro 1981) and even longer before its putative colonisation in the 1440s (Beier & Kramer 2018). Although the first maritime chart indicating São Miguel dates from 1339 CE (possibly 1331 – Callender & Henshall 1968), this is still some 50 years after the dated pollen zone, which implies that medieval navigators not only observed the island but also settled it possibly for a limited time.[8] Pollen records from Pico and Flores do not show any evidence of early vegetation disturbance.

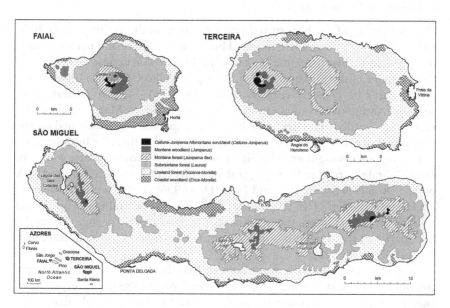

Figure 1.2 Potential natural vegetation (PNV) on São Miguel, Terceira and Faial. Based on Elias et al. 2016, Phytocoenolgia 46 (2), Figure 3, p. 117, Copyright permission Professor Rui Elias. Additional information from: Tutin 1953; Dias 1996 and field and air-photo research by the authors. The PNV was a concept developed by Reinhold Tüxen (1956), to describe the expected state of mature vegetation in the absence of human intervention.

The soils of the Azores have not been so intensively studied as has its vegetation, but the broad picture is well known. Using the system of classification first proposed by the United Nations Food and Agriculture Organisation (FAO 1974) and later developed into the *World Reference Base for Soil Resources* (FAO 1998), Portuguese soil scientists recognise a number of distinctive soil types (Ricardo et al. 1977, 1979; Madeira et al. 2007). The majority of cultivated and pastured soils are *andosols* (i.e. *andisols*) developed on pyroclastic and other volcanic materials under forest cover. Globally amongst the most productive of soil types, *andosols* are rich in organic matter; of low bulk density, with the potential for wind erosion when dry and removal by water and mass-movement on steep slopes when uncovered. Highly porous *andosols* may rapidly absorb and precipitate phosphorus. The latter feature is caused by high concentrations of Al (e.g. allophane) and Fe clay minerals that react with phosphorus (Gardiner & Miller 2008), and which mean that application of organic or mineral phosphorus is required to maintain yields. Other soils of lower agricultural potential include *lithosols* (thin fragmental soil) and *regosols* (weakly developed soils in unconsolidated sediments), both of which are associated with more sparsely vegetated and poorly weathered volcanic products.

The development of Azorean economy and society

Discovery and settlement

Detailed dating of the history and first settlement of the archipelago is difficult because many details are disputed but, as noted above and notwithstanding possible earlier maritime contacts, in 1431 a Portuguese explorer and Commander of the Order of Christ[9], Gonçalo Velho Cabral (*c*.1400–*c*.1460), was engaged by Prince Henry the Navigator (1394–1460) to undertake an Atlantic voyage of reconnaissance and discovery. Santa Maria was the first island to be settled and later a toe-hold of occupation was established on the south coast of São Miguel at a place that would eventually grow into the town of Povoação. Leaving domesticated farm animals to provide a later source of food, Cabral returned to Portugal, collected further colonists and arrived back on São Miguel between 1439 and 1443 (Dias 1936). Terceira was the next island to be settled in 1450, followed by the rest of the central island group and finally Flores and Corvo, the latter not being permanently occupied until the last quarter of the sixteenth century (Santos 1995a; Halikowski-Smith 2010; Beier & Kramer 2018).

Drawing settlers initially from both the northern and southern extremities of metropolitan Portugal, the Minho and Algarve regions, respectively, later immigrants were more cosmopolitan and included people from Madeira, Islamic prisoners, Jews from Morocco, African slaves; and Europeans from Spain, Flanders, France, Italy, Scotland and England; present-day place names and surnames bearing witness to the ethnic heterogeneity of the initial settlers (Cabral et al. 2005). Some years after settlement and because

Table 1.3 Population growth and density in the Azores

	Sixteenth century	Seventeenth century	1776	1878	1950	2011	2020
São Miguel	27,132 (36)	34,241	62,903 (84)	119,933	164,136 (220)	137,830 (185)	137,229 (184)
Santa Maria	2,600 (27)	4,235	4,871 (50)	6,378	11,788 (122)	5,552 (57)	5,620 (58)
Terceira	21,560 (54)	21,078	29,117 (73)	45,034	60,608 (152)	56,437 (141)	55,179 (138)
Graciosa	2,708 (44)	6,656	5,447 (89)	8,321	9,522 (156)	4,391 (71)	4,217 (70)
São Jorge	2,676 (11)	6,716	9,345 (38)	18,272	16,400 (67)	9,171 (38)	8,310 (34)
Faial	4,048 (23)	13,287	12, 027 (70)	24,963	23,944 (138)	14,994 (87)	14,532 (84)
Pico	3,508 (8)	10, 259	15, 444 (35)	26,396	22,336 (50)	14,148 (32)	13,644 (31)
Flores	632 (4)	3,235	3,190 (23)	9,687	7,812 (55)	3,793 (27)	3,628 (26)
Corvo	80 (5)	478	551 (32)	850	731 (43)	430 (25)	465 (27)
Total	64,955	100,185	142,895	259,834	317, 277	246,746	242,821

Based on information from: Rocha (1990); Costa et al. (2008) and Anon (2020). Population density (population km^2) is given in brackets for the sixteenth century, 1776 and 1950. In the late eighteenth century and for reasons which are not too clear population growth stagnated, but grew rapidly after 1800 (Axelsson 2015).

of fears of insurrection, slaves were relocated to Brazil and the Caribbean (Santos 1995a). As Table 1.3 shows, population grew rapidly due at first to immigration and later through natural increase, but periodically the number of people exceeded the productive capacity of the islands, because of a combination of issues associated with long-term development (see below) and shorter-run pressures brought about by major earthquakes, eruptions and epidemic diseases (see Chapter 3). It is only after 1800 that a transition to rapid and sustained demographic growth occurred, which according to Matos & Sousa (2015) was brought about by a better diet amongst the poorer sections of the population following the introduction of maize and potatoes. Several phases of emigration followed: in the eighteenth century, mainly to Brazil; in the second half of the nineteenth century and up to the First World War to the African colonies, Brazil and the USA, but after 1918 these countries imposed restrictions on immigration putting increased pressure on living standards. Relaxation of immigration controls to the USA followed the 1957/1958 Capelinhos eruption and earthquake (Coutinho et al. 2010—see Chapter 5), and people were also able to settle in Bermuda (Minder 2015). More recently, migrants have been free to re-locate to mainland Portugal, other countries of the European Union as well as to North America.

Long-term issues leading to poverty and outmigration have been a function, not just of isolation, but are an outcome of both the nature of economic development and certain features of the socio/political and cultural fabric of Azorean society (see below). From the beginning of the fifteenth century and notwithstanding profits from its empire in the first two centuries following settlement, in terms of both Gross Domestic Product (GDP) and GDP per capita, Portugal has been one of the poorest countries in Western Europe (Maddison 2007). This has placed a severe restriction on the aid that could be made available from the central government in the event of a disaster, notwithstanding the policies of particular governments towards state aid (see Chapter 5). Although settlement has contributed to the historical and present-day vulnerability of the Azores, further issues hinge on the nature of economic development, politics and society. There are countervailing aspects of Azorean society and its culture that have in the past and continue to enhance the islands' resilience, and these are introduced below and discussed in more detail in Chapters 5.[10]

Development of the Azorean economy

The agricultural economy of the Azores and its urban fabric quickly developed, and a land-use map of São Miguel compiled by Halikowski-Smith (2010, p. 84) shows that just a century after settlement there were elements in the landscape that are still visible today. Although the types of crop have changed, the broad distribution of cultivated and pasture land that is recognisable today was already discernible in outline in the sixteenth century. There was, for instance, intensive cultivation of wheat, vines, oranges, vegetables and woad (*Isatis tinctoria*) at low altitudes (below *c.*300 m) near to the coast and a more widespread distribution of pastoralism – especially of cattle – spreading from the coast inland, but avoiding the higher parts of central São Miguel which remained undeveloped. Many settlements present today were already established and included Ponta Delgada, Relva, Feteiras, Ginetes, Mosteiros, Bretanha, Capelas, Ribeira Grande, Maia, Achada, Nordeste, Faial da Terra, Povoação, Vila Franca do Campo, Água de Pau and Lagoa near to the coast; with Furnas being the only substantial inland village. Although the evidence is not so robust it seems likely that, with the exception of the smaller islands, other parts of the archipelago developed equally rapidly in the century following first contact, and this is reflected in population numbers (Table 1.3). It is because the islands of the archipelago were settled from the sea that even today there are few inland towns and villages, population density declining rapidly with increasing distance from the shoreline. As will be discussed in Chapters 5 and 6, this is a fortunate circumstance from a hazard management perspective because it greatly reduces the number of relatively isolated population clusters.

Based on a number of sources (Boid 1835; Admiralty 1945; Santos 1995a; Halikowski-Smith 2010; Newitt 2015; Beier & Kramer 2018), it is possible to trace stages in the development of the agricultural economy and landscapes of the islands from the time of first settlement to the present day. In the fifteenth century, the Portuguese government required the islands to become self-sufficient, provide provisions for passing ships and exports to mainland markets. Attracting settlers proved difficult, with much of the initial clearance of forest to produce pasture and croplands being entrusted to: slaves before their forced migration to Brazil and the Caribbean; Islamic prisoners and convicts (*degradados*) transported from Portugal.

The principal crops were sugar cane and wheat, but neither was successful in the long-term and the main exports were colouring agents: orchil (a red or violet dye made from lichens) and woad (a blue dye). Fishing was important, cleared timber was used to build ships, and by the sixteenth century, sweet potatoes, citrus fruits, grapes, maize and flax were being grown. Whaling arrived in the Azores in 1765, lasted until the 1980s, and from the seventeenth to the mid-nineteenth century, seaborne trade generated a boom in Azorean agriculture. After the 1850s, steam ships set new courses across the Atlantic and did not require provisioning in mid-Ocean (Callender & Henshall 1968). In 1834 the British, Captain Edward Boid writing of Faial and its capital Horta, noted that 'all the crops appeared abundant and luxuriant and were... most in demand amongst the shipping that frequented the port, – namely, potatoes, yams, onions, peas etc., all of which are extremely fine and taken in large quantities by American whalers' (Boid 1835, p. 273). The nineteenth century saw new crops of tobacco, tea and pineapples, the latter being especially geared to export markets, but in 1853 most of the vineyards were devastated by a fungal infestation *Uncinula necator* (i.e. *Oidium tucheri*) and, with the exception of the wines from Pico, Azores viticulture never fully recovered its overseas markets. Oranges were also affected by blight in 1877 which wiped out two-thirds of the crop (Santos 1995b). In addition to commercial crops, further exotic plants were introduced in the nineteenth century. Mention has already been made of *Cryptomeria japonica* (Japanese red cedar), and this was joined by *Pinus pinaster* (maritime pine) in commercial forests, with the ornamental plant *Hydrangea macrophylla* (French hydrangea) and other decorative species being sown in private gardens, public parks and as hedgerow plants (Rull et al. 2017a).

From 1893 to the late 1960s, isolation was reduced by growth of international telegraphy, with the city of Horta on Faial becoming a relay station for trans-Atlantic cables owned by companies from the UK, USA, France and Germany. By 1900, there were 300 expatriates living in Horta, and by 1928 15 cables were being serviced (Callender & Henshall 1968; Beier & Kramer 2018). Traffic declined from the 1950s.

In the twentieth century and until the Carnation Revolution of 1974 and the establishment of democratic government in Portugal, the most significant episode forcing landscape change occurred in the era of the *Estado Novo* (New State),[11] which was inaugurated in 1932 and lasted until 1974. The New State was effectively a dictatorship ruled by its Prime Minister, António de Oliveira Salazar. For reasons of self-sufficiency (i.e. autarchy), the islands introduced new crops (e.g. sugar beet), but more significantly the decision was made to capitalise on the archipelago's comparative advantage to become the nation's specialised producer of beef and dairy products, the number of cattle reaching 150,000 by the 1970s with much more land being placed under pasture. In 1945, a war-time intelligence assessment produced for the British Government (Admiralty 1945 see also Soeiro de Brito 1955) provides a 'snap shot' of the primary economic sector of the Azores after more than a decade of *Estado Novo* rule. The principal characteristics were as follows:

a   Three to four crops were cultivated each year usually without irrigation.
b   The principal crops were maize and broad beans, with wheat, barley, sugar beet, Irish potatoes, sweet potatoes and vines being of lesser importance.
c.   The islands were self-sufficient in bread but, despite the primary sector[12] accounting for two-thirds of male employment (in 1950), efficiency was low and not all foodstuffs could be supplied locally (Callender & Henshall 1968).
d   There was a strong and increasing emphasis on cattle rearing. The products were mostly for export to the mainland and the colonies, but because of poverty and with the exception of poultry, few Azoreans regularly had meat in their diets although dairy products especially cheese were plentiful and cheap.
e   Fishing and whaling were major employers of labour, but the Azores still relied heavily on imported fish, including dried and salted cod (*Bacalhau*) which was (and is) an important element in both Azorean and metropolitan Portuguese cuisine.

One consequence of the Second World War was that the Azores became of strategic importance because of their position with respect to sea lanes linking North America and Europe. Although initially sympathetic to the Axis powers, as the war progressed Salazar agreed to the Allies using the islands to pursue anti-submarine warfare through the provision of port and airport facilities. British forces, later joined by Americans, built airports on Santa Maria, São Miguel (Santana)[13] and Terceira (Lajes), and foreign troops influenced local lifestyles, diets and economic expectations, with overseas military investment stimulating the economy. From the 1950s, improved communications allowed links to be strengthened with diasporic communities located in such places as: Fall River, Rhode Island, Oakland and Boston in the USA; and Toronto and Montreal in Canada (Minder 2015). Following the Capelinhos eruption and earthquake in 1957/8 (see Chapters 3 and 5), many people flew to North America from the new international airport on Santa Maria (Coutinho et al. 2010).

Until this new wave of emigration was firmly established from the late 1950s, population continued to rise reaching an all-time peak of over

320,000 in 1960, and this era of over-population was associated with low labour productively and widespread poverty. In Callender and Henshall's (1968, p. 21) seminal study of Faial, they found both unemployment and underemployment 'amongst that majority of the labour force employed in agriculture. There is rarely more than 90 days work on the land available each year and those employed on a monthly wage are never absolutely fully employed.'

The demise the *Estado Novo* in 1974, the establishment of democracy and later accession into the European Union in 1986 brought great changes to the Azorean economy greatly affecting its landscape. Under the 1976 constitution, the Azores became an Autonomous Region with a regional government responsible, *inter alia*, for economic promotion, welfare policies and the environment. Investment from Portugal and the European Union under several initiatives impacted the economy.[14] Lower birth rates and emigration meant that population declined to 237,315 in 2001, but showed renewed growth up to the time of the 2011 census (Table 1.3 and SREA 2011; Anon 2020). These policies and their impacts have transformed the islands, raising living standards, increasing employment, further enhancing the pastoral specialisation of agriculture, with 57% of the meat being produced by beef cattle, followed by pigs, poultry, goats and sheep. Between 1980 and 2015, the primary sector shrunk from 30% to 10% of the Gross Added Value of the Azorean economy, the secondary sector from 26% to 16%, the major change being in services which rose from 44% to over 74%, much of the growth being accounted for by tourism and government employment. The figure for the primary sector in Portugal as a whole was only 2.4% in 2015 (Massot 2015; Beier & Kramer 2018). The area devoted to urban-based activities including housing has also grown. In 2013, living standards as measured by Gross Domestic Product (GDP) reached 91% of those recorded for Portugal as a whole, but only 72% of the average for the European Union (Massot 2015).

Present-day land use is captured in Figure 1.3, and changes since the fifteenth century may be assessed by comparing this figure with maps of the vegetation at the time of European contact (Figure 1.2). One feature of Azorean involvement with the European Union and other supra-national organisations has been greater environmental awareness, and this is expressed in policies which includes the following: the designation by the *United Nations Educational, Scientific and Cultural Organization* (UNESCO) of Corvo, Flores and Graciosa as biosphere reserves; the incorporation of the Azores into the European Network of *Geoparks* and the fact that *c.*16% of the archipelago now enjoys some form of environmental protection (Massot 2015). The Azores is marketed as a destination for ecologically focused tourism and the protection of 'green' assets is a prudent policy (Tiago et al. 2016). Greater wealth and better integration with international bodies are important contemporary themes which enhance the resilience of the islands in the event of a disaster (see Chapter 5).

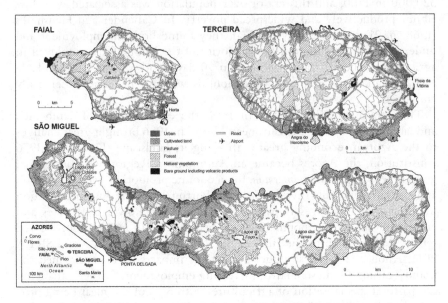

Figure 1.3 Land use of São Miguel, Terceira and Faial. Simplified from information in Anon 2007 and the field observations by the authors.

Politics and society

The development of administrative structures and frameworks of landholding are not arcane themes, but have had a bearing on the manner by which disasters in the Azores have been managed historically (see Chapter 5). Disasters may act as stimuli bringing into focus deep-seated political, economic and cultural concerns that are already present with a society (Pelling 2001; García-Acosta 2002; Wallenstein et al. 2015), with earthquakes and eruptions acting as 'highlighter(s) or amplifier(s) of daily hardship and everyday emergencies rather than… extreme and rare phenomena' (Gaillard & Texier 2007, p. 347). It may be argued that administrative structures and landholding have historically restricted the islands' agricultural potential, have placed a brake on wealth accumulation, encouraged emigration and exacerbated vulnerability. Both governance and landholding are themes that have been discussed in detail by a number of scholars (Admiralty 1945; Callender & Henshall 1968; Boxer 1969; Amaral 1987; Costa 2008; da Costa 2008; Beier & Kramer 2018), and it is upon these works that the following summary is based.

When first settled the Azores and Madeira were administered under a semi-feudal *donatário* (or donatary), system which was later adopted in other parts of the Portuguese empire. The monarch appointed Prince Henry the Navigator, and he in turn selected what became hereditary *capitães dos donatários* (captains of the donataries) to govern territories as his proxies, their principal role being to encourage settlement. By the sixteenth century, there were eight captaincies covering the islands and in later years the occupants of these offices were often absentee, delegating their powers to lieutenants, councillors and magistrates with the monarch also granting concessions to foreigners such as a Flemish settler, Jobst van Huerta, on Faial. In 1766, the administration was reformed by the king's first minister, the Marquis de Pombal (1699–1782), to form two administrative areas, and was modified again in 1835 to produce three districts: Angra (capital of Terceira, with responsibility for São Jorge and Graciosa); Ponta Delgada (on São Miguel, but also in charge of Santa Maria) and Horta (capital of Faial and administering the remaining islands within the archipelago). Each district was ruled by a governor appointed by the king who was assisted by a *Junta Geral* (general council), this system persisting until 1974. In common with the rest of Portugal, the Azores was sub-divided into *concelhos* (counties) and *freguesias* (parishes).

The *Capitães dos Donatários* and later administrators used a model of settlement that was first adopted during the Christian re-conquest of Portugal in the twelfth and thirteenth centuries. Territory was allocated to *sesmarias* (areas of uncultivated land) and, if the land was cleared and used for five years, it became the property of the settler. Sometimes settlers were established farmers, but often *sesmarias* were acquired by people of higher social and/or aristocratic status who frequently acquired several holdings on different islands. In addition, a number of *sesmarias* were acquired by artisans, many of whom were not farmers but speculators seeking economic advantage and with no intention of farming their land in the long term. By the mid-sixteenth century, all land in the Azores was either in private ownership or was deemed unattractive for investment, and these factors together conspired to cause only a proportion of the land to be cultivated. On São Miguel, this accounted for just *c*.1/3 of the total (Costa 2008).

It was legal stipulations governing inheritance, however, that had the most severe impacts on productivity and poverty. One-third of a deceased person's estate was reserved for the principal heir (usually the oldest male), and this system was disastrous for the following reasons (see Costa 2008):

a   It encouraged absentee ownership and inefficient management. Many owners had to manage numerous small scattered plots, dispersed over wide areas by means of agents. An extreme example was Pico where much land belonged to people who were actually resident on Faial.

b   Between the sixteenth and nineteenth centuries, profits were not generally re-invested into food production or land consolidation, but were converted into charitable donations to endow churches and/or to purchase indulgences.[15] The system left, not only a legacy of land hunger and retarded agricultural development, but was also a major 'push' factor in migration, so exacerbating economic polarisation.

c   Land comprising two-thirds of estates not inherited by principal heirs became fragmented as it passed down the generations through successive legatees.

The nature of vulnerability is dynamic and has changed through time and spatially. After more than three decades of *Estado Novo* government, in the late 1950s and early 1960s land ownership, the small size of holdings and poor employment prospects encouraged emigration and produced a residual economically disadvantaged and marginalised population.[16] In the 1950s on Faial, for example, there was: one substantial landowner; only 6% of the farmers employed labourers; 42% of proprietors or tenants had no paid workforce and 52% of agricultural personnel were either paid employees or family members. Sixty-one percent of farm workers were aged between 20 and 49 years (Cassola de Sousa 1957) and in 1950 illiteracy still averaged 39%, compared to 79% in 1900, with the rates in more isolated *freguesias* greatly exceeding this figure (Anon 2019a). In *pre-industrial* societies, high rates of illiteracy may wrongly imply an ignorant population. In 1894 in Ponta Delgada alone some 23 newspapers and periodicals were published and, although these were bought by elites, there was also a well-developed tradition of public reading in taverns, cafes and shops. In reviewing the progressive *Diário dos Açores,* Simões et al. (2012, p. 314) demonstrate that science was reported in some detail and included articles on hygiene and public health, as well geological topics including the 1908 Messina earthquake. From first settlement until at least two decades following the 1974 Revolution, vulnerability was expressed in a nexus of poverty, precarious employment, engagement with primary production and both educational and spatial isolation: within individual islands; amongst the islands of the archipelago and between the Azores and the mainland (see Chapter 5).

Since the advent of the democratic era in 1974, the nature of vulnerability has changed. Emigration is less significant, even in the recession of the last quarter of the decade after 2000, unemployment never approached the rates recorded historically and between the 2001 and 2011 censuses population increased by just 0.24%.[17] These features, combined with improvements in education, health care and a concomitant growth in alternative employment, have modified vulnerability (Wallenstein et al. 2015). By 2011, illiteracy rates had fallen to just 5% and were concentrated in the older population cohorts, whose numbers had increased – both absolutely and as a percentage – as a result of greater life expectancy, however levels of dependency[18] have increased (SREA 2011). In a study of São Miguel, Wallenstein et al. (2015) concluded that:

a   Illiteracy, poor pay and precarious employment in the primary sector has led to poverty and marginalisation amongst the older population cohort, especially in remote rural villages.

b   High dependency ratios have been a result of both rural/urban migration of younger people and commuting to centres of tertiary employment within larger urban centres, particularly to the island's capital, Ponta Delgada and its surrounding villages. This has left rural villages depleted in terms of literate and mature adults especially during the working day

(Chester et al. 1999). In these circumstances (Chester et al. 1995), a high proportion of the population will probably require some assistance, because of issues of mobility, responsibility for children and, amongst the very old, an inability to follow written instructions, should an eruption or earthquake-related emergency be declared (see Chapter 6).

c Today both São Miguel and other islands have transient populations, with many houses in rural areas only being used at weekends and/or for holidays.[19] Tourist numbers also vary over the course of the year, and in 2018 the number of visitors arriving in the Azores was over 840,000 (SREA 2019). From the perspective of hazard management, it is vital to determine: where the excess population, above that recorded in official statistics, is located; how many visitors are 'true' tourists and how many are Azorean expatriates returning to family homes that are either vacant or under-occupied for most of the year.[20] In an emergency, the demands of these disparate groups will be very different (see Chapters 5 and 6).

Culture and spirituality

There has only been one major sociological study of social class in the Azores and that is focused exclusively on São Miguel (Chapin 1989). Table 1.4 is based on Chapin's research, but has been updated and extensively modified using information from various sources, including census data and the results of our own research (Wallenstein et al. 2015). Self-explanatory overall there are two additional points that are not captured by the table and which require further discussion. First, the groups shown in the table reflect the socio/economic divisions that have already been discussed and have important implications for leadership. Leadership roles both in day-to-day administration and in emergencies have traditionally been filled by the *nobreza* at higher levels and, more locally, have been shared with *proprietários* and clergy, but today this responsibility falls heavily on people from an 'established educated' background. It is people from this group who fill posts as: politicians; regional and local government officials; teachers, doctors and other professionals within *concelhos* and *freguesisas* and provide academic and technical staff for the university. More specifically the university's multi-disciplinary research units: the *Instituto de Investigação em Vulcanologia e Avaliação de Riscos* (The Institute for Research in Volcanology and Risk Assessment – IVAR) and the *Centro de Informação e Vigilância Sismovulcânica dos Açores* (the Centre for Seismo-Volcanological Information and Surveillance of the Azores – CIVISA), both routinely advise the regional government and, in an emergency, would become its principal technical arm. Both units contain highly qualified staff many of whom have post-doctoral research experience and, although many are Azorean, others have been recruited from mainland Portugal and abroad. A second feature is that *trabalhadores* are not only concentrated in rural areas, but together with *proprietários* are also employed in agriculture, showing 'a

Table 1.4 Principal social grouping in the Azores

Socio/cultural group	Characteristics
Trabalhadores (manual workers)	Traditionally living in rural towns and villages and working as, either owners/renters of smallholdings (the minority), or as landless labourers (the majority), trabalhadores live in towns and villages and commute to work. Trabalhadores also work in fishing and as unskilled workers in urban centres. Even in the 2011, census illiteracy rates were over *c*.10% in some rural freguesias, although these figures have declined during the past decade as older members of this group have died (SREA 2012).
Proprietários (proprietors)	Owners of property. In the countryside, proprietários are small-scale village entrepreneurs such as shop keepers and bar/restaurant owners and/or own larger farms. Most proprietários have at least a primary education, but some older members of this group are illiterate. In rural areas local leaders have traditionally been drawn from this group, but many younger members are more upwardly mobile, being better educated and have often left their home villages to live in larger urban centres or abroad. Well represented in the Azorean diaspora, proprietors are sometimes known as 'educated proprietors' and have contacts access the social groups, both external and internal.
Established educated	This comprises two sub-groups. a. The educated 'middle class.' With the exception of doctors, local government officials and school teachers, few live in remote rural areas and are concentrated in the principal administrative centres especially: Ponta Delgada (São Miguel); Angra do Heroísmo (Terceira) and Horta (Faial). Members of this group often commute from villages within the larger urban conurbations. As tertiary sector employment has increased, so has this group which today contains some people born elsewhere in Portugal or abroad who are now resident in the Azores. b. Nobreza (nobility). Numerically very small, historically this group provided leaders in the four centuries following settlement of the archipelago. The nobreza still own large tracts of land.
New entrepreneurs	Again, highly concentrated in the larger urban centres, especially Ponta Delgada, the Azorean capital.

Based on information in: Chapin (1989); SREA (2012); Wallenstein et al. (2015).

strong attachment to land, community and livestock' that could mean that orders to evacuate could at best (be) ignored and at worst resisted. This feature, which is perceptual as well as social and cultural is more fully discussed in Chapters 5 and 6 (Wallenstein et al. 2015, p. 221).

Many literary writers, the foremost of whom was Vitorino Nemésio (1901–1978), have argued for the existence of distinctive Azorean identity. Based partly on a singular expression of Catholic spirituality (see below), this in addition involves features of identity that cut across social classes and groupings. Known as *Açorianidade* (Azoreanness – Almeida 1980; da Silva 2008), it encompasses attributes which, by increasing community solidarity, serve to boost resilience when society is faced with the unexpected. These features include the following:

a   mutual support from extended families, both living in the Azores and abroad, often invoking *saudade*[21], a nostalgic longing for such things as better times that were supposedly experienced in the past, and families not separated by wide ocean spaces;
b   *provincianismo* (provincialism), a strong identification with the archipelago, individual islands and villages, which is cemented by (mostly) friendly rivalry and
c   a pessimistic, stoical worldview that views Azorean history as survival under difficult circumstances, which includes the effects of disasters caused by eruptions and earthquakes together with poverty, isolation, emigration of family members and frequent outbreaks of epidemic disease.[22]

The religious *milieu* of the Azores is superficially similar to that found in other areas of southern Europe, with around 91% of the inhabitants declaring confessional adherence to Roman Catholicism (SREA 2012). Other similarities include a highly visible and colourful religious experience that is expressed in richly decorated church interiors and elaborate highly choreographed and sometimes tactile patronal festivals and processions which are attended by local people, tourists and returnees from the diaspora (Anon 2016b). The most famous festival is the *Festas do Senhor Santo Cristo dos Milagres* (Festival of the Christ of Miracles) held in Ponta Delgada (São Miguel) on the fifth Sunday after Easter. Also in common with southern Europe, churches are dedicated to a wide variety of saints with the Madonna being the most frequently encountered (da Silva 2008). Another marked similarity is with southern Italy where popular Catholicism has developed against a backdrop of major disasters caused by earthquakes and volcanic eruptions.

In spite of these similarities there are also significant differences. Whereas the catholic spirituality of southern Europe has evolved slowly since Roman Catholicism became the prescribed religion of the Roman Empire in 380 CE, in the Azores Christianity arrived with missionaries from Portugal who were predominantly members of religious orders: the Franciscans arriving in 1446 (Boehrer 1955); followed sometime later by the Jesuits. From the start, there was a strong emphasis on pastoral concern as well as worship with, monasteries, hospitals and schools being established (Chester et al. 2019). By the close of the eighteenth century, the Azores archipelago was home to around 1,000 priests – 1 for every 156 people – and 105 parishes (da Silva 2008, p. 76).

Another dissimilarity is that, whereas in southern Europe losses from disasters have in the past, and in the case of southern Italy frequently still are, explained as being the manifestations of divine wrath for sins supposedly committed, retribution is far from being sole model of theodicy[23] encountered in history and current practice of Catholicism in the Azores. Although between the fifteenth and eighteenth centuries, there are numerous examples of divine retribution being invoked (see Chapter 5) and, indeed, the term *mistério* (i.e. mystery of God) is used in the Azores for new land created by lava (Forjaz 2008, pp. 28–29), these forms of explanation were far from being universal, and from the sixteenth century educated priests such as Gaspar Frutuoso, many Jesuits (Udías 2009) and educated lay people began to question whether extreme geophysical events were under divine control by increasingly embracing an Aristotelian understanding that natural processes alone were responsible (Pinto 2003). Following the Lisbon earthquake of 1755 alternative models of theodicy, such as 'the best of all possible worlds,' became more popular in educated circles both clerical and lay. From the earliest times the Azorean clergy have focused, not only on theology but also on praxis, with examples including solidarity with congregations regardless of social background; self-sacrifice and attention to the pastoral, as well as the spiritual needs of the people (Chapter 5).

The major area of contrast with the rest of southern European Catholicism is the devotion of the people of the Azores and diasporic communities to the *Culto do Espírito Santo* (i.e. Cult of the Holy Spirit). This cult has a complex origin and heterodox theology (Chester et al. 2019), was condemned by the Pope

Table 1.5 The principal characteristics of the *Culto do Espírito Santo*

Characteristic	Detail
Império (empire or theatre)	This is a small building that looks like a chapel. Most date from the nineteenth century and before this time a wooden stump was used. Annual festivals are centred on the *Império* and their architectural style varies across the archipelago.
Egalitarianism of the Brotherhood of the Holy Spirit (*Irmandades do Divino Espírito Santo*)	The *mordomo* (leader or steward) is chosen by lot and all members are equal and all can be elected leaders or stewards, who are known as *mordomos*. The *mordomo* is chosen by ballot and a crown is placed on the head of an adult or child, to represent equality of status. Once installed, the social class of a new *mordomo* is irrelevant and all are shown equal respect. Ceremonies are organised by the *mordomo*, which take place between Easter and Pentecost and are sometimes extended a week until Trinity Sunday.
Hope (*Esperança*)	Human spiritual development requires time. The Holy Spirit is viewed as a font of knowledge, calmness and order.
Solidarity and charity	Every year there is a ritualised and symbolic distribution of alms and food, which is traditionally meat, soup and milk. Charity and solidarity are the central and guiding values of the cult. They go beyond the ceremonial and ideally influence a person's whole life.
Divine faith (*Fé no Divino*)	Members believe that the Holy Spirit is omnipresent, omniscient and all-knowing. Vows made to God and neighbour must be kept and the brother/sister hood virtues of: wisdom; understanding; counsel; fortitude; knowledge; piety and fear of the Lord are enjoined on all members.

Based on Chester et al. (2019).

in 1256 which led to its later suppression by the inquisition (Montez 2007). It was brought to the archipelago by Flemish settlers and Portuguese Franciscans and has a number of distinctive attributes which are summarised in Table 1.5. Three additional points need to be made. First, because the cult was declared anathema by the Pope in 1256 in the Azores, it flourished at arms length from the official church. The cult was led by lay people from the beginning, and it has never been under the control of the Catholic hierarchy, although the crowning ceremony may include a church parade and the involvement of a local priest. This means that the individual member is enjoined to live his or her life according to precepts of charitable giving, solidity with neighbours and concern for their well-being. A second theme is that, in the context of historical isolation and during times of trouble including those associated with earthquakes and eruptions, the values of cult were an important ingredient in community resilience. Members have often looked after people when they were hungry, homeless and required comfort to the body as well as succour for the soul. Finally, the sociologist of religion, Grace Davie (1990), makes a distinction between believing and belonging and, even though many Azoreans do not attend mass on a weekly or even occasional basis, there is a strong sense of belonging to a popular Catholicism, which is not defined merely by dogma but by the ethics of solidarity, mutual aid and praxis (Chester et al. 2019).

Concluding remarks

The Azores have never been completely remote from the rest of the world: the settlement of the archipelago occurring during the so-called European Age of Discovery; its prosperity fluctuating with the vicissitudes of world trade and the dynamics of trans-Atlantic shipping and the adjustments of population to the resource base being dependent on the willing of donor countries to receive Azorean migrants. Isolation shaped the relationships between people and environment in a more profound manner than in all but a small minority of areas within the European cultural orbit, to create a distinctive suite of landscapes. These were forged by the inhabitants making the most of the possibilities afforded by the environments they encountered to create distinctive *genres de vie* or ways of life (Vidal de la Blache 1903). The manner in which *genres de vie* have been maintained, developed and enhanced over nearly six centuries of settlement, against a background of major seismic and volcanic episodes, will be spelt out in detail in the chapters which follow.

Notes

1 The oldest rocks in the Azores are from Santa Maria and date from *c*.6 Ma and the islands become progressively younger towards the west. Flores is around 2.15 million years old and is located entirely on the North American Plate, while Pico is the youngest island in the archipelago with its oldest rocks dating from 0.27 Ma (Azevedo & Ferreira 2006).

2 Maximum heights are as follows: Flores 914 m, Corvo 720 m, Faial 1043 m, Pico 2351 m, São Jorge 1053 m, Graciosa 403 m, Terceira 1023 m, São Miguel 1103 m and Santa Maria 587 m.

3 The oldest materials in cores recovered from Pico show dates of *c.*6 cal. ka yrs BP and *c.*2.7 cal. ka yrs BP, respectively, from the Lagoa do Caveiro and 'Pico Bog' (a mire north east of Lagoa de Peixinho). The oldest dated sediments from Lagoa Rasa on Flores also date from *c.* 2.7 cal. ka yrs BP (Connor et al. 2012). In a study of a core from the Lagoa Azul (Sete Cidades volcano, São Miguel) only dated sediments from the last 700 years have been recovered (Rull et al. 2017a).

4 Macronesia is the collective name for the islands of the sub-tropical and tropical Atlantic Ocean. In addition to the Azores, they include Madeira, the Canary Islands and Cape Verde Islands.

5 Writers who are not biologists, often use the short-hand term 'laurel forest,' when referring to the vegetation before human contact. As argued in this chapter, this is an oversimplification.

6 Climatic modelling reported by Triantis et al. (2011) implies that in the late Pleistocene and Holocene the Azores was climatically less stable than either Madeira or the Canaries. Using palaeoenvironmental data from Flores and Pico, Connor et al. (2013) found little evidence for this and, in fact, noted a very weak relationship between palaeo-vegetational and palaeoclimatic changes. Even if palaeoclimate varied markedly, this is not reflected in vegetation history. Historically and pre-historically, the vegetation of the Azores has been much more strongly impacted by volcanic eruptions, processes of soil formation and human impact.

7 Born in the Azores, Gaspar Frutuoso was educated at the University of Salamanca, was associated with the Jesuits and, as well as theology, studied natural history. Whilst serving as a parish priest in Ribeira Grande (São Miguel) he produced a multi-volume study: *Saudades da Terra* (Nostalgic Longing for the Earth). Within its six volumes only: Book 3 - Santa Maria; Book 4 - São Miguel and Book 6 - mostly focused on Terceira, are concerned with the Azores. Held for centuries in manuscript form by Jesuits in Ponta Delgada, it was used by António Cordeiro and Ernesto do Canto. Later it fell into private hands and was finally donated to the Public Library and Archive in Ponta Delgada (São Miguel) in 1950 (Pinto 2003). *Saudades da Terra* did not become widely available until the twentieth century (Frutuoso 2005).

8 It is notable that Rando et al. (2014) has suggested that Madeira may have been settled some 400 years before the generally accepted date of European contact.

9 The Knights Templar were suppressed by the Pope in 1312 and many of its members were killed, and in 1317, King Dinis I of Portugal created the Order of Christ for surviving knights. The order gained great wealth during the Age of Discovery, in the eighteenth century it became secularised and today confers honours on prominent Portuguese citizens and foreigners (Olival 2018).

10 In Chapter 5 there is a detailed discussion of the concepts of vulnerability and resilience, but briefly vulnerability is the extent to which a society is impacted by a disaster this being influenced by the characteristics of the eruption or earthquake and the particularities of the society affected, whilst resilience is its ability to resist and recover. The balance between vulnerability and resilience governs the seriousness of a disaster (Chester et al. 2012, p. 77)

11 Ruling from 1928 to 1968, the Head of Government and effective dictator of Portugal was António de Oliveira Salazar (1889–1970). From 1928 to 1932 he was the most prominent minister of the *Ditadura Nacional* (National Dictatorship) and from 1932 Prime Minster of the *Estado Novo* government. Sometimes compared with ostensibly similar fascist regimes in Spain and Italy, the *Estado Novo*, also displayed dissimilarities. Its principal features were:

a   Salazar's adept 'juggling of the interests of the army, the urban middle class, the monarchists and the church, (the great ideals being) .... patriotism, paternalism and prudence' (Birmingham 1993 p. 159), helped him remain in power for four decades.

b   *Estado Novo* was neither capitalist nor communist. It was a highly managed state-controlled market economy.

c   The regime strictly limited government expenditure, ran a balanced budget and did not normally borrow money on the international markets.

d   The government was anti-parliamentary and placed power in the hands of the executive, the governing party (*União Nacional*) and Salazar more particularly.

e   Centralised control was exercised over all areas of Portugal including the Azores and the empire, the civil service and the press. In the Azores Salazar selected civil governors and military commanders (Baklanoff 1992; Guill 1993; Anderson 2000).

Following Salazar's incapacity in 1968, the *Estado Novo* continued under Marcello Caetano (1906–80) until 'Carnation Revolution' of 1974.

12   In contrast to the primary sector (agriculture, forestry and fishing), the secondary sector (construction, supply of energy/water manufacturing and the processing of agro-food and forest products) and the tertiary sector (mostly public administration, tourism and services) were very small in the 1940s.

13   Since 1969, the airport on São Miguel has been located near to Ponta Delgada.

14   Major European Union-based development initiatives include (and have included): the designation of the Azores's as an 'outermost region,' allowing the free-trade provisions of the *Common Agricultural Policy* (CAP) to apply only partially and support through the *Programme of Options Specifically Relating to Remoteness and Insularity* (POSEI), which has included a further modification of the CAP. Under these policies there have been three principal concerns: stabilising livestock production, promoting traditional cropping and reducing production costs. Between 2014 and 2020 a rural development programme (PRORURAL+) was in place. Funding is shared between the European Union and the Lisbon government and is focused on preserving agriculture and forest ecosystems, enhancing the viability and competitiveness of farms and educational extension thorough policies of knowledge transfer (Massot 2015).

15   In Catholic teaching an indulgence is a reduction in the time a person has to spend in purgatory for the sins they have committed during their lifetime. It often involved giving money to a church and/or endowing some religious good cause. Later it became a major grievance of the sixteenth century Protestant reformers.

16   The population was marginalised in three ways (Wisner et al. 2012, pp. 174–176): economically (i.e. it was poor); socially (i.e. people were removed from those sections of society with power and influence) and spatially (i.e. it was concentrated in remote parts of the archipelago and more isolated areas of the larger and generally more prosperous islands).

17   A population estimate gives a figure of 243,862 for 2018 (Anon 2019b). Whether this small decline from the 2011 census figure of 246,746 (Table 3) represents a trend or merely a fluctuation is not possible to determine until the next census in 2021.

18   The number of people under 15 and over 65, expressed as a percentage of the total population.

19   The census only provides a snapshot of resident population on a specific date. In Portugal March or April in the first year of the decade have been the census months, yet many houses are often only occupied at weekends and/or in summer.

Hence the number resident and who may have to be evacuated on, say, a Saturday in August would be far greater than on weekday in March or April.

20 In 2018 there were 427,310 foreigners visited the Azores and 403,213 Portuguese nationals (SREA 2019). The problems are that some 'foreigners' are part of the Azorean diaspora from North America and a proportion of the Portuguese have no connection with the Azores, except that they are using it as tourists.

21 *Saudade* is often translated into English as a nostalgic longing, but is almost impossible to translate adequately. It represents emotions of sadness, melancholia, nostalgia and longing. There are many aspects to *saudade*, but one of these is the sadness which relates to feelings for people - especially family members – who have emigrated, while longing is expressed by the migrants who miss the Azores, its culture and their relatives (Teletin and Manole 2015; Anon 2016a).

22 These included, not only infestations of crops that affected grapes and oranges in the nineteenth century and which have already been discussed, but also epidemics affecting the population. These have been discussed by de Matos & Silveira e Sousa (2015) and the population was severely effected by plague and other epidemics in the 1780s and by smallpox in 1806.

23 Theodicy is defined as any attempt to reconcile theistic belief and innocent suffering. Two major models are: retributive justice as punishment of sinful humanity and the 'best of all possible worlds.' The latter maintains that the laws of physics control the universe and not 'special laws' that result from a deity. The earth is the best of all possible worlds that could be created and suffering is a consequence of the need to achieve a greater good (e.g. without volcanic activity the Earth would have no atmosphere) (Chester 1998).

References

Admiralty 1945. *Spain and Portugal Volume IV: The Atlantic Islands.* Naval Intelligence Division, Geographical Handbook Series BR 502c, London.

Agostinho, J. 1938. Clima dos Açores – Parte I. *Açoriana (Angra do Heroísmo)* **2** (1), 35–65.

Agostinho, J. 1942. Clima dos Açores – Parte II. *Açoriana (Angra do Heroísmo)* **3** (1), 42p.

Agostinho, J. 1948. Clima dos Açores – contribuição para o estudo da sua variação secular. *Açoreana. Angra do Heroísmo* **4** (3), 263–266.

Almeida, O.T. 1980. A profile of the Azorean. *In:* Macedo, D.P. (ed) *Issues in Portuguese Bilingual Education. National Assessment Discrimination Center*, Cambridge, MA, 115–164.

Amaral, J.B.M. 1987. Development of the Azores: Problems and prospects. *Ekistics* **323/324**, 143–146.

Anderson, J.M., 2000. *History of Portugal.* Greenwood, Westport, CT.

Anon 2007. *Ocupação do Solo – Portal do Ordenamento do Território dos Açores.* Governo dos Açores, Ponta Delgada, http://ot.azores.gov.pt/Ocupacao-Solo.aspx

Anon 2016a. *Saudade* expresses two emotions – sadness and longing. Live Azores, January 9, http://liveazores.com/saudade-expresses-two-emotions-sadness-and-longing/

Anon 2016b. *Festivals.* Azores Web, http://www.azoresweb.com/festivals_azores.html.

Anon 2019a. Taxas de Analbabestismo: Censos de Portugal. Instituto Nacional de Estatística, Lisboa, http://censosdeportugal.blogspot.com/2016/09/10-taxas-de-analfabetismo-1900-2011.html

Anon 2019b. *Azores.* Population Data Net, https://en.populationdata.net/countries/azores/

Anon 2020. *Retrato dos Açores PORDATA*. Fundação Francisco Manuel dos Santos, Lisboa.

Antunes, P. and Carvalho, M.R. 2018. Surface and groundwater in volcanic islands: Water from Azores islands. *In*: Kueppers, U. and Beier, C. (eds) *Volcanoes of the Azores*. Springer, Berlin, 301–329.

Axelsson, P., de Matos, P.T. and Siveira e Sousa, P. 2015. The demography of the Portuguese Empire. Sources, methods and results, 1776–1822. A Demografia do Império Português. Fontes, métodos e resultados, 1776–1882. *Anais de História de além mar* **16**, 20–49.

Azevedo, J.M.M. and Ferreira, M.R.P. 1999. Volcanic gaps and subaerial records of palaeo-sea-levels on Flores Island (Azores): Tectonic and morphological implications. *Geodynamics* **28**, 117–129.

Azevedo, J.M.M. and Ferreira, M.R.P. 2006. The volcanotectonic evolution of Flores Island, Azores (Portugal). *Journal of Volcanology and Geothermal Research* **156**, 90–102.

Baklanoff, E.N. 1992. The political economy of Portugal's Estado Novo: A critique of the stagnation thesis. *Luso-Brazilian Review* **20** (1), 1–17.

Beier, R. and Kramer, J. 2018. A portrait of the Azores: From natural forces to cultural identity. *In*: Kueppers, U. and Beier, C. (eds) *Volcanoes of the Azores, Active Volcanoes of the World*. Springer-Verlag, Germany, 3–26.

Bettencourt, M.L. 1979. *O clima dos Açores como recurso natural, especialmente em agricultura e indústria de turismo*. O Clima de Portugal, 18, Instituto Nacional de Meteorologia e Geofísica (INMG), Lisboa.

Birmingham, D., 1993. *A Concise History of Portugal*. Cambridge University Press, Cambridge.

Boehrer, G.C.A. 1955. The Franciscans and Portuguese in Africa and the Atlantic Islands, 1415–1499. *The Americas* **11** (3), 389–403.

Boid, E. 1835. *A Description of the Azores or Western Islands from Personal Observation*. Churton, London.

Borges, P., 2003. *Ambientes litorais nos grupos Central e Oriental do arquipélago dos Açores. Conteúdos e dinâmica de microescala*. Ph.D. thesis, Departemento de Geociências, Universidade dos Açores.

Boxer, C.R. 1969. *The Portuguese Seaborne Empire 1415–1825*. Hutchinson, London.

Cabral, R., Branco, C.C., Costa, S., Caravello, G., Tasso, M., Peixoto, B.R. and Mota-Viera, L. 2005. Geogrpahy of surnames in the Azores. Specificity and spatial distribution analysis. *American Journal of Human Biology* **17**, 634–645.

Callender, J.M. and Henshall, J.D., 1968. The land use of faial in the Azores. *In*: Cook, A.N. (ed) *Four Island Studies. The World Land Use Survey*. Monograph 5, Geographical Publications, Bude, Chapter 11, 367–395.

Cassola de Sousa, A. 1957. *Plano de Povoamento Florestal do Distrito Autónomo da Horta*. Ministério da Economia, Lisboa.

Chapin, F.W. 1989. Tides of Migration: A study of migration decision making and social progress in São Miguel.

Chazarra, A., Barceló, A.M., Pires, V., Cunha, S., Silva, A., Marques, J., Carvalho, F, Mendes, M. and Neto, J. 2012. *Altas Climático de Los Archipiélagos de Canarias, Madeira y Azores. Atlas Climático dos Arquipélagos das Canárias, da Madeira e dos Açores. Climate Atlas of the Archiplelagos of the Canary Islands, Madeira and the Azores*. Agencia Estatal de Meterologia Ministerio de Agricultura, Alimentación y Medio Ambiente, Madrid and Instituto de Meteorologia de Portugal, Lisboa.

Chester, D.K. 1998. The theodicy of natural disasters. *Scottish Journal of Theology* **51** (4), 485–505.

Chester, D.K., Dibben, C. and Coutinho, R. 1995. Report of the evacuation of the *Furnas District,* São Miguel, Azores, in the event of a future eruption. CEC Environment ESF - Laboratory Volcano, Eruptive History and Hazard Open File Report 4, Planetary Image Centre, University College London and Universidade dos Açores.

Chester, D.K., Dibben, C., Coutinho, R., Duncan, A.M., Guest, J.E. and Baxter, P.J. 1999. Human adjustments and social vulnerability to volcanic hazards: The case of Furnas Volcano, Sao Miguel, Azores. *In*: Firth, C. and McGuire, W.J. (eds) *Volcanoes in the Quaternary*. Geological Society of London, Special Publication **161**, 189–207.

Chester, D.K., Duncan, A.M. and Sangster, H. 2012. Human responses to eruptions of Etna (Sicily) during the late-Pre-Industrial Era and their implications for present-day disaster planning. *Journal of Volcanology and Geothermal Research* **225**, 65–80.

Chester, D.K., Duncan, A.M., Coutinho, R. and Wallenstein, N. 2019. The role of religion in shaping responses to earthquake and volcanic eruptions: A comparison between Southern Italy and the Azores, Portugal. *Philosophy, Theology and the Sciences* **6**, 33–45.

Connor, S.E., Van Leuwen, J.F.N., Rittenour, T.M., van der Knaap, W.O., Ammann, B. and Björk, S. 2012. The ecological impact of oceanic island colonization - a palaeoecological perspective from the Azores. *Journal of Biogeography* **39**, 1007–1023.

Connor, S.E., Van der Knapp, W.O., Van Leeuwen, J.F.N. and Kuneš, P. 2013. Holocene palaeoclimate and palaeovegetation on the Islands of Flores and Pico. *In*: Fernández-Palacios, J.M., de Nascimento, L., Hernández, J.C., Clemente, S., González, A. and Días-González, P. (eds) *Climate Change Perspectives from the Atlantic: Past, Present and Future*. Servicio de Publicaciones, Universidad de La Laguma, 149–162.

Cordeiro, A. 1981. *História insulana das Ilhas a Portugal sugeytas no Oceano Occidental*. Região Autónoma dos Açores, Secretaria Regional da Educação e Cultura, Ponta Delgada.

Costa, C. 1950. Arvoredos dos Açores: algumas achegas para a sua história. *Comissão Reguladora dos Cereais do Arquipélago dos Açores* **11**, 45–60.

Costa, S.G. 2008. *Açores. Nove Ilhas, Uma História. Azores. Nine Islands, One History*. Institute of Government Studies Press, University of California, Berkeley.

Coutinho, R., Chester, D.K., Wallenstein, N. and Duncan, A.M. 2010. Responses to, and the short and long-term impacts of the 1957/1958 Capelinhos volcanic eruption and associated earthquake activity on Faial, Azores. *Journal of Volcanology and Geothermal Research* **196**, 265–280.

Cropper, T. 2013. The weather and climate of Macronesia: Past, present and future. *Weather* **68** (11), 300–307.

Cropper, T.E. and Hanna, E. 2014. An analysis of the climate of Macronesia, 1865–2012. *International Journal of Climatology* **34** (3), 604–622.

da Costa, R.M.M. 2008. A brief introduction to the settlement of the Azores and a historical outline of the island of Faial. *In*: Goulart, T (ed) *Capelinhos, a Volcano of Synergies. Azorean Emigration to America*. Portuguese Heritage Publications of California, San Jose, 3–16.

da Silva, H.G. 2008. Portuguese-Azorean cultural identity: Common traits of Acorianidade/Azoreanness. *In:* Goulart, T. (ed) *Capelinhos, A Volcano of Synergies. Azorean Emigration to America.* Portuguese Heritage Publications, San Jose, CA, 73–83.

Dansereau, P. 1970. Macaronesian studies IV. Natural ecosystems of the Azores. *Revue Canadienne de Géographie* **24**, 21–42.

Davie, G. 1990. Believing without belonging: Is this the future of religion in Britain? *Social Compass* **37** (4), 455–469.

de Matos, P. and Silveir e Sousa, P. 2015. Settlers for the empire: The demography of the Azores Islands (1766–1835). *Anais de História de Além* **Mar XVI**, 19–49.

Dias, E. 1996. Vegetação natural dos Açores. Ecologia e sintaxonomia das florestas naturais. PhD thesis, Universidade dos Açores.

Dias, U.M. 1936. *História do valle das Furnas.* Empresa Tip Lda, Vila Franca do Campo.

Duncan, A.M., Guest, J.E., Wallenstein, N. and Chester, D.K. 2015. The older volcanic complexes of São Miguel, Azores: Nordeste and Povoação. *In:* Gaspar, J.L., Guest, J.E., Duncan, A.M., Barriga, F.J.A.S. and Chester, D.K. (eds) *Volcanic Geology of São Miguel Island (Azores Archipelago).* Geological Society, London, Memoir 44, 147–153.

Elias, R.B., Gil, A., Silva, L., Fernández-Palacios, Azevado, E.B. and Reis, F. 2016. Natural zonal vegetation of the Azores Islands; characterization and potential distribution. *Phytocoenologia* **46** (2), 107–123.

Engelhart, S.E., Horton, B.P. and Kemp, A.C. 2011. Holocene sea level changes along the United States Atlantic coast. *Oceanography* **24** (2), 70–79.

FAO, 1974. *Soil Map of the World, 1:5,000,000.* Vol. 1. UNESCO, Paris.

FAO, 1998. *World Reference Base for Soil Resources.* World Soil Resources, Report 84. Food and Agriculture Organization of the United Nations, International Soil Reference and Information Centre, and International Society of Soil Science, Rome.

Fernándes-Palacios, J.M., de Nascimento, L., Otto, R., Delgado, J.D., García-del-Rey, E., Arévalo, J.R. and Whittaker, R.J. 2011. A reconstruction of Palaeo-Macaronesia, with particular reference to the long-term biogeography of the Atlantic island laurel forests. *Journal of Biogeography* **38**, 226–246.

Forjaz, V.H. 2008. Historical eruptions in the Azores. *In:* Goulart, T. (ed) *Capelinhos, a Volcano of Synergies. Azorean Emigration to America.* Portuguese Heritage Publications of California, San Jose, 25–43.

Frutuoso, G. 2005. *Saudades da Terra.* Instituto, Cultural de Ponta Delgada, Ponta Delgada, Açores (6 volumes) original manuscript 1586–1590.

Gaillard, J.-C. and Texier, P. 2007. Natural hazards and disasters in South-east Asia. *Disaster Prevention and Management* **16**, 346–349.

García-Acosta, V. 2002. Historical Disaster Research. *In:* Hoffman, S.M. and Oliver-Smith, A. (eds) *Catastrophe and Culture.* James Currey, Oxford, 49–66.

Gardiner, D.T. and Miller, R.W. 2008. *Soils on Our Environment.* Pearson-Prentice Hall, New Jersey.

Gaspar, J.L., Queiroz, G., Ferreira, T., Medeiros, A.R., Goulart, C. and Medeiros, J. 2015. Earthquakes and volcanic eruptions in the Azores region: Geodynamic implications from major historical events and instrumental seismicity. *In:* Gaspar, J.L., Guest, J.E., Duncan, A.M., Barriga, F.J.A.S. and Chester, D.K. (eds) *Volcanic Geology of São Miguel Island (Azores Archipelago).* Geological Society, London, Memoir **44**, 33–49.

Guill, M., 1993. *A history of the Azores Islands Handbook*. Gold Shield International, Tulare, CA, p. 662.

Guppy, H.B. 1917. *Plants, Seeds and Currents in the West Indies and the Azores*. Williams and Norgate, London.

Halikowski-Smith, S. 2010. The Mid-Atlantic Islands: A theatre of early modern ecocide. *International Review of Social History* **55** (suppl.), 51–77.

Madeira, M., Pinheiro, J., Madruga, J. and Monteiro, F. 2007. Soils of volcanic systems in Portugal. *In*: Arnalds, O., Óskarsson, H., Bartoli, F., Buurman, O, Stoops, G. and García-Rodeja, E. (eds) *Soils of Volcanic Regions in Europe*. Springer, Berlin, 69–81.

Maddison, A, 2007. *Contours of the World Economy, 1–2030 AD*. Oxford University Press, Oxford.

Malheiro, A. 2006. Geological hazards in the Azores archipelago: Volcanic terrain instability and human vulnerability. *Journal of Volcanology and Geothermal Research* **156**, 158–171

Marques, R., Zêzere, J., Trigo, R., Gaspar, J. and Trigo, I. 2008. Rainfall patterns and critical values associated with landslides in Povoação County (São Miguel Island, Azores): Relationships with the North Atlantic oscillation. *Hydrological Processes* **22**, 478–494.

Massot, A. 2015. *The Agriculture of the Azores Islands*. Director General for Internal Policies, European Parliament, Brussels.

Matos, P.T. and Sousa, P.S. 2015. Settlers for the empire: The demography of the Azores Islands. *Anais de História de Além-Mar* **16**, 19–49.

Meco, J., Petit-Maire, N., Guillou, H., Carracedo, J.C., Lomoschitz, A., Ramos, A.J.G. and Ballester, J. 2003. Climatic changes over the last 5,000,000 years as recorded in the Canary Islands. *Episodes* **6**, 133–134.

Minder, R. 2015. Azorean diaspora can't resist the powerful pull of home. *The New York Times*, June 4, https://www.nytimes.com/2015/06/05/world/europe/azores-diaspora-holy-christ-of-miracles.html

Montez, M.S. 2007. The Império of the Azores: The five senses of rituals to the Holy Spirit. *Traditiones* **36**, 169–176.

Newitt, M. 2015. *Emigration and the sea. An Alternative History of Portugal and the Portuguese*. Hurst and Company, London.

Olival, F. 2018. *The Military Orders and the Portuguese Expansion (15th to 17th Centuries)*. Portuguese Studies Review Monographs, Vol. 3, Baywolf Press and The Portuguese Studies Review, Peterborough.

Pelling, M. 2001. Natural disaster? *In*: Castree, N. and Braun, B. (eds) *Social Nature: Theory, Practice and Politics*. Blackwell, Oxford, 170–188.

Pinto, M.S. 2003. "Gaspar Frutuoso, a Portuguese volcanologist of the 16th century." *Açoreana* **10**, 207–226.

Rando, J.C., Harald, P. and Alcover, J.A. 2014. Radiocarbon evidence for the presence of mice on Madeira island (North Atlantic) one millennium ago. *Proceeding of the Royal Society (London)* **B 281**, 20133126, http://dx.doi.org/10.1098/rspb.2013.3126

Ricardo, R.P., Madeira, M.A.V., Medina, J.M.B., Marques, M.M. and Furtado, A.F.A.S. 1977. Esboço Pedológico da Ilha de S. Miguel (Açores). *Anais do Instituto Superior de Agronomia* **37**, 275–385.

Ricardo, R.P., Madeira, M.A.V. and Medina, J.M.B. 1979. Enquadramento taxonómico dos principais tipos de solos que se admite ocorrerem no Arquipélago dos Açores. *Anais do Instituto Superior de Agronomia* **38**, 167–180.

Rocha, G.P.N. 1990. A transição demográfica nos Açores. *Arqipélago (Revista da Universidade dos Açores, Ciências Sociais)* **5**, 125–168.

Rull, V., Lara, A., Rubio-Inglés, M.J., Giralt, S., Conçalves, V., Raposeiro, P., Hernández, A., Sábchez-López, G., Vázquez-Loureiro, D., Bao, R., Masqué, P. and Sáez, A. 2017a. Vegetation and landscape dynamics under natural and anthropogenic forcing on the Azores Islands: A 700 year pollen record from the São Miguel Island. *Quaternary Science Reviews* **159**, 155–168.

Rull, V., Connor, S.E. and Elias, R.B. 2017b. Potential natural vegetation and pre-anthropic pollen records on the Azores Islands in a Macaronesian context. *Journal of Biogeography* **44**, 2437–2440.

Santos, R.L. 1995a. Azores Islands. *In*: Santos, R.L. (ed) *Azoreans to California: A History of Migration and Settlement.* Alley-Cass Publications, Denair, CA, Section II, 3–33.

Santos, R.L. 1995b. Azorean Migration. *In*: Santos, R.L. (ed) *Azoreans to California: A History of Migration and Settlement.* Alley-Cass Publications, Denair, CA, Section III, 33–43.

Schaefer, H. 2005. *Flora of the Azores: A Field Guide.* Margraf Publishers, Weikersheim.

Schirone, B., Ferreira, R.C., Vessela, F., Schirone, A., Piredda, R. and Simeone, M.C. 2010. Taxus baccata in the Azores: A relict form at risk of imminent extinction. *Biodiversity and Conservation* **19**, 1547–1565.

Simões A., Carneiro A. and Diogo, M.P. 2012. Riding the wave to reach the masses: Natural events in the early twentieth century Portuguese daily press. *Science and Education* **21**, 311–333.

Soeiro de Brito, R. 1955. *I Ilha da São Miguel: Estudo Geográfico.* Instituto de Alta Cultura, Centro de Estudos Geográficos, Lisboa.

SREA 2011. *Censos 2011: Resultados Preliminares.* Serviço Regional de Estatística dos Açores, Açores.

SREA 2012. *Censos 2011.* Serviço Regional de Estatística dos Açores, Açores.

SREA 2019. *Turismo.* Serviço Regional de Estatística dos Açores, Açores, https://srea.azores.gov.pt/Conteudos/Relatorios/lista_relatorios.aspx?idc=6194&idsc=6712&lang_id=2

Teletin, A. and Manole, V. 2015. Expressing cultural Identity through saudade and dor: A Portuguese-Romanian comparative study. *In*: Baptista, M.M. (eds) *Identity: Concepts, Theories, History and Present Realities (A European Overview) Vol. 1* Grácio, Coimbra, 155–171.

Tiago, T., Faria, S.D., Cogumbreiro, J.L., Couto, J.P. and Tiago, F. 2016. Different shades of green on small islands. *Island Studies Journal* **11** (2), 601–618.

Triantis, K.A. Hortal, J., Amorim, I., Cardoso, P, Santos, A.M.C., Gabriel, R. and Borges, P.A.V. 2011. Resolving the Azorean knot: A response to Carine and Schaefer (2010). *Journal of Biogeography* **39**, 1179–1184.,

Tutin, T.G. 1953. The vegetation of the Azores. *Journal of Ecology* **41**, 53–61.

Tüxen, R. 1956. Die heutige potentielle natürliche Vegetation als Gegenstand der Vegetationskartierung. *Angewandte Pflanzensoziologie* **13**, 5–55.

Udías, A. 2009. Earthquakes as God's punishment in 17th and 18th century Spain. *In*: Kölbl-Edert, M. (ed) *Geology and Religion: A History of Harmony and Hostility.* The Geological Society, London, Special Publication **310**, 41–48.

Vidal de la Blache, P. 1903. *Tableau de la Géographie de la France.* Hachette, Paris.

Wallenstein, N., Duncan, A, Chester, D. and Marques, R. 2007. Fogo Volcano (São Miguel, Azores: A hazardous landform (*Le volcan Fogo, un paysage aux hazards dangereux). Géomorphologie: Relief, processues, environnement* **3**, 17–28.

Wallenstein, N., Chester, D., Coutinho, R., Duncan, A. and Dibben, C. 2015. Volcanic hazard vulnerability on São Miguel Island, Azores. *In*: Gaspar, J.L., Guest, J.E., Duncan, A.M., Barriga, F.J.A.S. and Chester, D.K. (eds) *Volcanic Geology of São Miguel Island (Azores Archipelago)*. Geological Society of London Memoir **44**, 213–225.

Wisner, B., Gaillard, J-C. and Kelman, I. 2012. Introduction to Part II. *In:* Wisner, B., Gaillard, J-C. and Kelman, I (eds) *The Routledge Handbook of Hazards and Disaster Risk Reduction*. Routledge, London, 171–176.

2 The tectonic and geological background

In terms of geological setting, it is of critical importance that the Azores archipelago straddles the Mid-Atlantic Ridge (MAR) and of the nine populated islands: Corvo and Flores, which lie to the west of the MAR; Faial, Pico, Terceira, Graciosa and São Jorge (the central group), which lie to the east of the MAR as do São Miguel and Santa Maria, which are located about 100 km to the SE of the central group (Figure 2.1). In this chapter, some reference is made to historic eruptions, but these are more fully considered in Chapter 3.

The tectonic background

The Azores are located at the triple point between the North American, Nubian and Eurasian plates (Figure 2.1). The islands sit on a region of thickened oceanic crust, referred to as the Azores Platform. The Azores Platform, delimited by the <2,000 m bathymetric contour (Needham & Francheteau

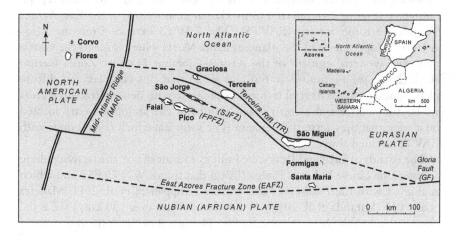

Figure 2.1 Azores: tectonic setting adapted from Carmo et al. 2015, Figure 6.1, p. 66. Copyright permission Geological Society of London, with additions and amendments.

DOI: 10.4324/9780429028007-2

1974), ranges in thickness from 8 to 20 km and formed over the last 20 Ma as a result of increased melt production (Beier et al. 2018; O'Neill & Sigloch 2018). The platform extends from Flores and Corvo, which are on its western margin, to Santa Maria in the east. O'Neill and Sigloch (2018) suggest that the Azores 'hot spot,' which marks the triple junction, is the result of a mantle plume. The nature of the plume is unclear, surface-wave models indicating that it is confined to the upper 250–300 km of the mantle, but it is of note that finite-frequency body-wave tomography suggests that the conduit may extend to the core-mantle boundary (see discussion in O'Neill & Sigloch 2018, pp. 83–84).

The East Azores Fracture Zone (EAFZ) formerly marked the boundary between the Nubian and Eurasian plates (Figure 2.1), but around 20 Ma ago this ceased to be the case and the EAFZ no longer acted as a transform fault (Miranda et al. 2018). Based on magnetic and geodetic data, Miranda et al. (2015) argue that much of the current differential motion between the Nubian and Eurasian plates is now accommodated by the Terceira Rift, an alignment of volcanic highs and tectonic basins that runs from Graciosa in the NW, through Terceira to the western part of São Miguel (Figure 2.1). McKenzie (1972) described the behaviour of this region as a 'leaky' transform fault, now commonly referred to as transtension (Mitchell et al. 2018). The Terceira Rift has a very slow spreading rate of around 3 mm/a (Miranda et al. 2015).

Although much of the differential movement occurs along the Terceira Rift, seismicity suggests that deformation is distributed widely across the region which includes Faial, Graciosa, Terceira and Pico (Mitchell et al. 2018), and in this area the margin between the Eurasian and Nubian plates is represented by a sheared zone rather than by a discrete boundary (Madeira et al. 2015). The Terceira Rift together with the *Faial-Pico* and *São Jorge Fracture Zones* form part of this sheared western segment of the Eurasian-Nubian plate boundary (Figure 2.1), referred to by Gaspar et al. (2015) as the East Azores Volcano-Tectonic System (EAVTS). The EAVTS extends from the Mid-Atlantic Ridge to east of the island of Santa Maria where it meets the Gloria fault, the western segment of the Azores-Gibraltar Fracture Zone. Carmo et al. (2015, see Figure 6.1 p. 66) also argue that this complex and diffuse plate boundary has a dextral transtensive regime with densely fractured areas where the central and eastern island groups of the archipelago are located on a zone that they term the Terceira Rift *sensu lato* which corresponds with EAVTS defined above.

The islands of Faial and Pico were built as a result of volcanic activity along a submarine Linear Volcanic Ridge (LVR) that trends WNW/ESE and which is aligned with the *Faial-Pico Fracture* Zone (Trippanera et al. 2014; Madeira et al. 2015; Miranda et al. 2018). At a distance of just over 130 km, Faial is the closest island to the Mid Atlantic Ridge. The island of São Jorge is the emergent portion of another WNW/ESE submarine LVR and is 27 km NE of, and parallel to, the Faial/Pico submarine LVR. Both these LVRs are parallel to the Terceira Rift (Hildenbrand et al. 2008). Whereas the distribution of the islands to the east of the MAR is controlled by the Terceira Rift and elongated volcanic ridges of the EAVTS, Corvo and Flores lie outside this system and are located on a ridge that runs parallel and to the west of the MAR.

Magma genesis

In the Azores magma is likely to have been generated adiabatically through the melting of ascending mantle material that had excess temperature (~35°C) relative to the ambient mantle, ascent being facilitated by presence of volatiles (H_2O, CO_2), i.e. it was a 'wet spot' (Beier et al. 2018). Sr-Nd-Pb-Hf isotope signatures indicate that magmas reflect source heterogeneity with mixing from variably enriched and depleted mantle components (Zanon 2015; Beier et al. 2018).

The volcanic rocks of Flores and Faial, both located only 120–150 km from the Mid Atlantic Ridge, show a less enriched isotope signature reflecting a significant input from depleted Mid Ocean Ridge Basalt (MORB) type mantle source that was mixed with a component from an enriched mantle. This is in contrast to São Jorge and Terceira, where volcanic products show a signature trending towards an HIMU (High μ) mantle component, which is suggestive of binary mixing between a common MORB mantle source and recycled oceanic crust (Zanon 2015). São Miguel shows a marked variation in its isotopic signature from west to east; with the volcanic products of Sete Cidades volcano in the west showing an isotope signature indicative of a mantle source typical of the Azores, whereas the products of Nordeste volcano in the east volcano reflect an unusually enriched, highly radiogenic source (Zanon 2015; Beier et al. 2018). Beier et al. suggest that the highly enriched source beneath the east of the island may be due to recycled oceanic crust combining with small amounts of evolved lavas probably from a sub-ducted seamount. The small scale of source mantle heterogeneity beneath São Miguel is demonstrated by the fact that Sete Cidades and Fogo volcanoes are only 30 km apart yet have differing source compositions (Beier et al. 2018).

Beneath Pico and Faial in the central group of islands, it is estimated by Zanon et al. (2013) that partial melting of the mantle occurred within a pressure range of 3–4 GPa (~95–130 km depth) and, as the lithosphere thickens eastwards, it is likely that initial melting would have occurred at higher pressures at São Miguel and Santa Maria. Magmatism in the Azores is characterised by low degrees of partial melting – typically in the order of 3%–5% – and with lower values of <4% for São Miguel and Santa Maria, and this generated primary magmas with MgO values ranging from 15% to 21% wt% (Larrea et al. 2018). These mafic magmas ascended and ponded at the Moho boundary, underwent fractional crystallisation and erupted predominantly basaltic lavas from fissure zones. In places where the tectonically controlled fissure zones are intersected by major faults, shallow magma reservoirs may have formed leading to the development of central volcanoes. As these central volcanoes mature, the shallow magma reservoirs may have undergone extensive fractional crystallisation and generated more silicic magmas of trachytic composition that gave rise to major explosive eruptions of sub-Plinian or Plinian scale and which often led to caldera collapse. These explosive eruptions deposited thick blankets of pumice lapilli and ash on the islands affected. For a fuller discussion see Zanon (2015) and Chapter 3.

The volcanic rocks of the Azores belong to the Alkalic Series (Na/K > 1) ranging from basalt to trachyte (see Table 2.1). On some islands, notably Terceira and Flores, the trend is peralkaline in character (Zanon 2015). Petrographically, the mafic rocks are often porphyritic with phenocrysts of olivine and clinopyroxene, and increasing amounts of plagioclase from alkali basalts to hawaiites. Megacrysts of xenocrystic origin also occur. Mugearites are typically porphyritic with plagioclase and scarce olivine and clinopyroxene, kaersutite amphibole is sometimes present as microphenocrysts. More evolved lavas, benmoreites and trachytes tend to be aphyric in character (for further details on petrography, see Zanon 2015, pp. 53–56). Plutonic xenoliths occur in the volcanic rocks of the Azores (Larrea et al. 2018). Ultramafic xenoliths (such as lherzolites, harzburgites, dunites, wherlites) are found on São Miguel, Terceira, Graciosa, Faial and Pico. These are mantle derived, with the exception of ultramafic nodules from Sete Cidades volcano, which are considered to have been formed by fractional crystallisation from mafic magma (Almeida 2001; Larrea et al. 2018). Gabbroic xenoliths are also found in São Miguel, Terceira, Graciosa, Faial, Pico, Flores and Corvo, these

Table 2.1 An explanation of volcanological terminology used in this and other chapters on styles and types of volcanic eruptions, their products and rock types

---

Eruptive products

A volcano forms when magma (molten rock material) is erupted at the Earth's surface. Compositionally mildly alkaline magmas range from: **mafic** (45–52 wt% SiO_2), that erupt as *basalts*; **intermediate** (52–63 wt% SiO_2) which erupt as *mugearites*, <57 wt% SiO_2, *benmoreites*, >57 wt% SiO_2, and **silicic** (> 63 wt% SiO_2), forming *trachytes*.

The nature of an eruption and the distribution of its products depend on the composition and volume of the magma involved. There are three types of volcanic product: (1) *lava* – the eruption of largely degassed magma as typically a silicate flow; *pyroclasts* – material produced by explosive discharge which comprises melt together with solid *lithics* material ripped from the walls of the conduit, *tephra* is often used as a term for pyroclastic deposits and (3) gas.

Lava Flows   Flows are commonly basaltic and there are two main types of morphology: *aa* and *pahoehoe*. These terms are derived from Hawaiian words (Harris and Roland 2015). *Aa* flows have a broken-up surface of lava fragments rafted on a mobile interior of fluid lava. As the lava progresses downslope, the active central portion tends to drain leaving a channel bounded by static levées of lava fragments. *Pahoehoe* lava flows have quite a different morphology with a coherent, smooth, glassy surface sometimes with a ropey texture. The fluid lava is transported internally often within tubes (Harris and Rowland 2015). Trachytic lava tends to be viscous and on eruption often forms a thick localised lava dome.

Pyroclasts These comprise fragmental material defined by clast size with: *ash* < 2 mm < *lapilli* < 64 mm < *blocks/bombs*. Bombs are discharged from the vent as projectiles, whereas pyroclastic fall material drops from the volcanic plume downwind of the vent mantling the topography as beds which become thinner and finer grained with distance. *Pyroclastic density currents* (PDCs), typically flows of hot gas and volcanic fragments, are formed by collapse of eruptive columns and by lateral blasts. Unconsolidated pyroclastic material deposited on the flanks of active volcanoes can often be reworked following heavy rainfall forming volcanic mudflows (often termed *lahars*).

Gas Gases discharged by volcanoes are often regarded as the volatile component as they are largely lost to the atmosphere. The gas comprises juvenile volatiles derived from melting of mantle material but also includes an input from the crust particularly some of the H_2O derived from interaction with groundwater. Volcanic gases include H_2O, CO_2, SO_2, H_2S and Radon (Rn), among others. Volcanic gases can react in the atmosphere to form aerosols such as sulphuric acid (H_2SO_4), hydrochloric acid (HCl) and hydrofluoric acid (HF). Gases such as CO_2 and Rn pose a hazard to human health and can seep from craters and the edifice of volcanoes even during non-eruptive periods (see Williams-Jones & Rymer 2015).

Styles of eruption and types of volcano

Classification of volcanic eruptions was traditionally based on eruptions at 'type' volcanoes (Macdonald 1972). The main types of eruptive styles are Hawaiian; Strombolian; Surtseyan; Vulcanian and Plinian (Walker 1973).

Hawaiian Eruptions of this style involve effusive discharge of mafic magma often with continuous jets of basaltic liquid generating lava fountains and feeding low-viscosity lava flows. This style of activity takes its name from Hawaii and has been well displayed by recent eruptions of Kilauea volcano (Houghton et al. 2016). Volcanoes with predominantly Hawaiian style eruptions typically build up shallow sloping edifices of lava called *shield volcanoes* (e.g. Kilauea and Mauna Loa volcanoes in Hawaii).

Strombolian Eruptions of a Strombolian style typically involve mafic magma with frequent small explosions (Houghton et al. 2016). This type of eruption is named after Stromboli volcano in the Eolian Islands in Italy. Strombolian eruptions are often 'single' eruption events which build up a monogenetic *cinder cone*. Lava flows are often erupted from vents at the foot of the cone.

Surtseyan Named after the birth of island of Surtsey, offshore the southern coast of Iceland on 1963–1964, this eruptive style was earlier described in Capelinhos 1957–1958 off Faial in the Azores, where this occurs through a basaltic eruption into shallow seawater. Its behaviour reflects magma coming into contact with an external source of water. This generates explosive eruptions described as *phreatomagmatic* or *hydromagmatic/hydrovolcanic*. Phreatomagmatic eruptions form *tuff cones* and *tuff rings* which have wider craters than cinder cones, and tuff rings tend to also have lower edifices with the outer slopes having a lower inclination (de Silva & Lindsay 2015).

Eruptive products

Vulcanian   Vulcanian eruptions are associated with intermediate to silicic
magmas. This style of eruption takes its name from the Italian
island of Vulcano, also one of the Eolian Islands and is based on
the classic 1888–1890 eruption of Fossa Volcano (Guest et al.
2003, pp. 129–133). This eruption was characterised by discrete
short-lived explosions, which during vigorous activity, occurred
every minute or so. Volcanic plumes up to 5000 m in height
were generated, and some of the larger paroxysmal explosions
created shock waves. They are also frequently associated with
phreatomagmatic activity.

Plinian   Eruptions of this type involve high-velocity discharge of typically
silicic magma (trachytic/rhyolitic) over a period of hours as
a jet of gas, solid fragments and liquid particles generating
a gas thrust phase during which air is entrained and heated.
When the jet reaches the apex of its trajectory the heated air
incorporated into the plume allows it to rise as a buoyant
convective phase until it reaches its maximum height (Cioni
et al. 2015). Pliny the Younger described this spectacle in his
observation of the 79 CE eruption of Vesuvius (Italy), the
classic Plinian eruption, and this style of activity is named after
him (Guest et al. 2003, pp. 35–45). As eruptions wane and the
diameter of the vent increases through erosion, the gas thrust
phase becomes less vigorous the eruptive column will then
begin to collapse leading to the formation of pyroclastic density
currents (PDCs). Included in the general category of Plinian
eruptions are two sub-categories: *sub-Plinian* and *Ultraplinian*.
Sub-Plinian eruptions are smaller in size and plumes do
not breach the tropopause, but otherwise share the general
characteristics as Plinian eruptions (Cioni et al. 2015), although
tending to be more unsteady and involving explosive pulses
followed by phases of collapse and the generation of PDCs. The
1630 eruption of Furnas volcano (Azores) is a good example of
a trachytic sub-Plinian eruption (Cole et al. 1995). The term
Ultraplinian is used for large Plinian eruptions.

Volcanoes that form through a mixture of major Plinian style
eruptions together with less explosive eruptions build up a
steep-sided composite volcanic edifice made up of lava and
tephra horizons termed a *stratovolcano*. Major explosive Plinian
style eruptions cause withdrawal of large volumes of magma
from high-level magma reservoirs leading to collapse of the
summit of the volcano forming large (>2 km) craters which are
termed *calderas*.

The relationships between VEI (Volcanic Explosivity Index), eruption type, volume of erupted material, and eruptive column height are shown in the diagram below (adapted from Siebert et al. 2010)

	0	1	2	3	4	5	6	7	8
VOLCANIC EXPLOSIVITY INDEX (VEI)	0	1	2	3	4	5	6	7	8
MAGNITUDE* (M)					4.4	5.3	6.3	7.7	8.2
GENERAL DESCRIPTION	Non-explosive	Small	Moderate	Moderate/ large	Large	Very large	———————————→		
VOLUME (m³)		10^4	10^6	10^7	10^8	10^9	10^{10}	10^{11}	10^{12}
ERUPTION COLUMN HEIGHT (km) Above the crater	< 0.1	0.1 - 1	1.5						
Above sea level				3 - 15	10 - 25	> 25			
ERUPTION TYPE	Hawaiian/Strombolian			Vulcanian	Sub-Plinian		Plinian	Ultra-plinian	

* Magnitude is based on the mean magnitude for eruptions for VEI's 4 - 8 for eruptions in the LaMEVE database (Crossweller *et al.* 2012).

The use of VEI is designed as a measure of explosivity and does not capture the impact of long-lasting basaltic effusive eruptions such as that of the 1672–1673 eruption of Faial. Also, the VEI diagram reflects eruptions governed by the composition of the magma (chemistry of the melt and gas content) and the volume involved. For consideration of magnitude see Crossweller et al. 2012.

VEI table copyright permission John Wiley & Sons, Newhall and Self (1982, Table 1, p. 1232).

gabbroic nodules probably deriving from cumulates formed by fractional crystallisation in magma reservoirs. Syenite nodules occur in volcanics from São Miguel, Terceira, Graciosa, Faial and Flores, with these nodules being derived from crystallisation of trachytic magma bodies within the plumbing system.

The volcanic islands

The islands are considered beginning first with the eastern group that contains Santa Maria, the oldest island, together with São Miguel, then the central group with Pico, Faial, São Jorge, Terceira and Graciosa and finally the western group to the east of the MAR, Flores and Corvo.

Santa Maria

Santa Maria is the oldest of the Azores islands and has not been volcanically active in historic times. The volcanic rocks of Santa Maria are the most alkaline of the Azores reflecting both a small degree of partial melting and the nature of the mantle source region. The volcanic edifice of Santa Maria emerged above sea level in the late Miocene around 6 Ma year ago, and the oldest exposed volcanics reflect both submarine Surtseyan and subaerial

Strombolian styles of eruptive activity (Ramalho et al. 2017; Ávila et al. 2018), for information on styles of volcanic activity see Table 2.1. This was followed by the development of a first shield volcano, the Anjos Volcanic Complex (5.8–5.3 Ma), building on what is now the western part of the island, and after this there was a period of quiescence. Subaerial and marine erosion, together with terrestrial plus marine sedimentation with some synchronous volcanic activity, gave rise to the Touril volcano-sedimentary complex which is of early Pliocene age and dates from 5.3 to 4.1 Ma (Ramalho et al. 2017; Larrea et al. 2018). Submarine activity then occurred to the east which subsequently broke the surface of the ocean to form a new island and built a shield volcano: the Pico Alto Complex. The Pico Alto Complex (4.1–3.5 Ma) is elongated NNW-SSE and sits astride the eastern portion of the older Anjos Volcanic Complex (Ramalho et al. 2017). Following the activity of the Pico Alto Complex, there was a period of subaerial and submarine erosion with incision producing coastal platforms and associated alluvial sediments. The island was also uplifted at this time (Ramalho et al. 2017; Larrea et al. 2018, pp. 198–203).

The final phase of volcanic activity involved the eruption of monogenetic subaerial and hydromagmatic cones, which together are known as the Feteiras Formation, with volcanic activity ending about 2.8 Ma (Ramalho et al. 2017). Marques et al. (2020) consider that the basaltic cones of the Feteiras Formation were continuing activity of the Pico Alto Volcano. After this time, Santa Maria entered a long phase of uplift and erosion with the development of a sequence of marine terraces with heights ranging from 7–11 to 210–230 m above present-day sea level (Larrea et al. 2018).

A marine geophysical investigation in 2016 undertook a high-resolution bathymetric survey of the marine shelf and the upper flanks of its submerged edifice (Ramalho et al. 2020). The survey revealed the presence of a volcanic cone on the seabed about 4 km offshore on the NE sector of the marine shelf. On the basis of a detailed analysis of the morphology of the feature, Ramalho et al. interpreted this cone as a tuff-ring which became emergent during eruption. They argue that eruption immediately preceded a rapid drop in sea-level allowing subaerial consolidation. Subsequently the rapid rise in sea-level during the Late Glacial Termination prevented the cone from being removed by marine erosion. Ramalho et al. (2020) suggest on this basis that the cone might be very late Pleistocene in age so implying that volcanism on Santa Maria is much younger than *c.*2.8 Ma, which is the generally accepted terminal date.

In terms of hazard, Santa Maria has not been volcanically active in historic times. If as indicated above Santa Maria has been active in the very late Pleistocene, then there is a possibility of Surtseyan type activity having hazard implications in the future (Ramalho et al. 2020). The island could also be inconvenienced by ash fall if there were a major explosive eruption on São Miguel. Ash could seriously affect the operation of the island's airport and the internationally vital *Santa Maria Oceanic Control Centre* which manages air traffic from Europe to Central and South America (Anon 2020)[1]. In terms of earthquakes, the island lies on the southern margin of the Azores epicentres zone and has not been the site of a destructive earthquake during historic time (Gaspar et al. 2015), but has suffered damage from off-shore earthquakes in 1937 and 1939 (Gaspar et al. 2015 – see Chapter 3).

São Miguel

São Miguel is made up of three active central volcanoes that from west to east are Sete Cidades, Fogo (also known as Água de Pau) and Furnas, and which are linked by fissure zones of predominantly basaltic volcanism (Figure 2.2). Each has undergone phases of trachytic Plinian-type eruptions (see Table 2.1) leading to caldera collapses and the deposition of pumice fall deposits which mantle much of the island. The eastern part of the island comprises two older volcanic systems, Povoação and Nordeste, which probably range in age from 100 ka to 4 Ma (Duncan et al. 2015). São Miguel is located towards the eastern end of the Terceira Rift and is characterised by two dominant tectonic structures (Carmo et al. 2015): the first ranging from NW–SE to WNW–ESE and, a second, NNW–SSE to N–S with central volcanoes located at the intersection of faults. The rift zones with monogenetic cones and fissure systems are aligned mainly on the NW–SE to WNW–ESE trending structures.

Sete Cidades volcano is situated at the western end of the island, has a subaerial diameter of about 12 km and rises to an altitude of 845 m asl at Pico da Cruz. There is a well-developed summit caldera (see Table 2.1) with a width of ~5 km and average depth of 350 m. Pumice cones, maars and domes occur within the caldera (Queiroz et al. 2015). The edifice is associated with tectonic structures that run NW–SE, and these are intersected

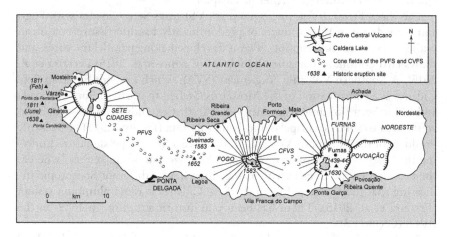

Figure 2.2 Volcanoes of São Miguel and the location of historic eruptions. This figure was published in Guest et al. 1999, Figure 1, p. 2, Copyright permission Elsevier, with additions and amendments.

by the N-S Feteiras Fault. The dominant structural feature is the Mosteiros Graben (aligned with the Terceira Rift), which extends SE from the coast at the village of Mosteiros, up the flank of the volcano to cut its caldera rim. The volcano emerged above sea level around 200 ka, and initially Sete Cidades was an island, that was subsequently joined to the volcanic edifice of Fogo some 20 km to the E by basalt fissure activity. This created a 'ridge' above sea level, that was formerly termed the 'waist zone' and which connects the two volcanoes (Booth et al. 1978), but is known today as the *Picos Fissural Volcanic System* (PFVS) (Ferreira et al. 2015). The subaerial history of Sete Cidades dates back, therefore, to around 200 ka (Moore 1991), but early products are poorly exposed and difficult to correlate. Major caldera collapse occurred about 36 ka and was associated with a large trachytic eruption, the Risco Formation (Queiroz et al. 2015). There was a further caldera forming eruption around 29 ka, and this was followed by the final phase of caldera collapse around 16 ka which was associated with the paroxysmal Santa Bárbara eruption (Queiroz et al. 2015). This was a highly explosive trachytic eruption triggered by injection of basaltic melt into a high-level trachytic magma reservoir (Porreca et al. 2018), with the climactic phase generating pyroclastic density currents that flowed down the SW and NE flanks of the volcano. Over the last 5 ka, Sete Cidades has been characterised by hydromagmatic explosive eruptions from the caldera, and Queiroz et al. (2015) have identified 17 such events, the youngest being dated by radiocarbon at around 600 years and just before the island was permanently settled (see Chapter 1). These eruptions, which have been relatively frequent in the last 5 ka, pose a threat to local communities (Cole et al. 2008; Kueppers et al. 2019 – see Chapter 4). Though there have been no historic eruptions of Sete Cidades, there have been eruptions just offshore in 1638 and 1811, and these are considered further in Chapter 3.

The Picos Fissural Volcanic System (PFVS) links Sete Cidades with Fogo and is made up of the products of predominantly basaltic fissure volcanism with scoria cones and eruption of lavas developed along parallel fractures, and shows two dominant volcanic alignments (Carmo et al. 2015; Ferreira et al. 2015). A Southern Volcanic Alignment (SVA), which extends SE from Sete Cidades as a continuation of the Mosteiros Graben to north of Ponta Delgada and a Northern Volcanic Alignment (NVA), that extends from Fogo Volcano westwards to south of Capelas and Fenais da Luz (Figure 2.2). The volcanic products of the Northern Volcanic Zone have a Sr-isotope signature implying that they were fed by magma from the Fogo plumbing system, whereas the products of the Southern Volcanic Zone have a different Sr-isotopic indicating that they were fed from the Sete Cidades magma plumbing system (Ferreira et al. 2015). There has been one historic eruption in 1652 on the PVFS (Chapter 3).

Fogo is the largest active volcano on São Miguel and dominates the island and its principal centres of population (Figure 2.2). The volcano is located where the Ribeira Grande graben, which is oriented NW-SE, and associated structures intersect the PFVS. The volcano rises to an altitude of 947 m asl at Pico do Barrosa. The volcano is truncated by a summit caldera about 3 km in diameter within which there is a lake: the Lagoa do Fogo (Wallenstein et al. 2015). The oldest dated rocks of Fogo are submarine basalts with an age of around 280 ka (with 50% uncertainty) which were recovered from a bore hole on the lower northern flank of the volcano; the oldest subaerial volcanic rock from the bore hole at a shallower depth was dated at around 117 ka (Muecke et al. 1974) and the emerging volcano must have built up above sea level in the north between these dates. Older subaerial volcanics with a date around 181 ka occur on the southern flank of Fogo (Gandino et al. 1985; Moore 1991), and it maybe that the graben on the northern flank delayed emergence of that sector of the volcano. The older subaerial volcanic products are poorly exposed and include trachytic ignimbrites and lavas, but the sequence of the last 40 ka is better understood (Wallenstein et al. 2015). For the period from 40 up to 5 ka, volcanic activity consisted of numerous explosive eruptions generating trachytic tephra fall and ignimbrites, and included the paroxysmal Ribeira Chã eruption (between 8 and 12 ka), that produced a thick ignimbrite and which includes a well-developed welded unit. The Fogo A (~ 4.6 ka) eruption was a classic Plinian event which gave rise to the most widespread recognisable fall deposit in the Azores making it a valuable marker horizon. The Fogo A deposits were first described in detail by Walker and Croasdale (1971) and since then have been the subject of numerous investigations (for details, see Pensa et al. 2015; Wallenstein et al. 2015). Four trachytic eruptions have occurred from the volcano since Fogo A, the most recent (and largest) of which was the historic sub-Plinian eruption of 1563 (see Chapter 3). There have also been basaltic (scoria cones and lava flows) and trachytic (domes) eruptions on the flanks of Fogo, including an historic eruption that also occurred in 1563 and which is associated with the sub-Plinian event. A basalt lava flow was generated and reached the north coast causing damage to the village of Ribeira Seca (see Chapter 3). On the northern flank of Fogo, there is an active hydrothermal system creating a geothermal reservoir with a liquid-dominated system with temperatures in excess of 240°C. This is exploited by two plants generating *c.*42% of São Miguel's electricity requirements (Rangel et al. 2015; Chester & Duncan 2018).

The *Congro Fissural Volcanic System* (CFVS), also known as the Achada das Furnas fissure zone (Figure 2.2), extends eastwards from Fogo some 7.5 km to the Furnas central volcano (Ferreira et al. 2015; Zanon 2015). The zone comprises mainly basaltic cones and lava flows together with lesser quantities of trachytic products forming pumice cones, domes and tephra mainly dating from 30 to 10 ka. There have been two more recent eruptions:

1   The Congro trachytic phreatomagmatic eruption, that formed a crater (now occupied by a lake) and a small trachyte dome. The dome was the last stage of the eruption and is dated between 3 and 4 ka (Booth et al. 1978).

2   Formation of a basaltic cone that erupted lavas and which lies on top of and is, therefore, younger than the Congro eruption products.

Since the beginning of the twenty-first century, the central zone of the CFVS has been one of the most seismically active regions of the Azores. Seismic crises have occurred in 2005 and in 2011–2012 and were associated with ground deformation (Ferreira et al. 2015).

Furnas central volcano (Figure 2.2) is located where two fault trends, north-south and NE–SW, intersect the trend of the CFVS (Carmo et al. 2015). Furnas, unlike Sete Cidades and Fogo volcanoes, does not have a well-defined edifice with a summit caldera but rather has a caldera complex which cuts into the western flank of the older Povoação volcano (Guest et al. 2015). Guest et al. argue that the lava pile, that extends from north of the Main Caldera to the coast, represents the Basal Lavas of the Furnas construct (~ 95 ka). The volcanic edifice was not only built through explosive activity producing trachytic pyroclastics but also included some trachyte lavas. At around 30 ka, there was a major caldera collapse forming the Main (outer) caldera and associated with this was an eruption that emplaced an extensive ignimbrite showing a distinctive welded facies and which is particularly well exposed in the Povoação caldera. This has been named the Povoação ignimbrite (Duncan et al. 1999). After the caldera collapsed a volcanic sequence, including the products of major explosive eruptions, formed the caldera fill. Following this, there was another episode of caldera collapse around 12 ka, which cut the chain of the distinctive Pico do Ferro trachyte domes to form the Inner Caldera (Guest et al. 2015). The last 5 ka of activity has ten trachytic eruptions from vents on the caldera floor, these eruptions typically involving magmatic explosive activity alternating with phases of hydromagmatic explosive activity and ending with the emplacement of a trachyte dome. The last two eruptions occurred in historic times, in 1439–1443 during initial settlement of the island and in 1630, and these are considered further in Chapter 3. There is an active hydrothermal system beneath the Inner Caldera, and fumarolic activity occurs both near to the shore of Furnas lake and on the edge of Furnas village, with diffuse discharge of CO_2 posing a threat to the local community (Viveiros et al. 2015 – see Chapters 3 and 4).

To the east of Furnas lies the inactive volcano of Povoação, which has an impressive caldera cut into a lava shield (Duncan et al. 2015). The basaltic lavas of the shield are well exposed in sea cliffs to the west of the village of Povoação. Post-caldera products are poorly exposed. The outer slopes of the Povoação edifice reach down to the north coast are less dissected than the slopes of the older volcano of Nordeste to the east (Figure 2.2). Nordeste is a highly eroded basaltic volcanic edifice with a small volume of trachytes and tristanites (a K-rich trachytic rock) generated in the last stage of activity (Fernandez 1980). There is no evidence of a summit caldera. Inland exposures are restricted, but the lava pile is well exposed in the high cliffs on the east coast of the island. Dating of the lava pile by Abdel-Monem et al. (1975)

gives an age range between 4 to 1.86 Ma, but more recent dating places the range between 0.88 and 0.80 Ma (Johnson et al. 1998), which would make Nordeste similar in age to the other central volcanoes of São Miguel. This dating conflicts with the morphological evidence, however, which implies that it is much older. For a discussion on this age discrepancy see Duncan et al. (2015). The NW–SE tectonic trend continues to the east of the island where the Água Retorta graben cuts the Nordeste edifice on the southern part of the east coast of the island.

São Miguel is affected by both tectonic and volcanic earthquakes. The whole island is situated within the zone of seismic epicentres that were recognised by Gaspar et al. (2015) between 1980 and 2013. Earthquakes exceeding an Intensity of IX have struck the island in 1713, 1811 and 1935, the largest earthquake with a maximum Intensity X destroying the then capital, Vila Franca do Campo, in 1522. Earthquake damage in association with volcanic eruptions occurred during the 1630 eruption of Furnas and eruptions off the coast of Sete Cidades in 1638 and 1811. The impact of earthquakes on São Miguel and its population is considered in more detail in Chapters 3 and 5.

Pico

The island of Pico is dominated by its distinctive stratovolcano that at 2351 m asl is the highest mountain in Portugal. Pico and Faial are both emergent islands on the submarine Linear Volcanic Ridge which is aligned with the Faial-Pico Fracture Zone. Pico is made up of three volcanic complexes: *Montanha*, the youngest forming the Pico central volcano; *São Roque-Piedade* and *Topo-Lajes*, the oldest which dates to at least 250 ka (Nunes et al. 1999; Cappello et al. 2015). The Pico stratovolcano, dominates the western part of the island, has a summit crater 500 m in width, is largely filled by pahoehoe lavas and has a lava cone in the NE corner which displays fumarolic activity (Woodhall 1974). The volcano comprises products of the Montanha Volcanic Complex and is covered by more recent lava flows. The basaltic products of the São Roque-Piedade Volcanic Complex were erupted along a WNW trending fissure system, the Planalto da Achada (Cappello et al. 2015), that extends for *c.*30 km along the eastern spine of the island and wraps around the northern side of the older Topo-Lajes Volcanic Complex (Woodhall 1974; França et al. 2006a). Most of the lavas on Pico are basaltic and more differentiated compositions are uncommon. It is of note that the early products of the 1718 (Santa Luzia) eruption on the NW flanks of the stratovolcano were mugearites and benmoreites (see Table 2.1) followed by basalt (Woodhall 1974; Cappello et al. 2015 – see Chapter 3). França et al. (2006a) argue that the genesis of these more evolved magma could reflect initial stages of high-level magma storage and differentiation within the volcanic system. Zanon et al. (2020) have carried out a detailed investigation of the magma plumbing system of Pico volcano using detailed petrological studies together with seismic monitoring and the role of three intersecting tectonic systems. They demonstrate that ascending magma ponds above the Moho Transition Zone at nearly 18 km depth. The Moho Transition Zone consists of ultramafic cumulates which overlie mantle rocks. Depending on the ascent path magma rising through the thickened oceanic crust can pond at several levels. Zanon et al. argue that ponding that occurs at a shallow depth of 4–7 km beneath the southern flank of Pico volcano may be the early stages of the development of a more complex magma reservoir (p. 15).

There are two main fault systems on Pico (Madeira 2005). The WNW-ESE structures define a graben which is aligned with the Pedro Miguel graben on Faial and cuts through the central and eastern parts of the island. The graben is buried in the western part of the island by volcanic products of the younger (<10 ka) Pico stratovolcano. The other NNW-SES trending faults are less abundant and typically reflect the alignment of volcanic vents and fissures (Madeira 2005). On Pico, in addition to tectonic movements, there is evidence of geomorphological instability. There have been flank failures on both the north and south sides of Pico, and Costa et al. (2014) describe submarine debris on the ocean floor of the São Jorge channel which they consider to have been derived from two scars on the north coast of the island. They consider this collapse to have had a minimum age of 70 ka and to have involved several km^3 of material, which generated a tsunami and impacted the south coast of São Jorge. Slumping has also been described in association with the old Topo-Lajes Volcanic Complex that is exposed on the south coast (Hildenbrand et al. 2012).

On Pico there have been four onshore eruptions during historic times (Figure 3.8):

1   From 1562 to 1564, on the spine of the São Roque-Piedade Volcanic Complex.
2   On the 1 February 1718 on the northern slopes of Pico Mountain, with lava moving towards Santa Luzia.
3   On 2 February, a third eruption began on the lower SE slopes of the mountain near to the village of São João.
4   The most recent eruption of 1720 occurred at Silveira at the coast just west of Lajes do Pico.

The eruptions of 1718 and 1720 occurred in association with the Montanha Volcanic Complex. The nature and impacts of these historic eruptions are considered in Chapter 3.

Sitting astride the Faial Pico Fracture Zone it is not surprising that Pico is seismically active. The October 1973–May 1974 earthquake swarm affected north Pico. The principal earthquake occurred on 23 November 1973 (Magnitude 5.5 and maximum Intensity VII) and had an epicentre 2 km SW of Santa António (Figure 3.2) on the north coast of Pico (Madeira 2005; Gaspar et al. 2015). The earthquake caused considerable damage on Pico with around 2,000 houses being destroyed, and 5,000 people were made homeless (Madeira 2005). In 1926 an earthquake with its epicentre in the sea between Faial and Pico caused damage on the west coast of the island. The 1757 earthquake on São Jorge (Maximum Intensity X), caused the church at Piedade to collapse and 11 people were killed on Pico (Madeira 2005; Gaspar et al. 2015). Pico was also affected by the 1998 earthquake off the NW coast of Faial. The impact of earthquakes on Pico is discussed in detail in Chapter 3 and responses to them in Chapter 5.

Faial

The subaerial portion of the Faial volcanic complex has been built up over the last 1 Ma (Hildenbrand et al. 2012) and a summary of its geological history has been published by Madeira (2005) and Trippanera et al. (2014). Recent interpretations of the geology of Faial are based largely on the geological map of Serralheiro et al. (1989). The oldest rocks on Faial formed a now eroded volcanic shield, the *Ribeirinha Volcanic Complex* (850–580 ka), which is exposed at the east end of the island and is cut by the Pedro Miguel graben. It remains active. The Pedro Miguel graben trends WNW-ESE aligning with the principal tectonic system of Faial. A second set of NNW-SSE conjugate faults intersects the principal tectonic system in the centre of the island, and it is here that the Caldeira Volcano has developed (Trippanera et al. 2014). The Caldeira Volcano (1043 m asl) is the highest point on the island and has a summit caldera 2 km wide and about 400 m deep.

The products of the Caldeira Volcano form the *Cedros Volcanic Complex* which ranges in age from 580 ka to the present and is divided into two units: the Lower Group and Upper Group. The Lower Group comprises basaltic to benmoreitic lavas which are exposed in the northern and southern cliffs of the island. There was then a change from effusive basaltic to trachytic explosive eruptions and this may reflect the development of high level (i.e. 3–5 km depth) ponding of magma leading to the evolution of a more silicic chemistry (Zanon and Frezzotti 2013; Zanon et al. 2013; Pimentel et al. 2015). The Upper Group is made up of trachytic pyroclastics derived from the 12 explosive eruptions (C1–C12), which have occurred during the last 16 ka of activity. The C11 event is notable, having a minimum eruptive volume of 0.1 km^3 (Dense Rock Equivalent), it is dated at *c.*1 ka and was one of the last major explosive eruptions in the Azores (Pimentel et al. 2015). The principal sub-Plinian phase of the eruption produced a column estimated at 14 km in height and thick pumice fall beds were deposited on the northern flank of the volcano. This phase was followed by prolonged pyroclastic fountaining which generated PDCs and deposited ignimbrites on the northern and eastern flanks (Pimentel et al. 2015). An eruption of this scale on a small island such as Faial would pose a significant threat. The *Almoxarife Formation* was a product of basaltic fissure activity which built up a basaltic platform on the SE corner of the island. With a poorly constrained age range 30 ± 20–11 ka, it is broadly contemporaneous with the boundary between the Upper and

Lower Groups of the *Cedros Formation* (Trippanera et al. 2014). To the west, the basaltic fissural activity of the *Capelo Formation*, which is aligned WNW-ESE and controlled by the main tectonic system, has constructed the north western peninsula of the island over the course of the last 8 ka (Di Chiara et al. 2014). There have been six eruptions on the Capelo peninsula in the last 2,000–3,000 years making this an area of significant volcanic activity and the site of the two historic eruptions (Di Chiara et al. 2014 – see Chapter 3).

Overall, Romer et al. (2018) argue that the periodicity of volcanicity on Faial with pulses of activity lasting less than 30 ka, followed by periods of quiescence has possibly been governed by tectonic movements (Hildenbrand et al. 2012), implying that different volcanic units have distinct primary magmas and plumbing systems. It appears that over time the degree of partial melting decreases progressively and, as melt supply declines, so volcanism becomes more localised (Romer et al. 2018).

Historic eruptions of Faial are considered in detail in Chapter 3 and occurred in 1672–1673, the *Cabeço do Fogo eruption* and in 1957–1958, the *Capelinhos eruption*, which is also associated with a small explosion event in the Caldeira Volcano caldera. The *Cabeço do Fogo eruption* was a subaerial effusive event and the *Capelinhos eruption* was initiated by Surtseyan activity offshore and subsequently forming an isthmus connecting it to the western end of Faial, later evolving to Strombolian and Hawaiian styles of activity.

In common with Pico, Faial is located on the *Faial-Pico Fracture Zone*, which is a zone of active seismicity, following settlement in the fifteenth century it was not, however, until the twentieth century that there are any major earthquakes were recorded. Major tectonic earthquakes occurred in 1920, 1924, 1926 and 1998 (Madeira 2005), with the 1926 and 1998 earthquakes causing significant damage and some fatalities (see Chapter 3).

São Jorge

São Jorge is a distinctive elongated island situated on a submarine Linear Volcanic Ridge which reflects tectonic alignment with the orientation of the Terceira Rift. Recent volcanic activity on São Jorge also shows the same tectonic control with a WNW-ESE trend along the *São Jorge Fracture Zone*. Forjaz and Fernandes (1975) identify three volcanic systems made up of basaltic to mugearitic products: *Serra do Topo* is the oldest and forms the eastern portion of the island; and this is separated from the younger *Rosais* and *Manadas* systems in the west by the NNW-SSE trending Ribeira Seca Fault. Further field work and K/Ar dating has led to a reinterpretation of the geological relationships (Hildenbrand et al. 2008; Marques et al. 2018). Marques and his colleagues argue that evidence for the occurrence of the Ribeira Seca fault is lacking, and it is not consistent with the volcanic stratigraphy (Marques et al. 2018, p. 43). Their dating reveals that the oldest exposed rocks in São Jorge range in age from 1.32 to 1.21 Ma, and following this there is a gap in the succession until 0.74 Ma, with the rocks of the Rosais Complex being erupted between 0.74 and 0.27 Ma. Marques et al. (2018) propose a revised stratigraphy: the *Old Volcanic Complex* (1.9–1.2 Ma) forms the eastern part of the island and is separated by a major fault from the *Intermediate Volcanic Complex* (0.8–0.2 Ma), which forms the bulk of the western two-thirds of São Jorge and the *Young Volcanic Complex* (< *c*.0.1 Ma), that includes the two historic eruptions and consists of Strombolian cones, lava and tephra which are unconformable with both the Old and Intermediate Volcanic Complexes (Marques et al. 2018, Figure 4, p. 45). The relationships between the three volcanic complexes are well illustrated in the coastal cliffs at Fajã de Santo Cristo on the north coast of São Jorge (Marques et al. 2018, Figure 4, p. 45).

The lava pile of the Old Volcanic Complex is tilted towards the east and Marques et al. (2018) argue that this may have been caused by uplift associated with the initiation of the Terceira Rift. They also argue that the former island constructed from the Old Volcanic Complex was truncated on its western end by a major landslide into the sea, and this explains the lateral discontinuity between the Old and Intermediate Volcanic Complexes. Volcanism then migrated westwards along the WNW/ESE São Jorge submarine volcanic ridge with the products of the Intermediate Volcanic Complex creating the remaining two-thirds of the current island.

There have been two onshore eruptions on São Jorge during historic times in 1580 near Velas and in 1808 above Urzelina (Figure 2.3). Both eruptions were basaltic producing lava flows which reached the coast. Explosive phreatomagmatic activity also occurred in both eruptions generating pyroclastic density currents (PDCs) that swept down the steep slopes to the sea causing a number of fatalities; around 15 in 1580 and more than 30 in 1808. These eruptions and their impact are considered in detail in Chapter 3. There is

Figure 2.3 Location map: São Jorge. This figure was published in Wallenstein et al. 2018, Figure 1, p. 166, Copyright permission Elsevier.

evidence of a probable submarine eruption in 1964 around 3 km offshore from Rosais, which is NW of Velas (Weston 1964 & Figure 2.3). On 15 February, numerous earthquakes occurred which were felt on Faial, Pico and Terceira as well as on São Jorge. In addition, in the W of São Jorge continuous tremor occurred. Poor weather prevented direct observation, there is no evidence that material was erupted at the surface, but on 18 February a sulphurous smell was noticed at Velas.

There have been a number of earthquakes that have impacted severely on the population of the island. The strongest earthquake reported in the Azores took place in 1757 (see above) just off the north coast of São Jorge. It had an estimated magnitude of 7.4 and killed over 1,000 people. An earthquake in 1964 associated with the submarine eruption caused severe damage in the NW of the island. São Jorge is a long, narrow island bounded by steep cliffs which are prone to collapse in the event of earthquakes or heavy rainfall. This provides a particular threat to communities living on *fajãs*, platforms of land formed by landslides or lava deltas at the foot of sea cliffs, which have been used both for settlement and agricultural production.

Terceira

Terceira is composed of four central volcanoes and a Basaltic Fissure Zone that broadly follows the trace of the Terceira Rift. The four central volcanoes are as follows: *Serra do Cume-Ribeirinha*, also known as *Cinco Picos*; *Guilherme Moniz*; *Pico Alto* and *Santa Bárbara* (Figure 2.4). Serra do Cume-Ribeirinha

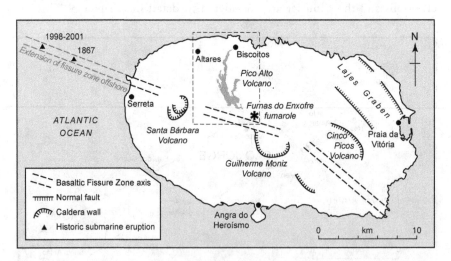

Figure 2.4 Location map: Terceira. This figure was published in Zanon and Pimentel 2015, Figure 1, p. 796, Copyright Mineralogical Society of America. Additions and amendments from Gertisser et al. 2010, Figure 1c, p. 135 and other sources.

(Cinco Picos volcano) erupted subaerial volcanics over 400,000 years ago and shows the remnants of the rim, of what is the largest caldera (7 × 9 km) in the Azores this caldera is considered to be younger than 370 ka (Calvert et al. 2006; Pimentel et al. 2016). Serra do Cume-Ribeirinha is heavily dissected and partly covered by volcanic deposits from younger centres and the Basaltic Fissure Zone. Guilherme Moniz, the next oldest volcano, is situated in the centre of the island. The oldest radiometric date from this volcano is *c.*270 ka (Calvert et al. 2006). The Pico Alto Centre, older than 141 ka (Gertisser et al. 2010), developed to the north of and partly overlapping the Guilherme Moniz edifice and is comprised predominantly of trachytic lava domes. Pico Alto has an 8 × 2 km caldera which formed from multiple collapses and is associated with eruption of trachytic ignimbrites including the Angra Ignimbrite, around 23 ka, and the Lajes Ignimbrite, around 21 ka (Calvert et al. 2006). The Lajes Ignimbrite is widely distributed across the island (Self 1976; Gertisser et al. 2010). The youngest date for the Pico Alto Centre is about 1,000 years BP, but this is not the youngest deposit, and Calvert et al. (2006) suggest that the centre was active up until almost the time when the island was first settled some 570 years ago. Santa Bárbara the youngest volcanic centre has built up over the last 100 ka and has been active historically. The summit of Santa Bárbara is truncated by two nested calderas about 2.5 × 2 km in size (Calvert et al. 2006). The outer caldera is considered by Self (1976) to have collapsed about 25 ka, and the inner caldera formed from a younger eruption possibly around 1–2 ka (Calvert et al. 2006). The inner caldera is partly filled by trachytic domes. There were two eruptions beginning in April 1761 (see Chapter 3). The first was a trachytic dome forming eruption which is the youngest event to have occurred on Santa Bárbara volcano.

The final volcanic unit, the Basaltic Fissure Zone, has been active for at least the last 50 ka (Zanon & Pimentel 2015), and Self (1976) has suggested that this fissure zone may be the subaerial expression of the Terceira Rift. The fissure zone extends from a WNW-ESE trending submarine ridge to the west of the island, the site of the 1998–2001 offshore Serreta eruption (see Chapter 3). On the west of the island, the fissure zone is obscured by the active Santa Bárbara volcano, presumably the presence of a crustal trachytic magma reservoir prevents access for the ascending denser basaltic magmas. A graben developed on the lower eastern flank of the volcano is an expression of the fissure zone and extends as a well-defined set of basaltic cones across the central part of the island to a zone between the Guilherme Moniz and Cinco Picos where it forms a junction. At this junction, the basaltic fissure zone is rotated in orientation to assume a NW-SE direction and then follows this trend to the SE of the island (Figure 2.4). Basaltic vents show a progressive trend by becoming younger towards the NW. Mungall and Martin (1995) refer to basaltic eruptions as being either *rift* basalts, with vents located on the Basaltic Fissure Zone, or *off-rift basalts* erupted elsewhere on the island. The *off-rift* basalts include those erupted by flank eruptions on the lower flanks of the active Santa Bárbara volcano. On the basis of petrogenetic modelling,

Mungall and Martin (1995) argue that the *rift* basalts show evidence of early crystallisation involving plagioclase in a low-pressure environment, whereas the *off-rift* basalts underwent initial crystallisation in a higher-pressure environment. In an investigation of fluid inclusions in mafic minerals in basaltic lavas from the fissure zone, Zanon and Pimentel (2015) propose that basalt magmas ponded at the Moho Transition Zone (20–21 km depth) and underwent fractional crystallisation. Magmas in the Serreta and SE segments of the fissure zone ascended directly through the crust to eruption. In the central segment of the Basaltic Fissure Zone – between the eastern foot of Santa Barbara volcano and the change in orientation of the fissure zone in the centre of the island – magmas rose by multi-stage ascent with further fractionation occurring between 17 and 7 km depth leading to eruption of more evolved trachybasalt magmas (Zanon & Pimentel 2015).

In historical times there has been one onshore eruption on Terceira which occurred in 1761 and consisted of two events (Pimentel et al. 2016, Chapter 3), first the trachytic dome forming eruption towards the base of the eastern flank of Santa Bárbara volcano and, second, an effusive basaltic eruption about 2.5 km away on the other side of Pico Gordo. This basaltic eruption, occurred on the margin of the Basaltic Fissure Zone between Santa Bárbara and Pico Alto volcanoes. There have been two offshore eruptions off the western coast of Terceira in 1867 and 1998–2001 on the Serreta submarine fissure zone (Chapter 3).

A number of damaging earthquakes have impacted Terceira. Sitting astride the Terceira Rift this island is in an active seismic zone and there have been particularly damaging earthquakes in: 1614 (at least 200 deaths, maximum Intensity X–XI); 1841, with around half the population in the affected area being rendered homeless, (maximum Intensity IX) and the devastating earthquake of 1980, with a maximum Intensity of IX and a magnitude 7.2 (Madeira 2005; Gaspar et al. 2015, Chapter 3).[2]

Graciosa

Graciosa is the only island in the Central Group not to have had a volcanic eruption in historic times. In common with Terceira, the tectonic structures of Graciosa are dominated, by its position sitting over the Terceira Rift and Gaspar (1996) has provided a detailed interpretation of the geology together with a geological map (Gaspar & Queiroz 1995). Three volcanic complexes are identified which are in order of age from the oldest to youngest: *Serra das Fontes; Serra Branca* and *Vitória-Vulcão Central.* Larrea et al. (2014) carried out a programme of $^{40}Ar/^{39}Ar$ radiometric dating of selected samples of the volcanic rocks of Graciosa and this has greatly enhanced the chronological framework first developed by Gaspar (1996) in his seminal study. The volcanic history as proposed by Larrea et al. (2014) dates the oldest subaerial rocks of the island to 1056.9 ± 28.0 ka, being basaltic lavas and pyroclastics that formed a shield volcano known as the Serra das Fontes volcanic complex.

This is now exposed in the centre of the island and is largely obscured by younger volcanics. Around 433 ka the magma plumbing system evolved to allow a high-level magma reservoir to form thus enabling more extensive fractional crystallisation to occur involving the eruption of trachytic magmas. These more silicic magmas built up the Serra Branca volcanic complex, a composite central volcano which remained active until about 270 ka. There then followed a gap in volcanic activity during which considerable erosion occurred. Volcanic activity recommenced and formed the Vitória-Vulcão Central volcanic complex, this complex being constructed from two units: the Vitória and the Vulcão Central (Larrea et al. 2014). The *Vitória* unit involved Strombolian activity with monogenetic basaltic eruptions building up the NW platform of the island. The oldest dated lava of the Vitória unit has an age of 96.3 ± 32.3 ka and activity continued with the eruption of what is likely to have been the most recent lava flow of Graciosa: the *Pico Timão,* which is dated at 3.9±1.4 ka (Larrea et al. 2014). The Vulcão Central stratovolcano rises to 402 m asl and is capped by an elliptical caldera (Jerram et al. 2017). Larrea et al. (2014) have dated lavas at the base of the pre-caldera sequence at 56.3±21.1 ka and 58.8±19.2 ka suggesting that the volcano was built around 50 ka. The stratovolcano was formed by basaltic volcanism, followed by explosive sub-Plinian and Plinian trachytic eruptions. An explosive hydromagmatic eruption led to caldera formation. Following caldera collapse, a basaltic effusion formed a lava lake and products from this lava spilled over the caldera rim in two places to feed an extensive lava flow field that spread around the western base of the volcano. This lava flow has been dated at 11.1±3.7 ka (Larrea et al. 2014).

There have been no historic eruptions on Graciosa since the island was settled in the late fifteenth century. With the last activity around 4,000 years ago future eruptions are a possibility. Indeed, even whilst inactive the Vulcão Central poses a threat. A fumarole sited in the Furna do Enxofre lava cave, located within the caldera where there are known discharges CO_2 asphyxiated two sailors from the Portuguese navy in 1992 (Gaspar 1996; Wallenstein et al. 2007).

The island is seismically active and an earthquake in 1837 (Intensity VII) caused three deaths and severe damage (Gaspar et al. 2015- see Chapter 3).

Flores

The islands of Flores and Corvo are located on the North American Plate to the west of the Mid Atlantic Ridge (MAR) and, in contrast to the other islands of the archipelago, are not located on the East Azores Volcano-Tectonic System (EAVTS). Most recent research into the geodynamics of the region has focused on the area to the east of the MAR (Azevedo & Ferreira 2006). Following a different tectonic alignment, Flores and Corvo are situated on a submarine ridge that trends NNE – SSW broadly parallel to the MAR. A westward displacement of Corvo relative to Flores at a rate of 1 cm/y is caused by transform faulting, which is reflected in the straight E–W coastlines on the north and south ends of Flores (Azevedo & Ferreira 2006).

Flores is composed of older submarine and emergent volcanic rocks overlain by subaerial volcanoes. Azevedo and Ferreira (2006) subdivide the volcanic products into the Base and Upper Complexes. The Base Complex (2.2–0.7 Ma) is composed of hydromagmatic breccias and tuffs with some pillow lavas and intercalated lava flows, and the stratigraphic sequence shows a progression from submarine to emergent volcanism. The products are mainly exposed around the coastline. Some outcrops of submarine volcanics do occur at an altitude of 400 m asl indicating considerable tectonic uplift.

The Upper Complex, which comprises the subaerial volcanoes, is subdivided into three units (Azevedo & Ferreira 2006). The Basal Unit dominates the subaerial part of Flores and Azevedo and Ferreira (2006) consider that this represents two phases of activity. Phase 1 (0.7–0.6 Ma) includes small to medium-sized central volcanoes which displayed basaltic Hawaiian and Strombolian activity. This was followed by Phase 2, when activity became localised in the centre of the island with activity evolving from Hawaiian/Strombolian activity to trachytic Plinian style eruptions with the formation of calderas. The Middle Unit (0.4–0.23 Ma) is formed of small-scale stratovolcanoes and monogenetic cones around the periphery of the island and was also associated with the Basal Unit caldera margins. Volcanism alternated between Hawaiian and Stromboli styles of activity. Activity of the Upper Unit (~ 3 ka) occurred after a long period of volcanic quiescence. Strombolian and phreatomagmatic activity formed scoria cones, generating pyroclastic surges and widespread tephra deposits (Azevedo & Ferreira 2006). The very young age of this activity suggests it may be associated with the current dormant volcanic system and, if this is the case, then it is possible that this system could erupt again. The geochemistry and mineralogy of the alkalic suite of volcanic rocks from Flores suggest that fractional crystallisation is the dominant differentiation process.

Corvo

Corvo is the smallest inhabited island of the Azores archipelago, 2 × 6 km in size, rising to a height of 270 m asl and is truncated by a small caldera 2 km in width. An outline description of the stratigraphy of Corvo based on Azevedo et al. (2003) is provided by França et al. (2006b) which identifies three main units: pre-caldera, syn-caldera and post-caldera. The base of the pre-caldera unit comprises hydromagmatic deposits which crop out in the coastal cliffs on the SW and north of the island; these early deposits being overlain by basaltic lava flows which built up a small shield volcano (Larrea et al. 2013). This was followed by eruption of progressively more evolved (silicic) magmas resulting in sub-Plinian/Plinian style trachytic explosive eruptions and caldera collapse. The products of this phase of activity make up the syn-caldera unit. Post-caldera activity involved monogenetic basaltic eruptions forming cones both within the caldera and on the flanks of the shield volcano.

The youngest eruption formed a lava flow near Vila Nova do Corvo at the south of the island. There is little information on the age of volcanism on the island. França et al. (2006b) note K/Ar dates of 0.7 and 0.4 Ma for younger basalt lavas of the shield volcano and they suggest that the youngest eruption forming the lava flow by Vila Nova do Corvo occurred around 80–100 ka, but no radiometric dates are available.

The lavas of Corvo are predominantly basaltic and fall into two petrographic categories: porphyritic and microporphyritic. The porphyritic lavas range from alkaline picrobasalts to alkali basalts and only occur in the pre-caldera and post-caldera units (Larrea et al. 2018). Larrea et al. (2013, 2018) argue that the porphyritic magmas form from accumulation of antecrysts that grew at depth from the crystallisation of primitive magmas which then ascended and were injected into the more evolved magma in the higher-level main fractionating reservoir. The microporphyritic lavas, which occur in all three units, represent near liquid compositions and their chemical variation reflects a single line of descent with differentiation caused by fractional crystallisation (Larrea et al. 2018). The trachytic magmas that caused the explosive eruptions associated with caldera collapse, probably evolved by fractionation in a shallow magma reservoir which developed at a late stage in the history of the shield volcano.

There is no evidence that Corvo has had an eruption in the last few thousand years, nor has it been subjected to significant seismic activity in historic times.

Notes

1 This would require winds from the north or NW. Historical meteorological data show that annually winds from these directions can be expected, respectively, and on average, *c.*13% and *c.*12% of the time (Chester et al. 1995, p. 11-12).
2 The conventions used in citing earthquake Intensity and Magnitude are discussed in Chapter 3.

References

Abdel-Monem, A.A., Fernandez, L.A. and Boone, G.M. 1975. K-Ar ages from the eastern Azores group (Santa Maria, São Miguel and the Formigas Islands). *Lithos* **8**, 247–254.

Almeida, M.H. 2001. *A fonte mantélica na região dos Açores. Constrangimentos impostos pelas características geoquímicas de rochas vulcânicas e de xenólitos ultramáficos.* PhD thesis, Universidade dos Açores, Portugal.

Anon 2020. *Santa Maria FIR (Oceanic Control Centre).* IVAO, Portugal. (https://pt.ivao.aero/portal/lppo/)

Ávila, S.P., Ramalho, R.S., Habermann, J.M. and Titschack, J. 2018. The marine fossil record at Santa Maria Island (Azores). *In:* Kueppers, L. and Beier, C. (eds) *Volcanoes of the Azores.* Springer, Berlin, 251–280.

Azevedo, J.M.M. and Ferreira, M.R.P. 2006. The volcanotectonic evolution of Flores Island, Azores (Portugal). *Journal of. Volcanology and Geothermal Research* **156**, 90–102.

Azevedo, J.M.M., Alves, E.I. and Dias, J.L. 2003. Contributo para a interpretação vulcanostrctural da ihla do Corvo, Açores. *Ciências da Terra (UNL)* Lisboa, no esp. V, A5–A8.

Beier, C., Haase, K.M. and Brandl, P.A. 2018. Melting and mantle sources in the Azores. *In:* Kueppers, L. and Beier, C. (eds) *Volcanoes of the Azores.* Springer, Berlin, 251–280.

Booth, B., Croasdale, R. and Walker, G.P.L. 1978. A quantitative study of five thousand years of volcanism on São Miguel, Azores. *Philosophical Transactions of the Royal Society of London* **228**, 271–319.

Calvert, A.T., Moore, R.B., McGeehin, J.P. and Rodrigues de Silva, A.M. 2006. Volcanic history and^{40}Ar/^{39}Ar and^{14}C geochronology of Terceira Island, Azores, Portugal. *Journal of Volcanology and Geothermal Research* **156**, 103–115.

Cappello, A., Zanon, V., Del Negro, C., Ferreira, T.J.L. and Queiroz, M.G.P.S. 2015. Exploring lava-flow hazards at Pico Island, Azores archipelago (Portugal). *Terra Nova* **27**, 156–161.

Carmo, R., Madeira, J., Ferreira, T., Queiroz, G. and Hipólito, A. 2015. Volcano-tectonic structures of São Miguel Island, Azores. *In:* Gaspar, J.L., Guest, J.E., Duncan, A.M., Chester, D.K. and Barriga, F. (eds) Volcano-tectonic structures of São Miguel Island, Azores. *Volcanic Geology of S. Miguel Island (Azores, Archipelago).* Memoir Geological Society of London **44**, 65–86.

Chester, D.K. and Duncan, A.M. 2018. Volcanic environments. *In:* Bobrowsky, B. and Marker, B. (eds) *Encyclopedia of Engineering Geology.* Springer International Publishing, 935–943.

Chester, D.K., Dibben, C. and Coutinho, R. 1995. *Report on the Evacuation of the Furnas District, São Miguel, Azores, in the Event of a Future Eruption.* CEC Environment: ESF Laboratory Volcano, Open File Report 4, University College London.

Cioni, R., Pistolesi M. and Rosi, M. 2015. Plinian and Subplinian eruptions. *In:* Sigurdsson, H. (ed) *The Encyclopedia of Volcanoes.* 2nd Edition. Academic Press, Amsterdam, 519–535.

Cole, P.D., Queiroz, G., Wallenstein, N., Gaspar, J., Duncan, A.M. and Guest, J.E. 1995. An historic subplinian/phreatomagmatic eruption: The 1630 AD eruption of Furnas Volcano, São Miguel, Azores. *Journal of Volcanology and Geothermal Research* **69**, 117–135.

Cole, P.D., Pacheco, J.M., Gunasekera, R., Queiroz, G., Gonçalves, P. and Gaspar, J.L. 2008. Contrasting styles of eruption at Sete Cidades, São Miguel, Azores, in the last 5000 years: Hazard implications from modelling. *Journal of Volcanology & Geothermal Research* **178**, 574–591.

Costa, A.C.G., Marques, F.O., Hildenbrand, A., Sibrant, A.L.R. and Catita, C.M.S. 2014. Large-scale catastrophic flank collapses in a steep volcanic ridge: The Pico-Faial Ridge, Azores Triple Junction. *Journal of Volcanology & Geothermal Research* **272**, 111–125.

Crossweller, H.S., Arora, B., Brown, S.K. et al. (2012) Global database on large magnitude explosive volcanic eruptions (LaMEVE). *Journal of Applied Volcanology* **1**, 1–13.

de Silva, S. and Lindsay, J.M. 2015. Primary volcanic landforms. *In:* Sigurdsson, H. (ed) *The Encyclopedia of Volcanoes.* 2nd Edition. Academic Press, Amsterdam, 273–297.

Di Chiara, A., Speranza, F., Porreca, M., Pimentel, A., Caraccioli, F. and Pacheco, J. 2014. Constraining chronology and space time evolution of Holocene volcanic activity on the Capelo Peninsula (Faial Island, Azores): The paleomagnetic contribution. *Geological Society of America Bulletin* **126**, 1164–1180.

Duncan, A.M., Queiroz G, Guest, J.E., Cole, P.D., Wallenstein, N. and Pacheco, J.M. 1999. The Povoação ignimbrite, Furnas volcano, Sao Miguel, Azores. *Journal of Volcanology and Geothermal Research* **92**, 55–65.

Duncan, A.M., Guest, J.E., Wallenstein, N. and Chester, D.K. 2015. The older volcanic complexes of São Miguel, Azores: Nordeste and Povoação. *In*: Gaspar, J.L., Guest, J.E., Duncan, A.M., Chester, D.K. and Barriga, F. (eds) *Volcanic Geology of S. Miguel Island (Azores, Archipelago)*. Memoir Geological Society of London **44**, 147–153.

Fernandez, L.A. 1980. Geology and petrology of the Nordeste volcanic complex, São Miguel: Summary. *Geological Society of America Bulletin*. Geological Society of America Bulletin **91**, 675–680.

Ferreira, T., Gomes, A., Gaspar, J.L. and Guest, J. 2015. Distribution and significance of basaltic centres: São Miguel, Azores. *In*: Gaspar, J.L., Guest, J.E., Duncan, A.M., Chester, D.K. and Barriga, F. (eds) *Volcanic Geology of S. Miguel Island (Azores, Archipelago)*. Memoir Geological Society of London **44**, 135–146.

Forjaz, V.H. and Fernandes, N.S.M. 1975. *Notícia explicativa das folhas "A" e "B" da ilha de S. Jorge (Açores) da Carta Geológica de Portugal na escala 1:50 000*. Serviços Geológicos De Portugal, Lisboa.

França, Z., Tassinari, C., Cruz, J., Aparicio, A., Araña, V. and Rodriques, B. 2006a. Petrology, geochemistry and Sr-Nd-Pb isotopes of the volcanic from Pico – Azores (Portugal). *Journal of Volcanology and Geothermal Research* **156**, 71–89.

França, Z., Lago, M., Nunes, J.C., Galé, C., Forjaz, V.H., Pueyo, O. and Arranz, E. 2006b. Geochemistry of alkaline basalts of Corvo Island (Azores, Portugal): Preliminary data. *Geogaceta* **40**, 87–90.

Gandino, A., Guidi, M., Merlo, C., Mete, L., Rossi, R. and Zan, L. 1985. Preliminary model of the Ribeira Grande geothermal field (Azores Islands). *Geothermics* **14**, 91–105.

Gaspar, J.L. 1996. *Ihla Graciosa (Açores): Historia vulcanológica e avaliação do hazard*. PhD thesis, Universidade dos Açores, Portugal.

Gaspar, J.L. and Queiroz, G. 1995. *Carta vulcanológica da Ilha Graciosa (escala 1:10000)*. Departament de Geociências, Universidade dos Açores, Portugal.

Gaspar, J.L., Queiroz, G., Ferreira, T., Medeiros, A.R., Goulart, C. and Medeiros, J. 2015. Earthquakes and volcanic eruptions in the Azores region: Geodynamic implications from major historical events and instrumental seismicity. *In*: Gaspar, J.L., Guest, J.E., Duncan, A.M., Chester, D.K. and Barriga, F. (eds) *Volcanic Geology of S. Miguel Island (Azores, Archipelago)*. Memoir Geological Society of London **44**, 33–49.

Gertisser, R., Self, S., Gaspar, J.L., Kelly, S.P., Pimentel, A., Eikenberg, J., Berry, T.L., Pacheco, J.M. and Vespa, M. 2010. Ignimbrite stratigraphy and chronology on Terceira Island, Azores. *In*: Groppelli, G. and Viereck-Goette, L. (eds) *Stratigraphy and Geology of Volcanic Areas*. Geological Society of America Special Paper **464**, 133–154.

Guest, J.E., Gaspar, J.L., Cole, P.D., Queiroz, G., Duncan, A.M., Wallenstein, N., Ferreira, T. and Pacheco, J-M. 1999. Volcanic geology of Furnas Volcano, São Miguel, Azores. *Journal of Volcanology and Geothermal Research* **92**, 1–29.

Guest, J., Cole, P., Duncan, A. and Chester, D. 2003. *Volcanoes of Southern Italy*. Geological Society of London, Bath.

Guest, J.E., Pacheco, J.M., Cole, P.D., Duncan, A.M., Wallenstein, N., Queiroz, G., Gaspar, J.L. and Ferreira, T. 2015. The volcanic history of Furnas Volcano, S.

Miguel, Azores. *In*: Gaspar, J.L., Guest, J.E., Duncan, A.M., Chester, D.K. and Barriga, F. (eds) *Volcanic Geology of S. Miguel Island (Azores, Archipelago)*. Memoir Geological Society of London **44**, 125–34.

Harris, A.J.L. and Rowland, S.K. 2015. Lava flows and rheology. *In*: Sigurdsson, H. (ed) *The Encyclopedia of Volcanoes*. Academic Press, Amsterdam, 321–342, 2nd Edition.

Hildenbrand, A., Madureira P., Ornelas Marques, F., Cruz, I., Henry, B. and Silva, P. 2008. Multi-stage evolution of a subaerial volcanic ridge over the last 1.3 Myr: S. Jorge Island, Azores Triple Junction. *Earth and Planetary Science Letters* **273**, 289–298.

Hildenbrand, A., Marques, F.O., Costa, A.C.G., Sibrant, A.L.R., Silva, P.M.F., Henry, B., Miranda, J.M. and Madureira, P. 2012. Reconstructing the architectural evolution of volcanic islands from combined K/Ar, morphologic, tectonic and magnetic data: The Faial Island example (Azores). *Journal of Volcanology & Geothermal Research* **241–242**, 39–48.

Houghton, B.F., Taddeucci, Adronico, D. et al. 2016. Stronger or longer: Discriminating between Hawaiian and Strombolian eruption styles. *Geology* **44**, 163–166.

Jerram, D., Scarth, A. and Tanguy, J-C. 2017. *Volcanoes of Europe*. Dunedin. Edinburgh, 2nd Edition.

Johnson, C.L., Wijbrans, J.R., Constable, C.G., Gee, J., Staudigal, H., Tauxe, L., Forjaz, V.-H. and Salgueiro, M. 1998. ^{40}Ar/^{39}Ar ages and Azores palaeomagnetism of São Miguel lavas. *Earth and Planetary Science Letters* **160**, 637–649.

Kueppers, U., Pimentel, A., Ellis, B., Forni, F., Neukampf, J., Pacheco, J., Perugini, D. and Queiroz, G. 2019. Biased volcanic hazard assessment due to incomplete eruption records on ocean islands: An example of Sete Cidades volcano, Azores. *Frontiers in Earth Science* **7**, 122.

Larrea, P., França, Z., Lago, M., Widom, E., Gale, C. and Ubide, T. 2013. Magmatic processes and the role of antecrysts in the genesis of Corvo Island (Azores Archipelago, Portugal). *Journal of Petrology* **54**, 769–793.

Larrea, P., Wijbrans, J.R., Gale, C., Ubide, T., Lago, M., França, Z. and Widom, E. 2014. ^{40}Ar/^{39}Ar constraints on the temporal evolution of Graciosa Island, Azores (Portugal). *Bulletin of Volcanology* **576**, 796.

Larrea, P., França, Z., Widom, E. and Lago, M. 2018. Petrology of the Azores Islands. *In*: Kueppers, L. and Beier, C. (eds) *Volcanoes of the Azores*. Springer, Berlin, 197–249.

Macdonald, G.A. 1972. *Volcanoes*. Prentice Hall, Eaglewood Cliffs.

Madeira, J. 2005. *The Volcanoes of the Azores Islands: World Class Heritage (Examples from Terceira, Pico)*. Field Trip Guide Book, IV International Symposium PROGEO on the Conservation of Geological Heritage. Univ. Lisbon, Portugal.

Madeira, J., Brum da Silveira, A, Hipólito, A. and Carmo, R. 2015. Active tectonics in the central and eastern Azores islands along the Eurasia-Nubia boundary: A review. *In*: Gaspar, J.L., Guest, J.E., Duncan, A.M., Chester, D.K. and Barriga, F. (eds) *Volcanic Geology of São Miguel Island (Azores, Archipelago)*. Memoir Geological Society of London **44**, 15–32.

Marques, F.O., Hildenbrand, A. and Hübscher, C. 2018. Evolution of a volcanic island on the shoulder of an oceanic rift and geodynamic implications: S. Jorge Island on the Terceira Rift, Azores Triple Junction. *Tectonophysics* **738–739**, 41–50.

Marques, F.O., Hildenbrand, A., Costa, A.C.G. and Sibrant, A.L.R. 2020. The evolution of Santa Maria Island in the context of the Azores Triple Junction. *Bulletin of Volcanology* **82**, 39.

McKenzie, D.P. 1972. Active tectonics of the Mediterranean region. *Geophysical Journal of the Royal Astronomical Society* **30**, 109–185.

Miranda, J.M., Luis, J.F., Lourenço, N. and Fernandes, R.M.S. 2015. The structure of the Azores Triple Junction: implications for São Miguel Island. *In*: Gaspar, J.L., Guest, J.E., Duncan, A.M., Chester, D.K. and Barriga, F. (eds) *Volcanic Geology of S. Miguel Island (Azores, Archipelago).* Memoir Geological Society of London **44**, 5–13.

Miranda, J.M., Luis, J.F. and Lourenço, N. 2018. The tectonic evolution of the Azores based on magnetic data. *In*: Kueppers, L. and Beier, C. (eds) *Volcanoes of the Azores*. Springer, Berlin, 89–100.

Mitchell, N., Stretch, R. and Tempera, F. 2018. Volcanism in the Azores: A marine geophysical perspective. *In*: Kueppers, L. and Beier, C. (eds) *Volcanoes of the Azores*. Springer, Berlin, 101–126.

Moore, R. 1991. *Geological Map of São Miguel, Azores. Map I-2007, Scale 1:50 000.* United States Geological Survey Miscellaneous Investigations Series, Washington, DC.

Muecke, G.K., Ade-Hall, J.M., Aumento, F., Macdonald, A., Reynolds, P.H., Hyndman, R.D., Quintino, J., Opdyke, N. and Lowrie, W. 1974. Deep drilling in an active geothermal area in the Azores. *Nature* **252**, 281–285.

Mungall, J.E. and Martin, R.F. 1995. Petrogenesis of basalt-comendite and basalt-pantellerite suites, Terceira, Azores, and some implications for the origin of ocean-island rhyolites. *Contributions to Mineralogy and Petrology* **119**, 43–55.

Needham, H. and Francheteau, J. 1974. Some characteristics of the rift valley in the Atlantic Ocean near 36° 48' north. *Earth and Planetary Science Letters* **22**, 29–43.

Newhall, C.G. and Self, S. 1982. The Volcanic Explosivity Index (VEI). An estimate of the explosive magnitude for historical volcanism. *Journal of Geophysical Research* **87** (C), 1231–8.

Nunes, J.C., França, Z., Cruz, J.V., Carvalho, M.R. and Serralheiro, A. 1999. *Carta Vulcanológica da Ilha do Pico (Açores) – Versão Preliminar. Escala 1:30,000.* Universidade dos Açores, Portugal.

O'Neill, C. and Sigloch, K. 2018. Crust and mantle structure beneath the Azores. *In*: Kueppers, L. and Beier, C. (eds) *Volcanoes of the Azores*. Springer, Berlin, 71–87.

Pensa, A., Cas, R., Giordano, G., Porreca, M. and Wallenstein, N. 2015. Transition from steady to unsteady Plinian eruption column: The VEI 5, 4.6 ka Fogo Plinian eruption, São Miguel, Azores. *Journal of Volcanology and Geothermal Research* **305**, 1–18.

Pimentel, A., Pacheco, J. and Self, S. 2015. The ~1000 years BP explosive eruption of Caldeira Volcano (Faial, Azores): The first stage of incremental caldera formation. *Bulletin of Volcanology* **77**, 42–68.

Pimentel, A, Zanon, V., de Groot, L.V., Hipólito, A., Di Chiara, A. and Self, S. 2016. Stress-induced comenditic trachyte effusion triggered by trachybasalt intrusion: Multidisciplinary study of the AD 1761 eruption at Terceira Island (Azores). *Bulletin of Volcanology* **78**, 22pp.

Porreca, M., Pimentel, A., Kueppers, U., Izquierdo, T., Pacheco, J. and Queiroz, G. 2018. Event stratigraphy and emplacement mechanisms of the last major caldera eruption on Sete Cidades Volcano (São Miguel, Azores): The 16 ka Santa Bárbara Formation. *Bulletin of Volcanology* **80**, 76.

Queiroz, G., Gaspar, J.L., Guest, J.E., Gomes, A. and, Almeida, M.H. 2015. Eruptive history and evolution of Sete Cidades Volcano, São Miguel island, Azores. *In*: Gaspar, J.L., Guest, J.E., Duncan, A.M., Chester, D.K. and Barriga, F. (eds)

Volcanic Geology of S. Miguel Island (Azores, Archipelago). Memoir Geological Society of London **44**, 87–104.

Ramalho, R.S., Helffrich, G., Madeira, J, Cosca, M., Quarto, R., Hipólito, A., Rovere, A., Hearty, P.J. and Ávila, S.P. 2017. Emergence and evolution of Santa Maria Island (Azores) – the conundrum of uplifted islands revisited. *Geological Society of America Bulletin* **129**, 372–390.

Ramalho, R.S., Quartau, R., Hóskuldsson, Á., Madeira, J., da Cruz, J.V. and Rodrigues, A. 2020. Evidence for late Pleistocene volcanism at Santa Maria Island, Azores? *Journal of Volcanology and Geothermal Research* **394**, 106829.

Rangel, G., Ponte, C. and Franco, A. 2015. Use of geothermal resources in the Azores Islands: A contribution to the energy self-sufficiency of a remote and isolated region. Azores. *In*: Gaspar, J.L., Guest, J.E., Duncan, A.M., Chester, D.K. and Barriga, F. (eds) *Volcanic Geology of S. Miguel Island (Azores, Archipelago)*. Memoir Geological Society of London **44**, 297–301.

Romer, R.H.W., Beier, C., Hasse, K.M. and Hübscher, C. 2018. Correlated changes between volcanic structures and magma composition in the Faial Volcanic System, Azores. *Frontiers in Earth Science* **6**, 78.

Self, S. 1976. The recent volcanology of Terceira, Azores. *Journal of the Geological Society of London* **132**, 228–240.

Serralheiro, A., Forjaz, V.H., Alves, C.A.M. and Rodrigues, E.B. 1989. *Carta Vulcanológica dos Açores, Ilha do Faial*. Centro de Vulanologia, Serviço Regional de Protecção Civile, University of the Azores.

Siebert, L., Simkin, T. and Kimberley, P. 2010. *Volcanoes of the World*. 3rd Edition., University of California Press.

Trippanera, D., Porreca, M., Ruch, J., Pimentel, A., Acocella, V., Pacheco, J., Salvatore, M. 2014. Relationships between tectonics and magmatism in a transtensive/transform setting: An example from Faial Island (Azores, Portugal). *Geological Society of America Bulletin* **126**, 164–181.

Viveiros, F., Ferreira, T., Silva, C., Gaspar, J.L., Virgili, G. and Amaral P. 2015. Permanent monitoring of soil CO_2 degassing at Furnas and Fogo volcanoes (São Miguel Island, Azores). *In*: Gaspar, J.L., Guest, J.E., Duncan, A.M., Chester, D.K. and Barriga, F. (eds) *Volcanic Geology of S. Miguel Island (Azores, Archipelago)*. Memoir Geological Society of London **44**, 271–288.

Walker, G.P.L. 1973. Explosive volcanic eruptions – a new classification. *Geologische Rundschau* **62**, 431–446.

Walker, G.P.L. and Croasdale, R. 1971. Two Plinian-type eruptions in the Azores. *Journal of the Geological Society London* **127**, 17–55.

Wallenstein, N., Duncan, A.M., Chester, D.K. and Marques, R. 2007. Fogo Volcano (São Miguel, Azores): A hazardous landform. *Géomorphologie: Relief, Processus, Environnement* **3**, 259–270.

Wallenstein, N., Duncan, A., Guest, J.E. and Almeida, M.H. 2015. Eruptive history of Fogo volcano São Miguel, Azores. *In*: Gaspar, J.L., Guest, J.E., Duncan, A.M., Chester, D.K. and Barriga, F. (eds) *Volcanic Geology of S. Miguel Island (Azores, Archipelago)*. Memoir Geological Society of London **44**, 105–23.

Wallenstein, N., Duncan, A., Coutinho, R. and Chester, D. 2018. Origin of the term *nuées ardentes* and the 1580 and 1808 eruptions on São Miguel, Azores. *Journal of Volcanology and Geothermal Research* **358**, 165–170.

Weston, F.S. 1964. List of recorded volcanic eruptions in the Azores with brief reports. *Boletim do Museo e Laboratório Mineralógico da Faculdade de Ciências de Lisboa* **10**, 3–18.

Williams-Jones, G. and Rymer, H. 2015. Hazards of volcanic gases. *In*: Sigurdsson, H. (ed) *The Encyclopedia of Volcanoes*. 2nd Edition. Academic Press, Amsterdam, 985–992.

Woodhall, D. 1974. Geology and volcanic history of Pico Island volcano Azores. *Nature* **248**, 663–665.

Zanon, V. 2015. The magmatism of the Azores Islands. Azores. *In*: Gaspar, J.L., Guest, J.E., Duncan, A.M., Chester, D.K. and Barriga, F. (eds) *Volcanic Geology of S. Miguel Island (Azores, Archipelago)*. Memoir Geological Society of London **44**, 51–64.

Zanon, V. and Frezzotti, M.L. 2013. Magma storage and ascent conditions beneath Pico and Faial islands (Azores archipelago): A study on fluid inclusions. *Geochemistry, Geophysics, Geosystems* **14** (9), 3494–3514.

Zanon, V. and Pimentel, A. 2015. Spatio-temporal constraints on magma storage and ascending conditions in a transtensional tectonic setting: The case of the Terceira Island (Azores). *American Mineralogist* **100**, 795–805.

Zanon, V., Kueppers, U., Pacheco, J.M. and Cruz, I. 2013. Volcanism from fissure zones and the Caldeira central volcano of Faial Island, Azores archipelago: Geochemical processes in multi feeding systems. *Geological Magazine* **150**, 536–555.

Zanon, V., Pimentel, A., Auxerre, M., Marchini, G. and Stuart, F.M. 2020. Unravelling the magma feeding system of young basaltic oceanic volcano. *Lithos* **352–353**, 105325.

3 Historical eruptions and earthquakes

Information on past earthquakes and eruptions in the Azores is available from three sources. The first two comprise general histories of the islands and more focused contemporary and near contemporary accounts of individual seismic and volcanic events. The third source of information is obtained via a different route. Informed by a knowledge of volcanic and seismic processes, field work is used both to test the veracity of historical narratives and to extend these through: the mapping, dating and interpretation of volcanic deposits; assessments of the impacts of earthquakes on people, buildings and infrastructure and, in the case of the 1522 earthquake on São Miguel, the archaeological investigations of settlements entombed by earthquake-triggered landslide sediments (Bento 1989, 1994; Marques et al. 2009; Anon 2014; Garrard et al. 2021).

Three early histories of the islands contain a wealth of information on eruptions, earthquakes and their impacts (see Chapter 1). These were authored by the cleric and scholar Gaspar Frutuoso (1522–1591), the Jesuit António Cordeiro (1641–1722) and the nineteenth-century literary figure Ernesto do Canto (1831–1900), who in his *Archivo dos Açores* compiled and edited many previously published and unpublished narratives of eruptions and earthquakes from earliest settlement to the close of the nineteenth century. For historic earthquakes additional information on their size[1] and impact are to be found in publications by Mendonça (1758), Drummond (1856–1859), Machado (1973), Madeira (1998), Madeira and Brum da Silveira (2003), Silveira et al. (2003), Silva (2005) and Malheiro (2006); and for volcanoes works by Agostinho (1931), Neumann van Padang et al. (1967), Weston (1964) and Madeira (1998), may be added. Many writers have produced narratives of individual events and recently Gaspar et al. (2015) have examined afresh all the historical materials, assessed in detail their reliability and have published two historical catalogues: one for earthquakes the other for eruptions (see below). These are accepted as definitive in the current volume.

History of research

In contrast to the situation in other volcanically and seismically active regions in Europe, research on eruptions and earthquakes in the Azores is not marked

DOI: 10.4324/9780429028007-3

by steady progress and even by the second half of the twentieth century, there were still gaps in historical catalogues and even some major destructive events had never been researched in detail. For instance, Gaspar Frutuoso's major work, *Saudades da Terra*, is a rich source of information on the earliest earthquakes and eruptions, but it remained in manuscript until the late nineteenth century, was only published in part from 1873 (Luz 1996) and did not become generally available internationally until well into the second half of the twentieth century (see Chapter 1). Accounts of seismic and volcanic events recorded in Ernesto do Canto's journal, *Archivo dos Açores*, had limited circulation within the archipelago and only reached an international audience when it was made available digitally by University of the Azores.[2] When the definitive recent historical catalogues of earthquakes and eruptions published by Gaspar et al. (2015) are compared with earlier compilations, the deficiencies of the latter become clear. With respect to eruptions, the Azores volume of the *Catalogue of Active Volcanoes of the World* (Neumann van Padang et al. 1967), was unaware of the *c*.1439 to 1443 eruption, while (Weston 1964), not only discounts this eruption, but also lists a 1439 eruption of Sete Cidades (on São Miguel) that did not occur, and notes a lava flowing in the direction of Rabo de Peixe (São Miguel) in 1652 which as discussed later is incorrect (Ferreira et al. 2015; Chester et al. 2017). Although pioneer studies of the geology of the Azores were published in the nineteenth century, most notably by the eminent French geologist Ferdinand Fouqué (1867, 1873), the first detailed geological maps (1: 50,000 scale) and associated memoirs were only issued by the *Serviços Geológicos de Portugal*[3] in the late 1950s and early 1960s (Zbyszewski et al. 1958, 1959; Zbyszewski 1961). These are incomplete in their treatment of historical eruptions and their impacts.

With respect to earthquakes, it was only from the time of the 1909 Benavente earthquake near to Lisbon that there was local lobbying for a seismic network to be established in the Azores, the archipelago having to rely on just two stations set up seven years earlier: one in Ponta Delgada (São Miguel) and the other at Horta (Faial). In 1932 a further seismograph was installed at Angra do Heroísmo (Terceira) (Senos et al. 2000; Chester et al. 2017), and the islands had to rely on these three stations until 1980 severally restricting the recording of small earthquakes and seismic swarms (Fontiela et al. 2018).

The reasons why seismology and volcanology were so undeveloped cannot be blamed solely on the physical isolation of the Azores, because in comparison with many mid-oceanic islands seaborne transport was well established, with Horta (Faial), Angra do Heroísmo (Terceira) and Ponta Delgada (São Miguel) being major ports on Atlantic trade routes where ships could be provisioned (Chapter 1). In addition from the 1890s the islands possessed telegraphic links to the rest of the world, Faial, in particular, being an important node on the international network and acted as host to expatriate personnel from Germany, the UK and the USA. Despite this degree of connectivity with the outside world there were few scientific or even international news reports of eruptions until well into the twentieth century, though it must be admitted that the only

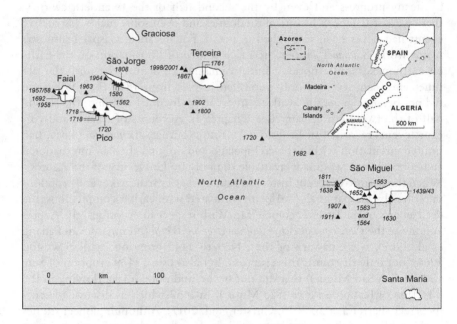

Figure 3.1 Location of historical eruptions. Based on various sources.

eruptions between 1808 and 1957/8 were submarine in origin (Figure 3.1). In contrast in the nineteenth century, earthquakes caused death and destruction to places visited by mariners, and events with an Intensity of VIII or greater affected Terceira in 1800, 1801 and 1841, and São Miguel in 1810, 1811 and 1852 (Figure 3.2). These earthquakes were reported in the international press, though not to any great extent in learned journals (Chester et al. 2017).

For centuries the Portuguese intellectual climate was not conducive to geological investigations. In a devoutly Catholic country in which the Inquisition only officially ended in 1821, though it was ineffective following the reforms initiated by the Marquis of Pombal in 1773–1774, it might be thought that religion constrained the development of the geological sciences especially following the publication of Lyell's *Principles of Geology* in 1830 and Darwin's *Origin of Species* in 1859, but this was not the case. As Carneiro et al. (2013) have argued, higher education had been outside Catholic control since the 1770s and there was a long-established tradition within political, social and academic elites of recognising that science and religion represented different spheres of intellectual enquiry. By the close of the second decade of the twentieth century, the *Estado Novo* was in power (see Chapter 1) and 'despite the close relationships between the newly established regime and the Roman Catholic Church, António Salazar maintained the separation of Church and State; in addition, the State held no official position regarding scientific matters' (Carneiro et al. 2013, p.333).

The problem lay elsewhere and, as historians of science have noted, 'despite the creation of a national geological survey being coeval to other European countries, the teaching of geology and geological research became effective only as late as the mid-twentieth century and the Portuguese Geological Society was only founded in 1940' (Mota & Caneiro 2013, p. 24). For centuries, there was one university at Coimbra, Oporto and Lisbon opened their doors in 1911 and the University of the Azores only dates from 1976. Until the twentieth century, students (almost always men) who wanted careers in science had three options 'medicine; the clergy and, in the nineteenth century, military engineering' (Carneiro et al. 2013, p. 332). A lack of trained personnel was clear during the 1957/8 eruption and earthquake on Faial when the most prominent scientists involved in recording the event and responding to it were not earth scientists by training (Coutinho et al. 2010). It is notable that neither of the two leading figures in Azorean earth science in the first six decades of the twentieth century was a geologist or geophysicist by training; José Agostinho (1888–1978) was an army officer and meteorologist, and Frederico Machado (1918–2000) a civil engineer (Chester et al. 2017).

The last half of the twentieth century marked a sea-change in Azorean earth science. From the late 1950s and as mentioned above, Georges Zbyszewski (1909–1999) of the *Serviços Geológicos de Portugal* led a programme that mapped the geology of the Azores. In the 1970s, the eminent volcanologist George Walker and his colleagues researched explosive silicic activity (Walker & Croasdale 1971; Booth et al. 1978) but with the exception of the Fogo 1563 eruption they mainly investigated pre-historic eruptions, and took the first tentative steps in exploring volcanic hazards (Booth et al. 1983, see Chapter 4). In the early 1990s, the United States Geological Survey produced a new geological map of São Miguel (Moore 1991a, 1991b) undertaken on behalf of the Geothermal Project on Fogo which also funded several seismic stations on the island (Rangel et al. 2015). While the high magnitude (M 7.2) earthquake in January 1 1980 on Terceira, not only devastated Angra and others settlements and also caused damage on other islands, but acted as a stimulus to Portuguese and foreign interest in the seismic risks faced by the Azoreans (e.g. Anon 1983; Oliveira et al. 1992; Silva 2005; Feio et al. 2015).

It was the establishment of the university which acted as a catalyst stimulating Azorean participation in major international projects, the establishment of specialised research centres and more effective administrative structures. These are more fully discussed in Chapter 6.

Historical catalogues

Table 3.1 lists the volcanic events that have occurred since the archipelago was settled in the fifteenth century, and Figure 3.1 shows their locations. Although the table is mostly self-explanatory several points require elaboration. First, of the 28 events listed in the table, 15 have occurred on land many of which have affected people and their activities sometimes very significantly. These are described below, as are the submarine eruptions of 1638, 1811 and 1867 which caused significant damage on land. A second feature of Table 3.1 is that the number of submarine eruptions increases over time, which implies that early events of this type may be missing from the record. Indeed even as late as the early twentieth century, submarine eruptions in 1902 and 1907 (Table 3.1) only became apparent because they severed submarine telegraph cables, having no impact on land and no affects on people or their activities (Chaves 1960).

Finally, Gaspar et al. (2015, p. 37) draw associations between eruptive styles (see Table 2.1) and the tectonic structure of the archipelago (see Chapter 2). The basaltic eruptions of: 1563 (São Miguel); 1761 (Terceira); 1580 and 1808 (São Jorge); 1562–1564 and 1720 (Pico) and 1672–1673 (Faial) are related to active fissures, their principal orientations reflecting the NW-SE general tectonic trend (see Chapter 2). The 1957–1958 Capelinhos eruption (Faial) also belongs to the category of basaltic eruptions associated with the NW-SE fissure zone, but began offshore as a submarine phreatomagmatic eruption. The edifice built up above sea level and eventually formed an isthmus connected to the NW end of Faial (see below). Explosive eruptions of trachyte magma occurred at central volcanoes in: 1439–1443 and 1630 (Furnas, São Miguel); 1563–1564 (Fogo, São Miguel) and included phreatomagmatic activity produced by water-magma interaction from crater lakes. Formation of lava domes were features of the latter stages of the 1439–1443 and 1630 (sub-Plinian) events at Furnas and the 1652 (Vulcanian) eruption on São Miguel, which was associated with the Picos Fissural Volcanic System (see Chapter 2). Gaspar et al. (2015, p. 37) using an account by Drummond (1856–1859), argue that a similar style of eruption may have characterised the 1761 *Mistérios Negros* eruption on Terceira, this is considered in the section on Historic Eruptions below.

In the Azores the epicentres of earthquakes are spatially concentrated in a zone running SE-NW from the Gloria Fault east of Santa Maria to the Mid Atlantic Ridge (MAR) between Faial and Flores, there having been no recorded earthquakes to the west of the MAR[4] (see Chapter 2 and Figure 3.2). The historical catalogue (Table 3.2) shows that since settlement in the fifteenth century the islands have been affected by 'about 15 major earthquakes and 16 seismic crises, of which 10 were related to observed or probable volcanic eruptions (and these) have been responsible for numerous deaths and significant damage' (Gaspar et al. 2015, p. 34). In many volcanic regions it is increasingly being recognised that risks posed by volcanic earthquakes have been underestimated (e.g. Branca et al. 2017) and in the Azores offshore volcanic eruptions and associated seismic activity, such as those that occurred in: 1638, west of São Miguel; 1800 and 1801, off the coast of Terceira; 1810 and 1811, off the west coast of São Miguel; 1867 near to Terceira and in 1964 close to São Jorge, were particularly problematic (Figure 3.1). Minimal or in some cases no damage was caused by volcanic activity directly, but earthquake losses were significant and in 1801 included two deaths (Gaspar et al. 2015, p. 35).

Table 3.1 Azores: list of volcanic eruptions since the time of settlement

Starting date	Duration	Location	Eruptive style	Volcanic products	Related phenomena
1439/1443	?	Pico do Gaspar, Furnas Volcano, São Miguel	Subplinian, Phreato-magmatic and dome forming	Trachytic pumice, ash & lapilli, lava dome	Seismic activity
September 21, 1562	At least two years	Prainha eruption, Pico	Hawaiian and Strombolian	Basaltic lava flows and pyroclasts	Seismic activity
June 28, 1563	Six days	Summit caldera, Fogo Volcano, São Miguel	Subplinian and phreatomagmatic	Trachytic pumice, ash & lapilli, surges	Seismic activity and secondary lahars
July 2, 1563	c. one month	Pico do Sapateiro, Fogo Volcano, São Miguel	Hawaiian	Basaltic lava flows, spatter and pyroclasts	Seismic activity
February 10, 1564	*Few hours?*	Summit caldera, Fogo Volcano	*Phreatic*		*Earthquake were simultaneous with the first explosion and lahars*
May 1, 1580	About four months	South of Velas, São Jorge	Hawaiian and Strombolian	Basaltic lava flows and pyroclastic and a nuée ardente	Seismic activity. Possible landslides associated with the slope collapse of a volcanic cone
September 3, 1630	At least until November	Furnas Volcano, São Miguel	Subplinian, phreatomagmatic and dome forming	Trachytic pumice, ash and lapilli. Pyroclastic density currents and a lava dome	Seismic activity and landslides
July 3, 1638	About 25 days	Submarine, off the west coast of S. Miguel, Candelária Submarine Volcano	Surtseyan	Submarine and subaerial (?) pyroclasts	Seismic activity

Starting date	Duration	Location	Eruptive style	Volcanic products	Related phenomena
September 19, 1652	About eight days	Pico do Fogo, Picos Fissural Volcanic System	Vulcanian and dome forming	Trachytic ashes and blocks, lava domes and flows	Seismic activity
April 24, 1672	Until March 1673	Fogo, Faial	Hawaiian and Strombolian	Basaltic lava flows and pyroclasts	Seismic activity
December 13, 1682	?	*Off the west coast of S. Miguel*	*Submarine*	*?*	*Seismic activity*
February 1, 1718	Few days in the case of Santa Luzia and until November in the case of S. João	Santa Luzia and Sao. João eruptions, Pico Volcano	Hawaiian, Strombolian and Surtseyan	Basaltic submarine and subaerial pyroclasts, and lava flows	
July 10, 1720	Until December 18	Silveira eruption, Pico	Hawaiian and Strombolian	Basaltic pyroclasts and lava flows	Seismic activity
December 1720	*Until September 1721*	*Banco D João de Castro*	*Surtseyan*	*Basaltic submarine and subaerial(?) pyroclasts*	*Seismic activity*
April 17, 1761	Few days	First Phase, Mistério dos Negros, Terceira	Vulcanian (?)	Trachytic ashes/ blocks and lava domes	Seismic activity
April 21, 1761	At least eight days	Second phase, 1761 lava flow, Terceira	Hawaiian/ Strombolian	Basaltic bombs, lapilli and ash, and lava flows	
June 24, 1800	?	*SSW of Terceira*	*Submarine*	?	*Seismic activity*
May 1 1808	Until 10 June (?)	South of Velas, São Jorge	Hawaiian/ Strombolian/ Phreatomagmatic	Basaltic lapilli and ash, lava flows and pyroclastic density currents	Seismic activity
February 1 and June 14, 1811	One week for both events	Submarine (Sabrina) eruption off the west coast of S. Miguel	Surtseyan	Basaltic submarine pyroclasts, blocks and ash	Seismic activity

Date	Duration	Location	Type	Products	Notes
June, 1867	Visible until June 8	Submarine west of Terceira	Submarine	Basaltic submarine ash and lava balloons (?)	Seismic activity, very intense in the previous six months. Onshore landslides noted
June 7 to June 8, 1902	*?*	*SW of Terceira*	*Submarine*	?	*?*
April 1, 1907	?	*SW of S. Miguel*	*Submarine*	?	*?*
March 7 1911	*About one hour*	*SW of S. Miguel*	*Submarine*	?	*Sea water jets*
September 27, 1957	Until October 24, 1958	Submarine, 800–1,200 m NW of Capelinhos lighthouse, Faial	Surtseyan/Hawaiian/Strombolian	Submarine and subaerial basaltic pyroclasts, volcanic bombs, surges and lava flows	Seismic activity. Highest intensities on May 12–13 1958
May 13–14, 1958	Until October	In caldera of Caldeira Volcano, Faial	Phreatic or Phreatomagmatic	Ash	
December 15, 1963	*?*	*Submarine, North of Pico*	*Submarine*	Lava balloons (?)	*Seismic activity December 13 to 14 and continuous volcanic tremor recorded on the seismograph in Horta*
February 18, 1964		*Offshore of Velas, S. Jorge*	*Submarine*	?	*Seismic crisis between February 15 and 24*
December 18, 1998	*Until August 2001 (?)*	*Submarine, 10 km west of Ponta da Serrata, Terceira*	*Submarine*	*Basaltic lava balloons, submarine ashes and volcanic gases*	*Low-magnitude seismic activity began on November 23*

Adapted from Gaspar et al. (2015). *Italic script indicates submarine eruptions which had little or no impact on people and their activities.*

Table 3.2 Destructive earthquakes and seismic crises that have occurred in the Azores in historical times

Date	M^1	Maximum intensity^2	Related phenomena	Island(s) affected
October 22, 1522		X	Landslides, debris flows and possible tsunami	S. Miguel
May 17, 1547		VII		Terceira
June 28, 1563		X	Volcanic eruption	S. Miguel
July 26, 1591		VIII–IX	Landslides and cracks in the ground	S. Miguel
May 24, 1614	6.3	X	Landslides, cracks in the ground and a tsunami	Terceira
September 2, 1630		VIII	Volcanic eruptions and landslides	S. Miguel
September 3, 1638		?	Submarine volcanic eruption	S. Miguel
October 10, 1652		VIII	Volcanic eruption	S. Miguel
November 14, 1713		IX–X	Landslides and secondary lahars (?)	S. Miguel
1717?				Graciosa
June 13, 1730		VII	Landslides on the inner caldera walls	Graciosa
July 9, 1757		X	Landslides, cracks in the ground, topographic changes and a tsunami affected Terceira, Graciosa and Faial	S. Jorge (also affected Pico and Terceira)
June 24, 1800		VII–VIII	Probably related to submarine volcanic activity	Terceira
January 26, 1801		VII–VIII	Probably related to submarine volcanic activity	Terceira
June 24, 1810		VII–VIII	Probably related to submarine volcanic activity	S. Miguel
June 13, 1811		IX	Volcanic activity	S. Miguel
January 21, 1837		VII		Graciosa
June 15, 1841		IX	Cracks in the ground	Terceira
April 6, 1852		VIII	Sete Cidades Volcano: widespread landslides on the slopes and sea cliffs of the volcano	S. Miguel
June 1, 1867			Related to submarine volcanic activity	Terceira
1868 (month and day unknown)			Landslides	Graciosa
August 31, 1926		IX	Landslides and small ground fissures	Faial (also affected Pico)

Date	M	Intensity	Description	Location
August 5, 1932		VIII	Landslides	S. Miguel
April 27, 1935		IX	Landslides	S. Miguel
November 21, 1937		VII (MM 56)		S. Maria
May 8, 1939	7	VII (MM56)	Small tsunami	S. Maria
December 29, 1950		VII		Terceira
June 26, 1952	5.6	VIII	Landslides	S. Miguel
May 13, 1958		VII–VIII	Volcanic eruption and landslides	Faial
February 21, 1964	4.8	VIII	Volcanic eruption	S. Jorge
November 23, 1973	5.5	VII	One injury and severe damage to the northern part of Pico	Pico (also affected Faial)
January 1, 1980	7.2	VIII	Small tsunami – Terceira and Faial	Terceira (also affected S. Jorge, Graciosa, Pico and Faial)
July 9, 1998	5.8	VIII	Landslides and ground cracks	Faial (also affected Pico and S. Jorge)

1 M = Magnitude
2 Maximum Intensity EMS 98 (Grünthal 1998), unless otherwise stated. MM 56 Modified Mercalli Scale 1956 (Richter 1958, p. 768).
Based on information in Gaspar et al. (2015).

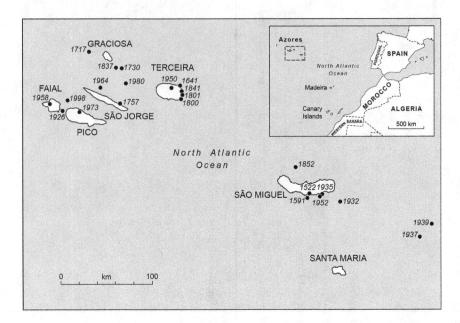

Figure 3.2 Location of the epicentres of destructive historical earthquakes. Based on information from Gaspar et al. 2015, Figure 4.3, p. 37. Copyright permission Geological Society of London.

Historic earthquakes may be sub-divided into three groups: those that occurred before 1900 and comprise a small number of mainly high magnitude events; those which are known instrumentally and occurred from the start of the twentieth century until the end of its seventh decade, so producing a more complete catalogue and post-1980 when the record becomes comprehensive (Nunes 1986). It is highly likely that there were small or even moderately sized earthquakes in the early years of settlement which were not recorded because of a lack of literate observers. There are, for instance, 13 recorded events in the twentieth century, but only 11 in the 278 years from 1522 to 1800. There are also local earthquake generated tsunamis in the historical record, but others are associated with distant fault ruptures and a detailed investigation, including a critical study of source materials has been carried out by Andrade et al. (2006, also see Table 3.3).

A feature that is apparent in Table 3.2 and in the discussion of the individual events that follows, is the shear scale of building losses that have occurred in even moderately sized historic earthquakes. Vulnerability of buildings is discussed in detail in later chapters, but briefly the European Macroseismic Scale (i.e. EMS-98 – Grünthal 1998) classifies buildings into six categories: A – being the least resistant and F – the most resistant. Most traditional buildings in the Azores fall into Classes A and B making them

Table 3.3 Historical tsunamis known to have affected the Azores

Date	Source	Reliability[1]	Cause[1]	Seismic intensity/ magnitude	Tsunami intensity (Papadopoulos and Imamura 2001)[2]	Run-up height (m)	Island(s) affected
22/10/1522	**Local**	**3**	**Seismic**	**X**	**?**	**?**	**S. Miguel**
??/7/1571	Local	−1	Seismic	VII?	III–IV?	1	S. Miguel, Terceira?, Graciosa?, S. Jorge, Pico? and Faial.
26/07/1591	**Local**	**−1**	**Seismic**	**VIII–IX**	**IV–V**	**1**	**S. Miguel and other islands?**
24/05/1614	**Local**	**3**	**Seismic and landslides**	**IX**	**IV–V**	**3**	**Terceira**
21/12/1641	Local	1	Seismic/ meteorological?	/	VI–VII	9	S. Jorge
??/??/1653	Local/ distal	1	Seismic/ meteorological?	/	V–VI	4	Terceira
23/11/1668	Local	1	Seismic/ meteorological?	/	V–VI	7	S. Jorge
??/??/1676	Local	1	Seismic/ Meteorological?	/	V–VI	4	Terceira
26/07/1691	Local	−1	?	/			Terceira
01/11/1755	Distal ('Lisbon' earthquake)	4	Seismic	8.5–8.8	VII–VIII	11–15	All islands
09/07/1757	**Local**	**4**	**Seismic and landslide**	**XI/7.4**	**IV–V**	**1**	**Terceira. Graciosa and Faial**
31/03/1761	Distal	4	Seismic		IV–V	1	Terceira
??/??1787	/	−1	/		III?	1	/

Date	Source	Reliability[1]	Cause	Seismic intensity/magnitude	Tsunami intensity (Papadopoulos and Imamura 2001)[2]	Run-up height (m)	Island(s) affected
23/01/1792	/	1	/	/		8	S. Jorge
09/07/1847	Local	4	Large scale landslides in sea cliffs. May have been seismically induced		V–VI	30?	Flores and also Corvo
1/02/1855	Local to distal	1		/	V?	5–10?	Terceira
06/01/1856	Local to distal	1	Seismic/submarine slide		V–VI	10	S. Jorge
03/02/1899	Local to distal	1	Seismic/submarine slide		V–VI	10	Faial and S. Jorge
31/08/1926	**Local**	**−1**	**Seismic**	**X/5.6–5.9**	**II–III?**	**0.5**	**Faial**
31/08/1931	Local	1	Seismic?		IV–V?	1	Faial
22/08/1935	Local	1	Seismic?		/	/	Faial
08/05/1939	**Distal**	**4**	**Seismic**	**VII/7.1**	**II**	**0.5**	**Santa Maria and Terceira**
25/09/1940	Local	4	Meteorological?		/	/	Flores and Terceira
25/11/1941	Distal	4	Seismic	8.2	I–II	0.5	Terceira, S. Miguel and Faial
28/2/1969	Distal	4	Seismic	7.3	II–III	1	S. Miguel, Terceira and Faial
26/05/1975	Distal	4	Seismic	IV–V/7.9	III–IV	1	S. Miguel and Faial
01/01/1980	**Local**	**4**	**Seismic**	**VIII/7.2**	**I–II**	**0.5**	**Terceira and Faial**
22/05/1980	Local	4	Landslide	/	/	/	Flores

1 Reliability is based on an assessment by Cabral (2020): −1, erroneous; 0, event only caused a seiche or disturbance in an inland river; 1, very doubtful; 2, questionable; 3, probable and 4, definite.

2 The Papadoulos/Imamura scale runs from I ('not felt') to XII ('complete devastation') Bold type indicates that the tsunami was associated with a destructive earthquake. Based on information in Andrade et al. (2006); Calado et al. (2011); Gaspar et al. (2015); Baptista et al. (2016) and Cabral (2020).

extremely vulnerable to seismic shaking (Neves et al. 2012). In more recent earthquakes the building stock has become more varied and some buildings show greater resilience.[5] For masonry buildings damage is classified into five 'grades' (Grünthal 1998): Grade 1 – negligible to slight damage (no structural damage, slight non-structural damage); Grade 2 – moderate damage (slight structural damage. moderate non-structural damage); Grade 3 – substantial to heavy damage (moderate structural damage, heavy non-structural damage); Grade 4 – very heavy damage (heavy structural damage, very heavy non-structural damage) and Grade 5 – destruction (very heavy structural damage). In historic earthquakes Azorean buildings have frequently shown damage equal to or greater than Grade 3.

Losses in earthquakes are not solely related to size and building vulnerability. Shallow focus earthquakes are, for instance, more destructive than deep-focus ones (Bolt 2004, pp. 39–41) and in recent years it has been recognised that a further consideration is *three-way harmonic interaction* which involves the resonance coupling of earthquake waves, surficial deposits and buildings (Degg 1995). First researched intensively following the Mexico City earthquake of 1985, where damage was enhanced on lake-bed deposits (Degg 1992), in Iberia generally and particularly in the Lisbon area (Mendes-Victor et al. 1994) and in the Algarve (Chester & Chester 2010), this is also highly relevant because of the wide diversity of rock types and associated soils which have varying seismic properties. In the latter area soils and sediments have been classified into several categories based on their seismic properties, particularly their average shear wave velocity (V_s): the lower the V_s value the higher the seismic enhancement (Roca et al. 2008; Carvalho et al. 2009) and *vice-versa*. Being almost exclusively volcanic, substrates in the Azores are less varied than those found on the mainland, but soil-amplification has been a factor in losses associated with the 1980 and 1998 earthquakes, the two most recent large events to have affected the archipelago (Table 3.2, and see below). This begs the question of whether the 'soil-factor' may have been of importance in earlier historical events for which relationships between buildings, soils and substrate geology are difficult and in some cases impossible to establish without detailed contemporaneous on-site surveys (Teves-Costa et al. 2007).

Calculations by Bebo (1992) and Oliveira and Ferreira (2008) indicate that the direct costs of the 1980 Terceira and the 1998 Faial earthquakes (at 1980 and 1998 costs) were, respectively, 125 million Euros and 300 million Euros. In 2015, costs of a similar catastrophe were estimated at over half the annual budget of the Regional Government (Gaspar et al. 2015, p. 37).[6]

Historic eruptions

São Miguel

There are five active volcanic systems on São Miguel (Gaspar et al. 2015) which are from west to east: Sete Cidades; the Picos Fissural Volcanic System (PFVS); Fogo; the Congro Fissural Volcanic System (CFVS) and Furnas. Sete

Cidades, Fogo and Furnas are trachytic central volcanoes linked by the two fissural volcanic systems (Figure 2.2). Fogo, Furnas and the Picos Fissural Volcanic System (PFVS) have all been active in historic times. Sete Cidades and CFVS have erupted in the last 5,000 years and, indeed, Sete Cidades is considered to have been the most active volcano in the Azores over this period (Queiroz et al. 2015). Weston (1964) and Zbyszewski (1963) describe tentative and uncertain historic accounts of an eruption of Sete Cidades in 1439, based on evidence of floating pumice and tree trunks on the sea and, when viewed from the sea, the disappearance of a peak, but there are no contemporary accounts of a volcanic eruption. Queiroz et al. (2015), in a detailed field investigation of Sete Cidades, identify the youngest eruption as occurring about 600 years ago and this would be just before the island was settled. This is supported by ^{14}C and U-series disequilibria work on these younger deposits of Sete Cidades volcano by Conte et al. (2019, p. 75), who consider the carbon ages associated with the youngest eruption to be robust. Queiroz et al. (1995) argue that the early accounts of an eruption at Sete Cidades actually relate to the 1439–1443 eruption of Furnas described below.

Furnas

Furnas volcano does not have a well-defined edifice and in terms of topography is the least impressive of the three trachytic central volcanoes of São Miguel (Guest et al. 1999). The volcano has a distinctive nested caldera complex (Figure 3.3) with the Main (outer) caldera elliptical in shape, 8 × 5 km in diameter and formed about 30 ka years ago. The Inner Caldera cuts the fill of the Main Caldera and is 6 × 3.5 km in diameter and about 12 ka in age (Guest et al. 2015). The formation of both calderas was probably associated with major Plinian scale eruptions.

In the last 5,000 years there have been ten trachytic eruptions at Furnas which from oldest to youngest are labelled A, B, C, D, E, F, G, H and I with the tenth and youngest being the 1630 eruption (Booth et al. 1978; Cole et al. 1995; Guest et al. 1999), the largest of these was Furnas C, 1870±120 BP (Guest et al. 2015). It is argued (Queiroz et al. 1995; Guest et al. 1999), that Furnas I is the eruption that was witnessed by the first settlers of São Miguel between 1439 and 1444 (Queiroz et al. 1995).

1439–1443 Eruption It is thought that the first settlement on São Miguel was at what is now the village of Povoação (Dias 1936; Bento 1994). In an analysis of historic documents, Queiroz et al. (1995) describe that during their first year these early settlers were terrified by loud noises, lightning and 'tongues of fire' coming from the Furnas caldera which lay around 5 km to the west. A priest climbed up to the edge of the Furnas caldera to see what was happening. He saw vapour rising from a barren depression covered in white material and the vapour sometimes glowed red. Guest et al. (1999) infer that the white material was fresh trachytic tephra and the red glow was a reflection from fresh lava in an active dome.

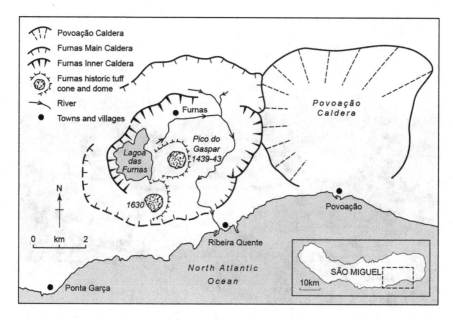

Figure 3.3 The nested calderas of Furnas Volcano, São Miguel, showing the sites of the 1439–1443 and 1630 eruptions. Based on information in Guest et al. 1999 and 2015, Figure 9.2, p. 126. Copyright permission Geological Society of London.

Queiroz et al. (1995) and Guest et al. (1999) argue that the priest was witnessing the activity of the Pico do Gaspar dome and this was the last phase of the Furnas I eruption. This interpretation accords with both the geographical and volcanological contexts. The Furnas I deposits sit on top of products of Furnas H with little evidence of any significant time gap in terms of a soil or weathering horizon (Booth et al. 1978). Furnas H is dated with a likely age range of between 1410 and 1440 by radiocarbon using charcoal fragments within the eruptive products (Queiroz et al. 1995; Guest et al. 1999). This supports the proposition that Furnas I, whose deposits are from the only eruption between Furnas H and the 1630 eruption is the eruption witnessed by the first settlers.

The 1439–1443 Furnas I eruption formed the Pico do Gaspar dome and surrounding tuff/pumice cone (Figure 3.4) on the east side of the current Furnas lake. The explosive eruption was sub-Plinian in scale (Booth et al. 1978) and produced alternating ash and lapilli layers interpreted by Cole et al. (1999) as switching between magmatic (pumice lapilli fall layers) and phreatomagmatic (tephra fall and PDC) activity. Relatively small PDCs would probably not have been able to overtop the high caldera wall to the east. Only tephra fall up to 20 cm in thickness was deposited in the village of Povoação the likely site of the first settlement. There are no historic records of this

Figure 3.4 View of Pico do Gaspar lava dome (marked a) and surrounding tuff
cone of 1439–1443 eruption Furnas Volcano. The edge of Furnas Lake is
marked b. Authors' photograph.

phase of the eruption, which must have preceded the first settlement, as the
tephra fall and vigorous eruptive column rising above the western rim of the
Povoação caldera would have had a most frightening impact if the settlers had
been present at this stage. It is likely that this first phase lasted no more than a
few days; the explosive phase of the similar 1630 eruption being about three
days in duration. The dome building phase which was witnessed by the priest
may have lasted several months.

The 1630 Eruption The explosive phase of the eruption started on 3 September at the southern end of Furnas lake. This corresponds to the position of two smaller lakes noted by the priest who observed the final stage
of the 1439–1443 eruption (Guest et al. 1999). One of these smaller lakes,
Escura, had dark and malodorous water and the other, *Barrenta,* was muddy
(Dias 1936). These small lakes were destroyed during the eruption (Cole
et al. 1995). The 1630 eruption was the most damaging to have occurred
in the Azores in historic times, the settlement of Furnas was destroyed, and
it is estimated that there were up to 195 fatalities (Corrêa 1924; Dias 1936;
Guest et al. 1999, p. 25). The stratigraphy of this eruption was first studied by
Booth et al. (1978). A detailed consideration of this eruption is provided by
Cole et al. (1995, pp. 118–122) and only a summary is presented here.

There were felt earthquakes for at least six hours before the start of the
eruption, and these were experienced as far away as 30 km at Ponta Delgada

(Jerónimo 1989; Cole et al. 1995). One source (Corrêa 1924) notes that earthquakes were felt, 18 hours before the start of the eruption. Earthquakes were strong and continuous causing church bells to ring (Cole et al. 1995). The earthquakes caused considerable damage and there were reports that most houses in Ponta Garça and Povoação were destroyed (Anonymous account extracted from Canto 1881a, pp. 527–547). These earthquakes caused landslides and a collapse of the sea cliffs to the south of Furnas, that generated a debris flow which covered part of the area now occupied by the village of Ribeira Quente (Figure 3.3), although there was no permanent habitation in the seventeenth century (Cole et al. 1999).

The eruption began with an explosive phase between 2 and 3 am on 3 September with descriptions (Cole et al. 1995) of: 'clouds of fire rising from the two lakes' described by Manoel Purificação (Guest et al. 1994, p. 15) and 'fire in the sky that lit up the whole island' (Corrêa 1924). Friars[7] who had a religious building in Furnas caldera were busy on the night of 2 September rescuing valuables from their church which had been damaged by the earthquakes. Following the loud explosions at the start of the eruption, they described a sight of a 'mountain rising through the air' (Purificação, cited in Canto 1881a, p. 528). The eruption cloud was blown towards the south and the friars fled north to the villages of Maia and Porto Formoso on the north coast claiming rather selfishly that 'God was very good to us: first we thought the hill would come to us, but instead it went the other way and killed the others' (Purificação, translated in Guest et al. 1994, p. 15). Thirty people who were ready to start picking fruit in the morning were spending the night near to where the eruption began and were killed by its first explosions (Dias 1936). As the eruption progressed, there are descriptions of people being burnt to death in the fields to the SW (Purificação, cited in Canto 1881a, pp. 527–547; Corrêa 1924) and this, together with the report by Jerónimo (1989) of 'a burning stream that spread through the woods,' enabled Cole et al. (1995, p. 121) to argue that these reports are consistent with low density PDCs (pyroclastic surges) flowing in the direction of Ponta Garça. After the initial explosive phase, which deposited a coarse lapilli layer to the SW, the eruption continued with a style of alternating magmatic and phreatomagmatic activity (Cole et al. 1995): with the magmatic phases depositing layers of lapilli fall and the phreatomagmatic phases producing ash with accretionary lapilli and widespread vesicular ash. Cole et al. (1995) note that these ash-rich layers show lateral thickness variations and this, together with cross-bedding and sandwave bedforms, are indicative of deposition by pyroclastic surges. It is argued by Cole et al. that massive poorly sorted flow deposits that occur on the south coast, were deposited by high concentration PDCs that travelled down steep-sided valleys trending away from the eruption site. By the morning of 4 September, the whole island was covered by an ash cloud and was in darkness (Cole et al. 1995, p. 121). Initially ash was deposited west towards Ponta Delgada, but by the final explosive magmatic phase of the eruption the wind changed direction and was blowing strongly from the SW depositing tephra to the NE. The explosive sub-Plinian phase of the eruption lasted only a few days, three days being suggested by Cole et al. (1995). This was followed by the emplacement of a trachytic lava dome which continued until early November (Guest et al. 1999).

It is estimated that up to 195 people were killed (Corrêa 1924; Dias 1936; Guest et al. 1999), although a minority of historical sources give a figure of up to 300 (Madeira 2005, p. 22) more than by any other eruption in the Azores. From the historic accounts there were the 30 deaths of the fruit pickers camping near the two small lakes killed in the initial explosions of the eruption and around 80 were burnt to death by the pyroclastic surges near Ponta Garça. It is not known how many deaths occurred from building collapse caused by seismic activity or from roof collapse as a result of tephra accumulation. In Ponta Garça it was reported that only one house and the chapel remained (Canto 1881a, pp. 527–547) and in Povoação only one out of 250 houses survived the earthquake damage (Jerónimo 1989). Survivors from Ponta Garça were forced to evacuate, as also were the inhabitants of Vila Franca do Campo, a town on the coast about 10 km west of the eruption site (see Chapter 5). The southern half of the Furnas caldera was buried in over one metre of tephra (Booth et al. 1978). The thickness of ash in Ponta Garça was reported to have been around 6 m (Canto 1881a, pp. 527–547), but Cole et al. (1995) argue that this was probably due to the ash banking up against buildings and that the estimate of Corrêa (1924) of less than 3 m is more likely to be the average thickness. There was 1.5 m of ash at Vila Franca do Campo and 10 cm at Ponta Delgada, some 30 km to the west. In addition to the devastation in Furnas, Ponta Garça and Povoação, the eruption must have had an impact across the island with ashfall affecting livestock grazing and crops (Canto 1881a, p. 539). The inhabitants of Vila Franca do Campo returned after about seven days (Guest et al. 1999), but recovery across the affected area especially in Furnas was slow in the immediate vicinity of the eruption and is more fully discussed in Chapter 5.

Fogo

Fogo volcano (also known as Água de Pau) is the largest of the trachytic central volcanoes of São Miguel and dominates the centre of the island (Figure 3.5). In the last 10 ka, there have been a number of major explosive eruptions including two of Plinian scale: Ribeira Chã, 8–12 ka. and Fogo A at 4.6 ka (Wallenstein 1999; Wallenstein et al. 2015). Fogo was active in 1563–1564 with two linked eruptions: a sub-Plinian trachytic eruption from the summit caldera, followed by a basaltic eruption from low on the NW flank (Wallenstein et al. 2005, 2015). Even when not active Fogo displays several hazards including earthquakes, fumarolic activity, landslides and floods (Wallenstein et al. 2007).

1563 Trachytic Summit Caldera Eruption This eruption was preceded by felt seismic activity that started on 23 June 1563, the earthquakes causing damage

Figure 3.5 Fogo Volcano, São Miguel. Based on information in Wallenstein et al. 2015, Figure 8.4, p. 108. Copyright permission Geological Society of London.

to many buildings in the settlements of Ribeira Seca and Ribeira Grande which are located on the northern coast 7.5 km from the summit lake of Fogo (Wallenstein et al. 2015). The eruption began on the evening of 28 June from a vent in the middle of the caldera lake. The eruption of trachytic magma was explosive and of sub-Plinian style and, during the first 24 hours, deposited ash and lapilli on the eastern side of the island. On 29 June, the wind changed to the NE causing 10 cm of ash to fall on Ponta Delgada. For the next three days the wind was westerly depositing tephra on the eastern half of the island particularly affecting towns and villages such as Ribeira Grande, Porto Formoso, Maia, Achada, Nordeste, Povoação, Furnas and Vila Franca do Campo (Wallenstein et al. 2015). Explosive activity ended on Saturday, 3 July, with the eruption having lasted 6 days. The eruption deposited a thick mantle of tephra across the island to the east (Walker & Croasdale 1971), with *c.*0.5 m being found on the east coast at a distance of *c.*30 km (Figure 3.6). This caused roof collapse as far away as Nordeste (Wallenstein et al. 2005) and must have destroyed crops particularly on slopes leading down to the north coast. It would have rendered pastures and grazing land sterile for many months Wallenstein (1999). Further damage resulted from debris flows which

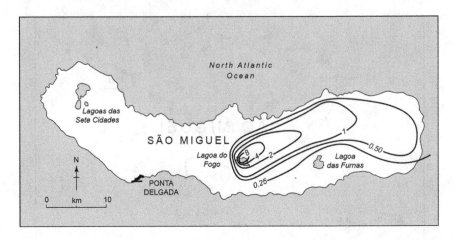

Figure 3.6 Isopachs (m) of the tephra fall of the 1563 sub-plinian eruption of Fogo
  Volcano, São Miguel. Based on Booth et al. 1978. Copyright permission
  Geological Society of London.

swept down valleys to the south coast and there was severe flooding in Villa
Franca do Campo (Wallenstein et al. 2005). Earthquakes in the five days
before and during the eruption, caused destruction of houses in Água de Pau
and Lagoa in addition to damage in Ribeira Seca and Ribeira Grande. There
were no deaths recorded during the eruption, but two people who visited
the caldera after the activity ended died after becoming overcome by gases
(Wallenstein 1999).

On 10 February of the following year, a large explosion was described by
Frutuoso (Canto 1880a, p.186), as having taken place in the caldera near the
site of the 1563 summit eruption. Frutuoso records that a crater opened, from
which thick smoke emerged and a *liqueur* flowed down the mountain towards
the south coast. It is likely that this was a phreatic explosion caused by water
(from winter rainfall?) coming into contact with the still hot material below the
vent of the 1653 eruption: the *liqueur* flow being a lahar (Wallenstein et al. 2015).

1563: The Pico do Sapateiro (Pico Queimado) Basaltic Flank Eruption on Friday
2 July, the day before the end of the 1563 summit activity, a basaltic fissure
eruption began on the lower NW flank of Fogo (Figure 3.5). The eruption
occurred along the line of a NW-SE fault that cuts the Pico do Sapateiro trachyte dome, but following the eruption this was renamed Pico Queimado[8]
(Wallenstein et al. 1998, 2015). It is probable that the trachytic summit eruption and the basaltic flank eruption are linked, and it is likely that the eruption from the zoned trachytic magma reservoir beneath Fogo was triggered
by the injection of basaltic magma into the base of the reservoir. Some of the
ascending basaltic magma was then injected laterally as a dyke along the line
of the NW-SE trending fault intersecting the surface at Pico Quiemado. It
should be noted that this lateral injection is along the line of the Picos Fissual

Volcanic System where it meets the flanks of Fogo Volcano (Ferreira et al. 2015). Although the eruption site is now covered in dense vegetation, the line of craters with spatter ramparts may be clearly observed.

The initial stage of the eruption was of Hawaiian style with eruption of ballistic material and a dark eruptive column developed as described by Frutuoso (Canto 1879a, pp. 458–466, 536–541) which was visible from Terceira according to an account by M. Menezes (Canto 1879a, pp. 456–458). After two days of such activity along the fissure and again according to Frutuoso, activity increased and a lava flow was erupted which flowed in a direction just east of north for about 1.25 km and reached Santa Bárbara. It then flowed for a further 3 km north down the river channel to the village of Ribeira Seca and then went on to reach the sea during the night of 6 July (Figure 3.5). It took the lava two days to reach Ribeira Seca which it partly destroyed. In the centre of the village the site where the lava buried the fountain has been preserved. The lava just missed the church and to commemorate this safe deliverance there is an annual ritual involving a group of horseman (the *Cavalhadas de São Pedro*) riding round the church (Chester et al. 2019, see Chapter 5). A second lava flow was erupted on 9 July, and this flowed in a NW direction across wheat fields for *c*.1.5 km. The date of the end of the eruption is not known, but there are records of activity still occurring on 28 July.

Sete Cidades

As discussed above there are no authenticated eruptions on Sete Cidades during historical times, but eruptions have occurred in the sea close (<3 km) to the coastline of the volcano. These offshore eruptions date from: 1638 and two events occurred in 1811, but tephra from these eruptions have not been identified at any onshore location (Queiroz et al. 2015).

1638 Submarine Eruption This eruption broke surface on 23 July and activity continued for 25 days (Weston 1964). An island emerged about 3 km west of Ponta da Candelária (Figure 2.2) and built up to height of around 120 m before being destroyed by marine erosion (Gaspar et al. 2015). Seismic activity associated with the eruption destroyed some houses in Várzea (Gomes et al. 2006).

1811 (February) Submarine Eruption During July and August 1810 the western part of São Miguel was affected by earthquakes of an Intensity VII–VIII (Gomes et al. 2006). Smaller earthquakes continued to be felt infrequently until January 1811 and on the last day of the month a much larger shock occurred (Webster 1821), to be followed by a number of earthquakes, of which 31 were counted in Ponta Delgada. On the next day, 1 February, 'a strong sulphurous smell confirmed the report of fire and smoke that issued from the sea opposite the village of Ginettes (*sic*), at the distance of about two miles from the shore' (Webster 1821, pp. 139–140). This was the start of the eruption which lasted 8 days (Figure 2.2). According to Webster an eruption column was formed and with a westerly wind ash was deposited on Ponta Delgada. When the eruption ended it left 'a shoal on which the sea broke' (Webster 1821, p. 140). Earthquakes associated with the eruption caused damage to many houses on the western coast of São Miguel.

1811 (June) Submarine Eruption Later that year a new submarine eruption took place on 13 June, *c*.1.5 km SW of Ponta da Ferraria (Queiroz et al. 2015). The eruption site was just under 3 km south of that of the February eruption (Figure 2.2). The first signs were violent earthquakes and a sulphurous smell followed by 'flames' and ashes erupting from the surface of the sea (Webster 1821, p.141). On 14 June, Commander James Tillard, captain of the British sloop H.M.S. Sabrina which had arrived at Ponta Delgada, travelled over-land to witness the eruption from the coast. A description of the eruption by Tillard was published in the *Philosophical Transactions of the Royal Society of London* (1812). He describes the eruptive column rising from the sea 'when suddenly a column of the blackest cinders, ashes, and stones would shoot up in the form of a spire at an angle from 10 to 20 degrees from a perpendicular line' (Tillard 1812, p.153) which suggests Surtseyan style tephra jets. The eruptive column displayed flashes of lightning. Tillard notes that the sea was more than 30 fathoms (*c*.55 m) deep at the site of the eruption. After Tillard and his companions arrived at the cliff overlooking the eruption site they observed a 'peak' developing out of the sea and after about three hours a complete crater had formed to a height of about 7 m asl, the diameter of the crater being more than 100 m. Earthquakes were noted by Tillard including a large shock which caused part of the cliff to collapse. On the next day, 15 June, Tillard took H.M.S. Sabrina along the coast to witness the eruption from the sea but was frustrated by poor visibility and when the sloop sailed under the eruptive plume about 5–6 km from the volcano, fine black ash covered the ship. The Sabrina departed from São Miguel on 16 June, and after a week of activity the eruption subsided. The seismic activity caused considerable damage to buildings in the vicinities of Ginetes, Várzea and Mosteiros (Figure 2.2).

Commander Tillard returned to São Miguel on 4 July after the eruption had finished and with a party of officers visited the new volcanic island. He described the circumference of the island as just less than a mile (1.61 km). There was a well-developed crater which was breached on the side facing São Miguel. The island, which was named Sabrina after Tillard's ship, was washed away by marine erosion in October.

Picos fissural volcanic system (PFVS)

The Picos Fissural Volcanic System (PFVS), previously called the Waist Zone by Booth et al. (1978), forms land that stretches for *c*.23 km between the Sete Cidades and Fogo central volcanoes (see Chapter 2). Alignments of cones and eruptive fissures form two main volcanic ridges: the North-ern Volcanic Alignment that develops westwards from Fogo and the sub-parallel Southern Volcanic Alignment, that extends eastwards from Sete Cidades (Ferreira et al. 2015). In 1652 there was an eruption from the Pico do Fogo area on the Northern Volcanic Alignment about 6 km north of Lagoa (Figure 2.2).

1652 Eruption The 1652 eruption was distinctive in that the style of activity was Vulcanian and erupted three trachytic domes (Ferreira et al. 2015). Previous workers (Webster 1821; Agostinho 1936; Weston 1964) considered that the 1652 eruption was basaltic and generated a lava flow field that extended to the north coast. Moore (1991a) shows this interpretation on his geological map of São Miguel, but also included a lava that flowed to the south coast. Booth et al. (1978) argued that the 1652 eruption was a small explosive Vulcanian style event of trachybasaltic composition, however, that occurred on the side of the Fogo cone and which they considered to be of prehistoric date. Ferreira et al. (2015) support the interpretation of Booth et al. and further confirmation is provided by palaeomagnetic data, with Di Chiara et al. (2014) showing that the southern lava flow yielded an inclination *c.*15° lower than the Historic Geomagnetic Field (*gufml*) predictions for the mid-seventeenth century and supporting an older date of around 800 CE.

Ferreira et al. (2015) provide a reconstruction of the eruption based on a detailed analysis of the recent literature, historic documents and field investigation. The area in which the 1652 eruption occurred has been known from the time of the eruption as Pico do Fogo, with two cones being named Fogo 1 and Fogo 2 (Figure 3.7). Ferreira et al. (2015, pp. 140–141) from an analysis of historical documents, consider that these cones pre-date the eruption, are pre-historic and relate to cones which before 1652 were known as the Pico de João Ramos (Fogo 1) and the Pico do Paio (Fogo 2). Booth et al. (1978) did not consider that the eruption was associated with Fogo 1, but occurred from a crater on the side of Pico do Paio (Fogo 2) which is nearby.

The eruption began just before sunset on 19 October and was preceded by several days of localised intense seismic activity (Gaspar et al. 2015). A translation of a contemporary description by Luiz Mendes de Vasconcellos (1652) by Ferreira et al. (2015, p. 140) 'describes the onset of the eruption as being marked by the appearance of small clouds of smoke that soon rose at a higher speed and, that by one hour after sunset, flames of fire began to be seen along with huge roars and hideous rumblings, with black dense clouds projecting large blocks into the night sky.' During the night ash was deposited to the west on vineyards at Atalhada and Rosto do Cão, which lie between Lagoa and Ponta Delgada (Ferreira et al. 2015). Following their investigation of historic sources and field study, Ferreira et al. *op cit.* consider that, in association with explosive Vulcanian explosions, three small trachytic lava domes were erupted along a NW-SE fissure on the flanks of the Pico do Paio and Pico de João Ramos scoria cones (Figure 3.7). The eruption reached its peak on 20 October and further explosive activity lasted for eight days.

The impact of the eruption was localised. The associated seismic activity caused damage in Lagoa and nearby villages (Gaspar et al. 2015). The start of the eruption which was explosive and developed through the night of 19 October, caused much concern with priests seeking support from the local

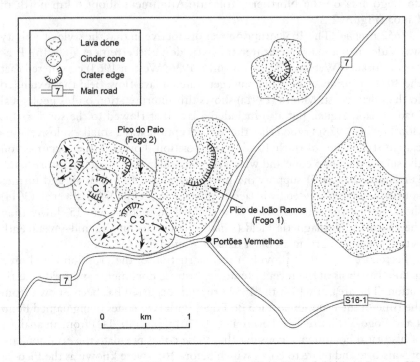

Figure 3.7 Features of the 1652 eruption in the *Picos Fissural Volcanic System*, São Miguel. Adapted from Ferreira et al. 2015, Figure 10.7, p. 142. Copyright permission Geological Society of London.

Captain-Major (Ferreira et al. 2015 citing Monte Alverne 1695). Ash fall was the main cause of damage particularly to agriculture in the coastal region between Lagoa and Ponta Delgada. It was fortunate that the wind was from the north and carried most of the tephra out to sea. This eruption was unusual for the PFVZ in being trachytic and explosive rather than basaltic and effusive which is the usual style of event in this region. The 1652 eruption should not be seen as a typical exemplar, therefore, of eruptions and their associated impact in the PFVS.

Pico

On Pico there have been four onshore eruptions during historic times (Figure 3.8). The first eruption (Prainha) occurred from 1562 to 1564 on the spine of the São Roque-Piedade Volcanic Complex. The second eruption (Santa Luzia) took place starting on 1 February 1718 on the northern slopes of Pico Mountains, sending lava towards Santa Luzia and this was followed

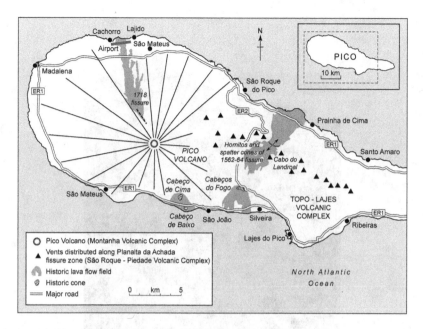

Figure 3.8 Pico island: the location of historic volcanic eruptions. Based on França et al. 2006, Figure 2, p. 73 and other sources. Copyright permission Elsevier.

by the third eruption (São João) a day later on the lower SE slopes of the mountain near to the village of São João. There was also a vent offshore. The fourth and most recent eruption occurred in 1720 at Silveira by the coast just west of Lajes do Pico. The 1718 and 1720 eruptions are the most recent events of the Montanha Volcanic Complex. The 1562 to 1564, 1718 (São João) and 1720 eruptions generated basalt lava flows. In contrast, the 1718 (Santa Luzia) erupted lavas ranging in composition from mugearite to just into the benmoreite field of the total alkalis/silica plot of Le Bas et al. (1986) and are among the most evolved lavas exposed on Pico (Woodhall 1974; França et al. 2006). As discussed in Chapter 2, this could represent the early stages in the development of a high-level magma reservoir in Pico volcano (França et al. 2006; Zanon et al. 2020).

An investigation of the hazard posed by lava flows on Pico has been undertaken by Cappello et al. (2015). Cappello et al. compared the spatial probability of future vent locations from an analysis of the distribution of eruptive fissures on Pico and, using eruption data from the four historic eruptions and simulation of likely lava flow paths, used the MAGFLOW model to generate a lava flow hazard map (see Chapter 4). The impacts of these eruptions from the *São Roque-Piedade* and *Montanha* volcanic complexes are considered below.

Figure 3.9 The lava delta formed on the north coast of Pico during the Prainha eruption of 1562–1564. Note Pico Volcano in the background. Authors' photograph.

São Roque-Piedade complex

1562–64 (Prainha) Eruption The eruption is generally considered to have started around midnight on 20/21 September 1562 (Madeira 1998; França et al. 2014; Gaspar et al. 2015), though some earlier sources (Frutuoso, quoted by Canto 1879b, p. 366) state that the eruption began on 22 September. This was the longest duration historic eruption of the Azores and lasted around two years, though a precise end-date is unknown. It is probable that the end of this eruption would have been difficult to recognise at that time, because these were tube fed lavas building a delta into the sea (Figure 3.9).

Activity was preceded by about a month of felt seismic activity which ceased 4 to 5 days before the onset of eruption (França et al. 2006), after which a fissure about 500 m in length opened on the central spine of the island about 0.7 km north of Cabo Do Landroal (Figure 3.8). On 23 September, a series of seven spatter cones (Figure 3.10) developed along the line of a WNW-ESE trending fissure (Nunes 1999). During the initial two days, there was explosive activity, and on 23 September activity was such that lapilli were being deposited on Pico and ash on São Jorge (Weston 1964). On 24 September, lava was erupted from vents along the fissure and flowed north advancing quickly to the coast. Contemporary accounts from a Spanish ship sailing from the West Indies to Horta in Faial describe 12 rivers of fire entering the sea (Canto 1879b, pp. 363–365). Lava also flowed to the south covering flat ground between older cones (França et al. 2006). On the night of 26 September, two "large fires" were observed on Serra Ventosa and there was significant ash fall on São Jorge (Nunes 1999). It is possible that the description by Gaspar Frutuoso of two *large fires* refers to Hawaiian style fountaining and (Canto, 1879b, p. 366 our translation) records '*The great fire rose illuminating the land all round and could be seen from São Miguel*' (a distance of about 250 km). As Nunes (1999) comments, historic records provide only scanty information about the details of the eruption but it is likely, that in common with many long-lasting effusive eruptions, the more vigorous explosive activity occurred in its very early stages.

A series of spatter cones and hornitos developed along the fissure (Figure 3.10). The lavas which flowed to the south formed a spread of thin *aa* lava that filled a shallow depression between older prehistoric cones. In places low tumuli formed from which small *pahoehoe* lava breakouts occurred. Most of the lava flowed down the steep slope to the NE initially at least with several *aa* channels, as witnessed by Spanish sailors in September 1562. A likely initial high effusion rate and steep slope would both favour open channel *aa* flows. Where recent eruptions have been observed on Etna 1983 (Guest et al. 1987; Duncan et al. 2004) and Pu'u 'Ō'ō, Kilauea (Kauahikaua et al. 2003), *aa* channels on steep slopes tend to roof over forming lava tubes. It was noted at the time that lava built out into the sea as described by Frutuoso (Canto 1879b, p. 366 our translation) '*...large stream of lava flowed to the north, for a league*[9] *and a half to dropdown a steep slope and form a great platform of lava rock.*' Lava flowed over the sea cliff

Figure 3.10 Eruptive fissure of the 1562–1564 Prainha eruption of Pico showing the line of well-developed spatter cones. Authors' photograph.

and built a delta into the sea to the west of Prainha de Cima (Figure 3.9), the original sea cliff can be clearly seen in this photograph), and tumuli can be observed on the surface of the delta. In a study of the 1792 lava flow field on Etna, Guest et al. (2012) demonstrated that tumuli formed at the foot of breaks of slope where tube-fed lava slowed as it encountered a shallower gradient. This led to an increase in pressure and the formation of ephemeral vents that erupted *pahoehoe* lava. Mitchell et al. (2013) used bathymetric data to analyse the submarine morphology of lava flows on Pico that had entered the sea and observed that *pahoehoe* flows, such as those of 1562–1564, formed lava deltas similar to the situation on Kilauea (Hawaii) in 1969–1971 (Moore et al. 1973). Poland and Orr (2014) describe how such lava deltas are unstable, can collapse causing explosions and represent a hazard to onlookers. Mitchell et al. (2013, Figure 14, p. 15) imaged the morphology of the 1562–1564 lava flow field from the shore down to a depth of 200 m below sea level.

The lava field covered an area of 12.9 km^2 (Cruz et al. 1995) and destroyed agricultural land and some farms mainly between the vent area and the coast. Macedo (1871) describes the devastation of land, houses and the parish church in the *freguesia* (i.e. parish), known as Nossa Senhora d'Ajuda da Pranha do Norte. Today this small settlement is called Prainha de Cima and the church of Nossa Senhora d'Ajuda is located in Prainha de Baixo, some 3 km to the SW and near the coast. Both Prainha de Cima and Prainha de Baixo form part of the *freguesia* of Prainha. The lava flow also affected the western part of the freguesia of São Roque do Pico in the place of Nossa Senhora da Piedade. In addition to direct damage by lava, Weston (1964, p.5) refers to lava setting 'alight thousands of bushes and tall trees that were half dry in the autumn, and forming a sea of fire hundreds of acres (i.e. *c.*202 ha) in extent.' When lava is erupted, wild fires are clearly a major category of hazard especially with regards to certain vegetation types under dry weather conditions and at certain times of the year.

The area of the 1562–1564 flows remains unproductive as far as agriculture is concerned, and the slopes leading down to the sea are largely forested with the area being known as the *Mistério da Prainha*. Even though the direct physical effect of the eruption on the population of Pico was limited, it seems to have had a major social impact. The eruption and associated seismic activity caused great fear amongst the inhabitants of Pico and many people fled to São Jorge, Faial and Terceira (see Chapter 5). Frutuoso (Canto 1879b, p. 366) states that the island became depopulated, though the actual reduction in numbers is unknown. It may be that this effusive eruption caused so much consternation because the initial vigorous fire fountaining was visible across much of the island and this, taken together with the extensive precursory seismic activity, would be alarming because this was the first eruption to be properly experienced in the Azores since they were settled, though this eruption was followed shortly after by the sub-Plinian eruption in 1563 of Fogo volcano on São Miguel (see above).

Montanha Volcanic complex

1718 (Santa Luzia) Eruption There were three phases in the Pico eruptions of 1718. First, a fissure trending NNW-SSE cut across Montanha do Pico (Pico Mountain) on 1 February, and the first eruption took place on the northern flank of the volcano sending a lava flow to the north towards Santa Luzia. Second, on the following day, vents opened near São João on the lower southern flank and produced another flow field. Finally, on 11 February, the third a submarine eruption occurred just offshore from São João.

The *Santa Luzia* eruption was unusual because it erupted relatively evolved lavas of mugearitic/benmoreitic composition, whereas the other historic eruptions were basaltic (Cappello et al. 2015). As discussed in Chapter 2, França et al. (2006) argue that the mugearitic/benmoreitic magma formed at shallow depth as a result of fractionation of magma stored within the plumbing system. Ascent of mafic magma from a depth of $c.15$ km (Zanon & Frezzotti 2013) through the plumbing system of mafic magma, triggered an eruption of the mugearitic/benmoreitic magma high on the NW flank of the Pico volcano. This mafic magma gave rise to the subsequent basaltic eruptions at São João and contemporaneous offshore eruptions.

The Santa Luzia eruption began at 6 am on 1 February and was preceded by a few hours of strong earthquakes (Macedo 1871, vol. 1, p. 21). Initially the eruption was mildly explosive and occurred along the line of the fissure at an altitude of around 1000 m on the flank of Pico Volcano (Figure 3.8). Tephra was deposited to the south causing disruption in the *Freguesia* de São Mateus leading the community to seek religious intervention to save them from what they assumed would be their fate (França 2002, see Chapter 5). The fissure is characterised by spatter cones, there being seven small craters and a larger crater that mark the highest altitude from which lava was erupted. By the evening of the first day the lava had travelled more than 7 km reaching the north coast between Lajido and Porto do Cachorro in less than six hours (Canto 1882, p. 506; Machado 1962; França 2002). The lava flow field shows *aa* morphology, which is what might be expected given the initial high effusion rate that may be inferred from the speed with which the lava reached the coast and the relatively steep slopes ($c.7–8°$) over which it flowed. On entering the sea *aa* lava fed dendritic submarine flows (Mitchell et al. 2013). The date of the end this eruption is unknown although Weston (1964) suggests it lasted around two to three weeks, whereas França (2002) considers it to have ended after a few days.

The lava flow field covered an area of 8.62 km^2 (Cruz et al. 1995) destroyed a number of vineyards and wine making establishments but no village was affected (Nunes 1999). Though nothing is explicitly stated in the archives or printed sources we have examined, it may be safely inferred that as the lava reached the coast, it must have disrupted communications on the northern side of the island.

1718 (São João) Eruption The development of fracturing and seismicity associated with the formation of the NNW-SSE fissure led to the opening of cracks in the ground between São Mateus and São João on the south side of Pico mountain. This was followed on 2 February by an eruption with a vent opening on the south coast at an altitude of about 300 m near to Cabeço de Cima from which six lava flows were erupted, with two reaching the sea (Figure 3.8). An extensive *mistério*[10] of lava covering 2.04 km^2 (Cruz et al. 1995) was formed. The lavas destroyed part of the town of São João and its parish church (França 2002). The eruption continued and there were further explosions and earthquakes on 24 and 27 February, which caused people to flee east and towards Silveira (Macedo 1871). Activity declined until 15 August when the eruption ceased, though it restarted in early September and was accompanied by earthquakes (Nunes 1999). This re-activation must have been short-lived because by the beginning of November the inhabitants had returned religious items from Silveira to the hermitage of Santo António (Zbyszewski 1963, citing Canto 1882).

The 1718 São João eruption also involved a submarine vent which began erupting with Surtseyan activity on 11 February and occurred at a depth of about 88 m, some 100 m offshore from the Church of St João Baptista in São João (França 2002). According to Macedo (1871, vol. I, p. 476), the eruptive vent built up above sea level as a pyroclastic cone and formed an island. This cone was called Cabeço de Baixo and as the eruption continued, the cone was eventually connected to the land by lava (Zbyszewski 1963, citing Canto 1882). Phreatomagmatic ash and the likely shallow, proximal seismic activity must have had a major impact on São João because the Surtseyan eruption was so close (Macedo 1871, Vol. III, p. 185). Macedo reports that three deaths occurred: one man went out to sea to fish and was 'buried'; a slave woman, who was collecting water from a source close to the sea was suffocated by 'horrible smells' and a further person was suffocated. There are reports of noxious fumes (sulphurous?) being released, these deaths being the result of 'inattention' with at least one being caused by asphyxiation (França 2002). Apart from the damage to São João and the two deaths, the main impact was the destruction of farmland and vineyards from tephra fall which probably came from both the onshore São João and the offshore Surtseyan eruptions.

As the eruption progressed, lavas from Cabeço de Cima extended offshore burying the original coastline (Canto 1882, p. 506), and it appears that these lavas surrounded the Cabeço de Baixo cone. According to Weston (1964), the onshore eruption of Cabeço de Cima and the eruption of Cabeço de Baixo were active at the same time. Both eruptions ended in December 1718.

1720 (Silveira) Eruption Twenty months after the end of the 1718 eruptions a new eruption occurred 5 km to the east of São João at Soldão near to Silveira (Figure 3.8). According to Weston (1964), this eruption was preceded by five months of earthquakes that were felt across the island, but he provides no source for this information (Madeira 1998). There is general agreement in historic sources, however, that there were precursory earthquakes for a month before the eruption began (Macedo 1871, p. 478). The eruption started on 10 July (Macedo 1871, p. 478; Nunes 1999) with four or five craters opening at 9 pm and forming the Cabeços do Fogo. Lava flowing from the vents, reached the coast at Silveira and formed an extensive flow field with an area of 4.94 km^2 (Cruz et al. 1995), called the Mistério de Silveira (França 2002).

The lava destroyed 30 houses in Soldão. There were no casualties (Nunes 1999), but many inhabitants sought refuge in Lajes do Pico. Farmland, including cereal crops, vineyards and orchards, was overwhelmed by lava between Soldão and the coast according to a description by Padre Francisco de S. Maria (Canto 1883a, pp. 343–44). Many animals were suffocated and 'people went around stunned by sulphurous fumes' (Macedo 1871, vol. I, p. 478). During the more explosive phase of the eruption, which generated an abundance of ash, the winds must have been southerly, as Padre Francisco de S. Maria reports damaging ash fall on São Jorge over 30 km to the north. The eruption ended on 18 December.

Faial

The two historic eruptions of Faial both occurred on the fissural system of the Capelo Formation (see Chapter 2) in: 1672–1673, the Cabeço do Fogo eruption and in 1957–1958 the Capelinhos eruption, which was also associated with a small explosive event in the Caldeira Volcano caldera. The Cabeço do Fogo eruption was a subaerial effusive event and the Capelinhos eruption was Surtseyan, being initiated offshore and subsequently forming an isthmus connecting with the western end of Faial.

1672–1673 Cabeço do Fogo Eruption The onset of the 1672–1673 eruption followed an extensive period of seismic activity. Felt earthquakes first occurred in September 1671 and grew in intensity during early 1672 with a large earthquake taking place on 17 April. This earthquake was of Intensity VII to VIII in the Capelo peninsula, causing the collapse of many buildings, and around Intensity IV in the vicinity of Horta at the eastern end of the island (Machado 1959). Seismic activity caused inhabitants from the villages of Praia do Norte and Capelo (Figure 3.11) to abandon their homes and live in huts made of straw (Madeira 1998, see Chapter 5).

The eruption began in the early hours of the morning of 24 April with a vent opening at the Cabeço do Silva or Cabeço do Rilha Boi (now called Cabeço do Fogo). It is documented that all the island was covered in ash, there was a stench of sulphur and the sun appeared yellow (Canto 1881b, pp. 346–351; Canto 1888, pp. 425–432). Lava flowed to the south travelling about 1.5 km before the flow halted at about 17.00 hours. At dawn on 25 April, a new vent opened in the same area which erupted lava that travelled south reaching the coast at Ponta Ruiva later that morning, having travelled 3 km in about six hours (Machado 1959). The village of Ribeira Brava, located between Capelo and Arieiro, was completely destroyed by this flow and was never reconstructed (Madeira 1998). The active lava entering the sea caused the water to boil and generated violent phreatic explosions heard for a distance of up to 10 km (Machado 1959). Lava also flowed to the north. On 25

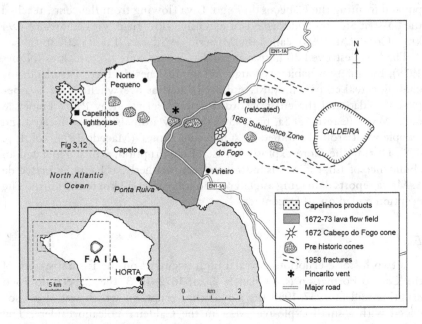

Figure 3.11 Characteristics of the 1672–1673 eruption on Faial. Based on Cole et al., 2001, Figure 1, Page 204. Copyright permission Springer Nature.

April, a new vent called Pincarito opened up with explosive activity and was located (Figure 3.11) about 2 km to the west of Cabeço do Fogo (Machado 1959; Serralheiro et al. 1989).

Following initial explosive activity, lava was erupted which added to the northern flow field with lava reaching the coast and also extending the flow field to the west. The village of Praia do Norte was destroyed. The western extremity of the island was now cut off from the main part of the island by the active lava flow fields (Figure 3.11). Some inhabitants of Praia do Norte spontaneously left the parish of Santa Bárbara dos Cedros on the north coast of the island, but the people of Capelo were trapped on the western side of the flow field with no access by land to other centres of population. These stranded people were supplied by boat, most did not evacuate but remained in this sealed enclave of the island until travel overland was possible once more (Weston 1964). The Cabeço do Fogo vent, which fed the lava flow field to the south coast, was active for three months (Weston 1964) so it is unlikely that foot travel would have been possible. The Pincarito vent area was active for about ten months (Weston 1964) and lava was still active in December and probably halted in February 1673. The eruption continued until the end of February or March (Machado 1959; Gaspar et al. 2015) the date being understandably imprecise because, unless effusive eruptions are closely monitored, it is often difficult to identify a precise end date from the historical record.

Machado (1959) has calculated that the volume of lava material erupted (including that discharged in the form of tephra) was $360 \times 10^6 \, \text{m}^3$.

Activity did much damage to the livelihoods of the affected inhabitants. As described above, Ribeira Brava was totally destroyed as was Praia do Norte. Father Diogo Soares Sarrão, Parish Priest of Capelo, notes 'it was the fire that destroyed the entire village of Praia do Norte along with the parish church... lava burned the most fertile lands of the island' (quoted in Madeira 1998, p.111 our translation). The ash and lava also had a deleterious impact on pastures used for grazing livestock. After the end of the eruption 10.7 km^2 of fertile land was rendered sterile by lava. The 1672–1673 flow field remains a *mistério*, a rough wooded terrain of lava unexploited for agriculture and classified as *scrub* on Callender and Henshall's (1968) land-use map. During the course of the eruption, ash was deposited over all the island which would have had a serious impact on the grazing of livestock. A contemporary document records the deaths of at least three people 'among the people who died were the lay friar, Brother Manuel da Luz, his brother and a youth from the convent who finding a place to watch the eruption were caught between two branches of the moving lava flow ... and perished miserably' (Madeira 1998, p. 112, our translation). The wording of this report suggests that there may have been more deaths (Madeira 1998). In the eruption 307 houses were destroyed and around 1,200 people were made homeless (Machado 1959).

Responses to the 1672–1673 *Cabeço do Fogo* eruption are well documented and exemplify aspects of the ways in which Azorean society coped with disaster during the *pre-industrial* era. These aspects are discussed in detail in Chapter 5.

Capelinhos 1957–1958 Eruption This is the best known of the Azorean historic eruptions. The eruption which began offshore from the Capelinhos lighthouse, formed at least three tuff cones which built up in a progressively easterly direction (Cole et al. 1996). The first two tuff cones were relatively short-lived, collapsed and were eroded below sea level before the eruption ended. The third cone became linked to the coast of Faial at the lighthouse by an isthmus formed from tephra produced by the eruption. The Capelinhos volcano is currently being eroded away by the Atlantic Ocean (Figure 3.12) and much of what remains was probably erupted in 1958 (Cole et al. 1996). The eruption was phreatomagmatic with the rising magma interacting explosively with seawater generating ash, lapilli and steam. At times when seawater was prevented from accessing the vent Strombolian activity occurred. This was a classic Surtseyan eruption (Camus et al. 1981) but, in contrast to Surtsey (Iceland), the products are currently well exposed and the proximity of the shore of Faial means that subaerial deposits that were laid down beyond the cone are preserved (Cole et al. 2001). Towards the end of the eruption in May 1958 Surtseyan activity declined and Strombolian activity formed a spatter cone within the crater of the tuff cone and subaerial lava flows were erupted. The eruption has been extensively studied and many of the papers are incorporated in a compilation volume entitled, *Vulcão dos*

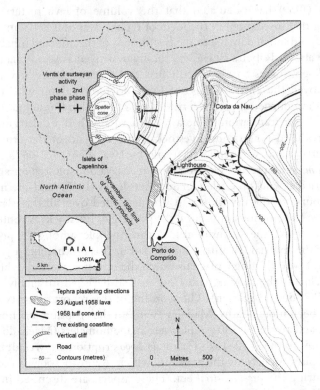

Figure 3.12 Capelinhos during and after the 1957/1958 eruption, Faial. Modified
after Cole et al., 2001, p. 205, Figure 2. Copyright permission Springer
Nature.

Capelinhos memórias 1957–2007 (Forjaz 2007); recent papers not included in
the volume include Cole et al. (1996, 2001) and Coutinho et al. (2010).

In contrast to the 1672–1673 eruption, the Capelinhos eruption was preceded
by only a short period of felt seismic activity but there was, however, no seismograph operating on Faial at that time (Machado et al. 1962). Between 16 and 27
September, more than 200 earthquakes were felt on Faial which did not exceed
an Intensity of V (Zbyszewski 1960). The eruption began on 27 September with
boiling of the sea about 1 km offshore NW from the Capelinhos lighthouse with
4 vents aligned WNW at a depth of about 50 m (Machado et al. 1962; Cole et al.
2001). More vigorous activity commenced on 29 September with tephra finger
jets and large amounts of steam (Cole et al. 2001). Within two weeks a tuff cone
100 m high and with a diameter of 800 m had been built above sea level. Activity
was phreatomagmatic with tephra jets and development of sustained eruption
columns, which when they collapsed produced pyroclastic surges. Analysis of
photographs taken from aircraft passing Faial led to the recognition of eruptions
of this style and, more particularly of a type of PDC, known as a 'base' surge
as described in a classic paper authored by Waters and Fisher (1971). Brown and

Andrews (2015) demonstrated the effectiveness of phreatomagmatic activity in fragmenting magma and generating fine ash, but the incorporation of water cools the erupting mixture, and these wet and relatively cool pyroclastic density currents (PDCs) can plaster deposits against vertical surfaces such as walls and cliffs (Cole et al. 2001) (Figure 3.12). Cas and Wright (1987) comment on the spectacular pyroclastic surges filmed by the US Airforce in October 1957. The deposits of the pyroclastic surges extend for up to 1.2 km from the vent and, to reach these distances, needed to surmount pre-eruption cliffs which led to their deceleration and distal surge deposits grade into fall deposits (Cole et al. 2001). Cole et al. (2001) note that, in addition to the low (typically < 1 km in height) eruption columns described by Waters and Fisher (1971) which were associated with the generation of the pyroclastic surges, there were more frequent and higher (> 1 km) eruption columns that lasted several hours. These continuous uprush columns were partly incandescent when seen at night (Castello Branco et al. 1959). During October much of Faial was covered by ash and lapilli fall from the eruption, and Pico and São Jorge islands were also impacted (Cole et al. 2001). Towards the end of October, the first phase of the eruption finished, the cone foundered and collapsed (Machado et al. 1962).

A new cone formed in November 100 m east of the position of the first cone and during the month an isthmus formed linking this cone to the coast of Faial (Cole et al. 2001). During early December, there was continuing explosive activity with sustained discharge lasting up to 3 hours and on the night of 10/11 December 10.5 cm of ash fell on Castel Branco. Partial collapses on the seaward size reduced the cone height. On 17 December, there was the first effusion of lava and magmatic eruptive activity, with Hawaiian lava fountaining and Strombolian explosive activity (Machado et al. 1962; Cole et al. 2001). The vents for this new magmatic activity were located around 300 m east of the crater of the second cone. Following a day or so of quiescence at the end of December, Surtseyan activity increased in intensity in January 1958 with the vent now centred on the position of the December magmatic activity and this is the position of the third cone that remained active till the end of the eruption. Between January and March 1958 magmatic and phreatomagmatic activity occurred, and there was switching between the two styles of activity. Surtseyan activity decreased at the end of April to the beginning of May which was followed by eruption of lava flows on the west side of the cone (Machado et al. 1962; Cole et al. 2001).

A seismic crisis began in the early evening of 12 May and continued throughout the day of 13 May. There were around 450 felt earthquakes with many of Intensity X on the Modified Mercalli Scale (Machado et al. 1962). A seismograph had recently been installed on the island (Figure 3.13) and the focal depths of the earthquakes were determined to be about 1 km (Machado et al. 1962). Associated with this seismic crisis, a zone of subsidence (up to 1.2 m in depth) opened (Figure 3.11) from Capelinhos parallel to the Capelo fissural system cinder cone belt to the summit caldera (Machado et al. 1962) and fractures developed along this zone. Seismic activity caused much damage on

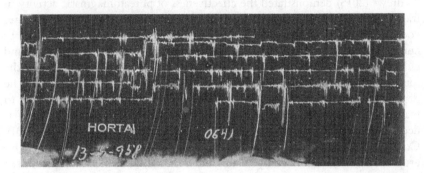

Figure 3.13 Seismogram of earthquakes during the seismic crisis on 12 and 13 May of the Capelinhos eruption by Professor Machado. Courtesy of Sr Renato Lemos, *Instituto Português do Mar e da Atmosfera* **(IPMA).**

the western part of the island, with 500 houses destroyed and many people being evacuated, but fortunately there was no loss of life (Machado et al. 1962). Coutinho et al. (2010) using documentary sources give a higher number of 1,037 homes destroyed and 3,023 people displaced. The traditional single-story rubble-stone construction of houses on Faial were at the time extremely susceptible to earthquake shaking (Coutinho et al. 2010 and see endnote 5). Following the seismic crisis, there was a phreatic explosion on 14 May in the summit caldera of Faial.

Volcanic activity changed after the seismic crisis, sea water was no longer able to gain access to the crater and Surtseyan volcanism ended (Machado et al. 1962). Magmatic eruptive activity involving Strombolian and Hawaiian styles of volcanism built a spatter cone in the crater and this style of activity continued and included effusion of lava. A lava flow was erupted on 23 August about 250 m west of the lighthouse. By 1994/1995 erosion by the sea of the tuff cliffs on the western beach had exposed the feeder dyke for this flow (Cole et al. 1996), see Figure 3.14. The dyke has a sinuous form as it has cut through incompetent tephra layers. In places laterally injected toes and fingers may be observed. The eruption ended on 24 August. Figure 3.15 shows an aerial view of Capelinhos taken shortly after the eruption in 1960. Since the eruption considerable marine erosion of the promontory, which links the main cone of the Capelinhos volcano to the western end of Faial island, has occurred and this has created a submarine platform (Zhao et al. 2019).

São Jorge

The two historic onshore basaltic eruptions on São Jorge in 1580 and in 1808 are significant because they both generated pyroclastic density currents (PDCs). Basaltic PDCs as discussed below are relatively unusual and the ones on São Jorge are of particular interest as they were described by contemporary observers as *nuvem ardente*, a burning cloud and subsequently

Figure 3.14 View of dyke feeding August 1958 lava flow of the Capelinhos eruption, Faial. Authors' photograph.

Figure 3.15 Aerial view of Capelinhos after end of eruption (United States Air Force Photograph – in the public domain).

the French volcanologist Ferdinand Fouqué (1828–1904), who visited the Azores (Fouqué 1873), first translated the term into French as *nuée ardente*. His son-in-law was Alfred Lacroix (1863–1948), who following the 1902 eruption of Mount Pelée (Martinique), introduced the term more widely into the volcanological literature (Wallenstein et al. 2018)[11]. In Wallenstein et al. we provided an account of these two eruptions which we have expanded in this chapter. Eruptions of this type present a serious hazard on São Jorge where eruptions occur on steep slopes.

Zanon and Viveiros (2019) undertook a petrological investigation of these eruptions which they consider to be associated with the same magmatic system. The products of the 1580 eruption range from basanite to hawaiite and for the 1808 eruption from hawaiite to mugearite in composition, the chemical variation in this suite of magmas being caused by crystal fractionation.

1580 Eruption Most knowledge of this eruption comes from the account by Gaspar Frutuoso recorded in *Saudades de Terra* and was extracted and published by Canto (1880b, pp. 188–190). Frutuoso did not directly witness the eruption but his account appears to be based on direct testimonials (Madeira 1998). In the latter part of the nineteenth century, there were further accounts by João Teixeira Soares (extracted by Canto 1880b, pp. 190–193) and Fouqué (1873) who give different dates and sequencing of the volcanic activity to that provided by Frutuoso (see below). They do not indicate the source of this information (Madeira 1998) and for the purposes of this account the dates provided by Frutuoso are used. The following section is largely derived from Frutuoso's account.

On the night of 28 April, 30 earthquakes were felt, on 29 April a further 50 and 50 more on 30 April, on which day two vents opened above the Fajãs de Estevam da Silveira forming scoria cones by the Ribeira do Almeida about 1 km SE of Velas. This first eruption locality is termed the Ribeira do Almeida eruption site (Figure 3.16). From Frutuoso's account, these vents opened on 30 April and this is the date given by Madeira (1998). There was intense ground shaking and tephra fall was generated from an eruption column above the two cones that rose to such a height that the top was not visible from the ground. People and cattle were reported killed, pregnant women scared of death and people either fled to Velas, or sought refuge in the church where the priest prayed and sought the mercy of God. On the morning of 1 May, two lava flows travelling from the scoria cones entered the sea near to Velas. Some accounts state that the eruption began on 1 May, but this is probably based on the record of this lava activity (Scarth & Tanguy 2001). The Ribeira do Almeida vents are situated at 250 m asl and formed two small scoria cones. The lavas travelled down a steep slope to the sea which was only 0.7 km away. On reaching the coast, the lava was accreted on to the western end of the Ponta da Queimada lava platform (Figure 3.16), a lava delta built up from recent basaltic eruptions of which 1580 is the youngest. To restore agriculture the steep slopes of the lava flow had to be terraced.

After a further six hours and later on 1 May, another area of volcanic activity opened 1.5 km further to the SE among vineyards near Queimada (Figure 3.16). This second event affected an important agricultural area which sold at the time 1500 barrels of wine every year. The majority of the vines

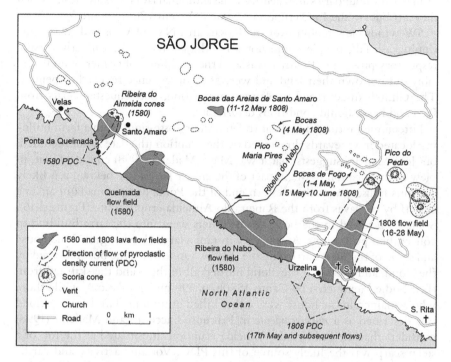

Figure 3.16 Map of the eruptive activity of 1580 and 1808 eruptions, São Jorge. Adapted from Wallenstein et al. 2018, Figure 2, p. 167. Copyright permission Elsevier. Additional information from Madeira, 1998 and 2017 personal communication.

were destroyed by lava and the flows continued for two more days adding the most recent contribution to the Ponta da Queimada delta on the coast. Today the western end of the runway of São Jorge's airport is built on this flow field. The flow field is *aa* lava and this fresh terrain of basaltic clinker renders the land sterile for agriculture and is referred to as the *Mistério da Queimada*.

Eruptive activity then began to the east, the precise date is not recorded, some 6 km from Velas at the Ribeira do Nabo (Figure 3.16). The Ribeira do Nabo basaltic lava flow field overwhelmed vineyards and the vast majority of around 300 *adegas* (i.e. wine making facilities) were destroyed and subsequently not rebuilt. In addition to grazing land being overwhelmed by lava, pastures were also covered by tephra and livestock could not be fed, with more than 4000 cattle being killed together with numerous sheep and goats. Large numbers of beehives were also destroyed.

Zanon and Viveiros (2019) have recently undertaken a re-investigation of the 1580 eruption with detailed mapping of the vents and lava flows aligned with close analysis of the historical records. On the basis of this research, they have refined flow boundaries and identify some additional lava flows and vents which they attribute to the 1580 eruption (Zanon & Viveiros 2019, Figure 3, p. 55).

SW winds blew tephra over the Rosais area NW of Velas and, with continuing ash fall, people could not leave the churches, the fine ash causing respiratory problems. Tephra devastated the livelihood of farmers who invariably did not own their land and were as a consequence reduced to penury. The Church discouraged unplanned evacuation, but requested boats from Pico and Faial should conditions deteriorate.

Fifteen men returned by boat to fetch their possessions from a farm building located in vineyards threatened by the eruption (the date is not recorded, but José Madeira suggests that it was May). Madeira (1998) considers that, in view of the location of the impact of the *ardente nuvem* (see below), it is likely that this house was at the western end of the Ponta da Queimada in an area cut off by the lavas from the Ribeira do Almeida eruption site (Figure 3.16). Some men stayed in the boat, whilst others went into the farm building to collect their possessions. At this very instant a large cloud engulfed the house and some of the men startled by the shadow ran towards the boat followed by the cloud 'and the air of the cloud burned all of them and the skin fell from their bodies' (extracted by Canto 1880b, p. 190 our translation), while those who remained in the house were either killed or injured. This burning cloud was described as a *nuvem ardente* in Frutuoso's account and Madeira (1998) considers that the Ribeira do Almeida eruption site situated at the top of a steep scarp was the likely source of this PDC. Volcanic activity and earthquakes continued for four months and in Velas there was widespread civil disturbance with fights breaking out. The visit of a priest from the Jesuit Order was welcomed and he sought to give reassurance and provided Communion to members of surrounding parishes. It is estimated that damage from the eruption and seismic activity, which included damage to buildings, the loss of vineyards, livestock and crops reached a value in excess of 150,000 crowns. In addition, parish records refer to at least ten fatalities (Avellar 1902, p. 414). Responses to this singular eruption are discussed in Chapter 5.

The 1808 Eruption According to Madeira (1998) the fullest account of the 1808 eruption is by Father João Ignácio da Silveira, Santo Amaro's parish priest, and this was published in the journal, *Jorgense,* on 6 May 1871 (republished by Canto 1884a, pp. 437–441) with notes by Dr João Teixeira Soares. This forms the framework for the account of the eruption which follows, with other sources being cited where appropriate.

Following a week during which several earthquakes occurred, on the night of 30 April/1 May eight earthquakes were felt and one before sunset was so large that it shook the people of Urzelina from their beds. During Mass on the morning of Sunday 1 May, a loud noise was heard accompanied by a strong earthquake and the congregation fled from the church. For some minutes before this large earthquake, animals in a nearby farm become agitated and stampeded. The congregation and other people who had fled their houses observed a dark cloud rising to a great height from the ridge above the town and many thought that this was divine punishment for sins committed by individuals in the parishes of Urzelina and Manadas (see Chapter 5). The large explosion was followed by a continuous roar marking the onset of the 1808 eruption. Dark clouds rose to a great height and ash fell over the parishes of Urzelina and Manadas. From the vents on the ridge came a continuous roaring noise, people were frightened, regarding this as a terrible punishment for their sins and abandoned the church at Urzelina. The fields were covered in ash to around a metre in depth, vines at Santa Rita were damaged and the roofs of houses were in danger of collapsing from the weight of ash. Seven vents were aligned along a fissure (Figure 3.16) and were extremely active in ejecting incandescent bombs for a distance of around 1 km. J.B. Dabney, the Consul of the United States of America[12], witnessed the start of the eruption from the city of Horta on Faial. He described a noise like a cannon shot from São Jorge followed by a dense column of smoke rising to a great height above the island (Webster 1821). On 3 May Dabney travelled to Velas on São Jorge to observe the eruption. He learned that the eruption started in the middle of a lake in a fertile pasture on the ridge above Urzelina and formed a crater of 24 acres (c.10 ha) and that a NE wind had deposited tephra up to more than 1 m in depth to the south in the area between the vent and the coast. Ash was also deposited on the eastern end of Pico. On the evening of 2 May Dabney's party were told that a new small crater had opened near Velas (Figure 3.16). The inhabitants of Velas were frightened and joined in religious ceremonies and devotions aimed at appeasing an angry deity. Dabney visited the second crater which opened on 2 May and found that within about a kilometre and a half the 'earth was rent... and as we approached nearer, some of the chasms were 6 feet (almost 2 m) wide' (Webster 1821, p. 238). The crater was about 50 m in circumference and there was a continuous tremor. Late on 3 May, Dabney returned to Velas and discovered that the inhabitants had mostly left their homes and were to be found either in the open or in tents (Webster 1821). It seems likely that it was earthquakes that caused people to abandon their homes.

On the night of 3 May, activity at the seven vents subsided, but in the early morning of 4 May a new eruption started about 1.5 km to the west (Figure 3.16), forming an eruption column and generating much tephra. The account by Father João Ignácio da Silveira records that two new vents opened on 4 May along a line (fissure) in the fractured area described by Webster above. The Vicar of Urzelina led a procession carrying the Senhora Santo Christo and Senhora das Dores images and, as they proceeded, they joined a second religious procession led by Padre José Machado. Some participants in the procession suffered breathing difficulties after inhaling 'sulphurous' fumes which made them feel dizzy, and progress was also inhibited by trees which had fallen under the weight of ash.

On 11 May, a new eruption began 1.5 km further to the west (Figure 3.16) in the parish of Santo Amaro (Canto 1884a, pp. 437–447; Madeira 1998). Two lava flows travelled west in the direction of Velas and were moving towards houses which had been evacuated. The Vicar of Velas, Father Antonio Teixeira, fearing the lava would disrupt the mass, sent word to change the venue to Beira, an act that produced widespread general consternation and caused many people to flee from Velas: nuns re-located to Rosais; Father Terceira and other clergy to Faial; the resident judge to Pico and many local people to the villages of Beira and Rosais. In the event, the lava stopped after a short distance and the threat to Velas was reduced (Madeira 1998).

Following a few days of relative quiet, vigorous eruption returned on 16 May with new activity just to the east of the original fissure which had opened on 1 May (Figure 3.16). For more detailed information on the location of the fissure see Zanon and Viveiros (2019, p. 61). Tephra driven by winds from the NE ruined much pasture land and many people fled from Urzelina to the villages of Manadas, Calheta and Rosais. Only the parish priest remained at his post at the church of S. Mateus in Urzelina. 'As the tide rose the volcano became more furious,' the vents began ejecting bombs and the vicar saw lava coming down the hill, he rang the church bell as a warning and some people came and helped remove valuable items to the Ermida do Senhor Jesus.

On 17 May, the Vicar of Manadas arrived with his parishioners at the church of S. Mateus in Urzelina to help in the recovery and, whilst they were working, they saw 'a huge flame rises from the volcano and a burning cloud, (a) "nuvem ardente," flowed down the slopes burning the vineyards, bushes and trees and scorched around 30 people as it approached the church. The cloud covered the sunlight as if it were night' (Canto 1884a, p. 440, our translation). This is likely to have been some form of relatively dilute PDC, possibly a pyroclastic surge. An account of this event is included by Dr Soares as a footnote in which he notes: 'After the eruption of the lavas and associated damage there was a short suspension of activity at the craters, and this was followed by the explosion of a dark cloud that came down the hill close to the ground reaching to the sea and destroying everything in its path and anyone in its way was killed. Afterwards the area affected looked like the desert of a volcano' (Canto 1884a, p. 440, our translation).

The vicar of Urzelina thinking it was the last hour of his life was seeking to remove the reserved sacrament when the cloud cleared and the sun shone. He left the church and met his clerical colleague from Manadas together and the congregation many of whom had been burnt. Those who were less severely hurt returned home, whilst the badly injured were treated either in the church or in nearby houses. The badly burned had skin hanging from their limbs, their bodies were blackened, their legs swollen and some were sneezing which is suggestive of respiratory damage. A proportion of injured died and a list of some of the dead is provided as footnote 2 in Canto (1884a, p. 440). The vicar of Urzelina, concerned about the risk to his church, went to the parish of Manadas to give the sacrament to the injured at Santa Rita. On the following day, he travelled to Ribeira do Nabo by a circuitous route to minimise further dangers so that he could give assistance to his parishioners who had escaped the 'nuvem ardente' of 17 May.

Dabney provides an interesting account of the impact of 17 May dilute pyroclastic density current 'most of the inhabitants fled; some, however, remained in the vicinity too long, endeavouring to save their furniture and effects, and were scalded by flashes of steam, which, without injuring their clothes, took off not only their skin, but also their flesh[13]. About 60 persons were thus miserably scalded, some of whom died on the spot or in a few days after. Numbers of cattle shared the same fate. The judge, and principal inhabitants, left the island very early' (Webster 1821, p. 239, see Chapter 5).

Fouqué (1873) describes the cloud of 17 May. It was a dilute pyroclastic density current, loaded with humid ash and from which witnesses did not observe incandescence and, from this evidence, he argued that the flows were not as hot as those produced by the 1580 eruption. The locals used the same term *nuvem ardente* to describe these bulbous clouds which flowed down the hill killing plants, animals and people. The asphyxiation of those caught by the cloud was noted by Fouqué who drew a similarity to the fate of some of the inhabitants of Pompeii in the eruption of Vesuvius in 79 CE by arguing that this would explain how the Roman inhabitants including robust males clearly died in agony. Fouqué (1873) considered that the asphyxiation was caused by inhalation of toxic gases, but it is more likely that this relatively cool pyroclastic density current which was, nevertheless, hot enough to scald victims would have asphyxiated the victims as they inhaled the hot gases burning and causing swelling of their throats.

Subsequently, there was an eruption of lava from the eastern-most vent that reached the coast at Urzelina (Figure 3.16). According to Father Silveira, the lava destroyed 67 houses together with agricultural land and formed a lava flow field, a *mistério,* and a list of the landowners who lost property was compiled and is recorded in Footnote 1 by Canto (1884a, p. 441). The Church of S. Mateus was partly buried by lava with just its tower remaining visible (Figure 3.17). The vents remained active until 5 July, and ash was deposited causing damage to fields and pastures, with farmers attempting to clear the ash so that their livestock could be fed, but in the event many cattle died.

After the eruption had ended there were still threats and vents continued to emit gas for many years. On 8 July 1810, some people were trying to clear ash from wells and died of asphyxiation (Canto 1884a, footnote 4, pp. 441–442), and it is likely that this was caused by CO_2.

The 1580 and 1808 eruptions on São Jorge are important, not only for their relevance to volcanism on the Azores, but also as examples of pyroclastic density currents (PDCs) which are relatively unusual when generated by basaltic eruptions (Wallenstein et al. 2018). PDCs from basaltic eruptions can be generated by phreatomagmatic explosions, or by gravitational collapse of scoria cones, or by accumulations of lava at the top of a steep slope (Cole et al. 2015). Behncke et al. (2008) describe PDCs generated by explosive activity during eruption of basaltic magmas from the SE Crater of Etna volcano in 2006, these PDCs travelled for a distance of up to 1 km. It is argued that explosive activity was triggered by the interaction of magma with wet rock

Figure 3.17 Tower of S. Mateus Church, Urzelina, protruding from lava erupted in
1808, São Jorge. Authors' photographs.

on the flank of the cone. Alternatively, Ferlito et al. (2010) suggest that this
paroxysmal event was caused by rapid ground fracturing leading to a drop
in confining pressure which was exploited by an injection of primitive, gas-
rich magma. In the case of the São Jorge eruptions, documented records and
the nature of their impacts suggest that the 1580 PDCs were hotter, whereas
the 1808 PDCs were both cooler and more moist. The hotter incandescent
nature of the 1580 eruption and the local topography might suggest collapse
of an active lava flow from the top of the scarp. This is supported by the interpretation of Madeira et al. (1998) who consider that the 1580 PDC erupted
from the Ribeira do Almeida site was a block and ash flow. In the case of
the 1808 PDC the observation of the moist ash laden nature of the turbulent
flow, its relative coolness and the observation of a dark cloud blotting out the
sun light supports the proposition that this was a phreatomagmatic eruption.
The presence of standing water at the top of the ridge described by Dabney
(Webster 1821) indicates the availability of water. Deposits from the 1580 and
1808 PDCs have not been identified. A prehistoric block and ash flow deposit
(radiocarbon date 2880±60 BP), however, has been located in the Ribeira da
Cancela above Manadas (Madeira et al. 1998).

The fact that two historic eruptions generated PDCs, caused loss of life, tephra fall that damaged agricultural land and lava flows that sterilised significant tracts of land for at least a hundred years, emphasises the significant hazard such events pose for the people of the Azores. Sao Jorge's airport is built in part on the 1580 lava flow field and the area between Velas and Urzelina is particularly at risk from future basaltic eruptions. The deaths in 1810 from asphyxiation of workers trying to clear wells of ash demonstrates the hazard posed by CO_2 fumarolic activity after eruptions have ceased.

Terceira

In historical times there has been one onshore eruption on Terceira which occurred in 1761 and consisted of two events (Pimentel et al. 2016), first a trachytic dome forming eruption towards the base of the eastern flank of Santa Barbara volcano and, second, an effusive basaltic eruption about 2.5 km away on the other side of Pico Gordo. This basaltic eruption, occurred on the margin of the Basaltic Fissure Zone between Santa Bárbara and Pico Alto volcanoes (Figure 2.4). There have been two offshore eruptions off the western coast of Terceira in 1867 and 1998–2001 on the Serreta submarine fissure zone.

1761 Eruption Santa Bárbara Volcano/Basaltic Fissure Zone In the literature there has been lack of clarity over whether the 1761 eruption consisted of one event, the eruption of basaltic lava from a line of vents to the east of Pico Gordo with the lava flow reaching the village of Biscoitos near the north coast, or two events including the trachytic dome forming eruption of Mistérios Negros near to the foot of Santa Bárbara volcano (Figure 3.18). Self (1976) identifies the Mistérios Negros domes as being recent but not historic, while Weston (1964) recognises a precursory explosive event before the 1761 eruption, that did not erupt lava but does not give a precise location. Zbyszewski et al. (1971), Scarth and Tanguy (2001) and Calvert et al. (2006) suggest that the Mistérios Negros domes are the products of this precursory explosive phase. Pimentel et al. (2016) carried out palaeomagnetic dating of measured samples from the Mistérios Negros domes and the 1761 basaltic lava and this supports the proposition that the two events were coeval. Assigning both events to an eruptive episode in 1761 is accepted by Gaspar et al. (2015, Table 4.2, pp. 39–41) in their catalogue, *Volcanic eruptions occurred in the Azores in historical times*.

The first event involved the eruption of the Mistérios Negros Domes and, according to an early account by José Acúrcio das Neves (Canto 1883b, pp. 362–365) the first sign of the impending eruption was when the fumarolic activity in the Furnas do Enxofre, a notable local phenomenon located about 5 km east of the Mistérios Negros (Figure 2.4), 'ceased smoking' in November 1760. On 17 April 1761, the eruption began following 3 or 4 days of felt earthquakes (Pimentel et al. 2016). José Acúrcio das Neves refers to 'underground explosions like artillery discharges … and fire being thrown up from

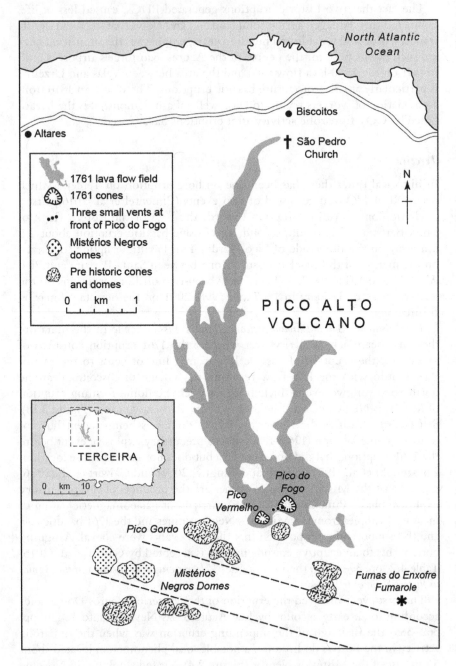

Figure 3.18 Features of the 1761 eruption, Terceira. This figure is based on Zanon and Pimentel 2015, Figure 1, p. 796, Copyright Mineralogical Society of America. Additional information from Pimentel et al. 2016.

the earth' (Canto 1883b, p. 363). The records are vague and provide little in the way of specific data. Pimentel et al. (2016) in an analysis of the records, note that the local priest describes the rise of small hills in the area and they consider that these are the Mistérios Negros Domes which are youthful features and not covered in tephra. The local name of Mistérios Negros they argue supports the possibility that this event was witnessed by locals as the name *mistérios*, as already discussed, refers to recent lava terrain. The Mistérios Negros Domes stretch for about 900 m trending in a WNW-ESE direction. The four domes range in height from 20 to 68 m, are composed of mildly peralkaline comenditic trachyte, which is porphyritic, with 30%–41% alkali feldspar being present and with very small volumes of clinopyroxene, olivine together and with occasional aenigmatite, apatite and eudyalite in the groundmass reflecting the peralkaline character of these rocks (Pimentel et al. 2016).

The second event was a basaltic eruption, which began on 21 April just over a kilometre east of Pico Gordo (Figure 3.18). A series of vents opened along a fissure aligned just north of east at about 80°. Scoria cones were formed: the largest is named Pico do Fogo, with a height of 53 m; Pico Vermelho, 52 m in height and three smaller vents (Pimentel et al. 2016). The lava just north of Pico Vermelho shows a surface of fine clinker covered in moss and lichens with abundant growth of *Erica azoria* (see Chapter 1). A road section provides a good exposure of the lava. Near one of the smaller vents there is a good exposure of toothpaste lava. Historic records presented by José Acúrcio das Neves (Canto 1883b, pp. 362–365) indicate that the lava first flowed to the south and east for a few hundred metres. Subsequently the main flow of lavas was to the north, with the main branch travelling 5.5 km and reaching the village of Biscoitos (Figure 3.18). Historical accounts describe the advance of the lavas as tranquil and inhabitants had time to escape with many of their possessions and there were no casualties, but 27 homes were destroyed. People now felt they were safe as the lava came to a halt on 29 April (Drummond 1856, extracted by Canto 1883b, pp. 364–365; Pimentel et al. 2016). In addition to the damage to houses, agricultural land south of the village was rendered derelict (Pimentel et al. 2016). Strombolian/Hawaiian activity from the vents generated ash that was deposited over most of the island (Canto 1883b).

The lava flow field covers an area of 3.5×10^6 m^2 (Pimentel et al. 2016, p. 13), and these authors calculate an estimated dense rock equivalent (DRE) of $11{\pm}4 \times 10^6$ m^3. The trachybasalt is strictly an hawaiite ($Na_2O - 2 \geq K_2O$ Middlemost 1997, p. 43), and the 1761 products are slightly more evolved than other basaltic lavas erupted from the fissure zone and have experienced greater olivine and clinopyroxene fractionation (Pimentel et al. 2016).

The relationship between the two eruptive events of 1761 Associated eruptions of basaltic and trachytic magma occur in the Azores. On the day before the end of the sub-Plinian trachytic eruption of Fogo volcano (São Miguel) in 1563, a basaltic eruption began on the lower NW flank of the volcano (see above).

Wallenstein et al. (2015) suggested that a pulse of ascending basaltic magma reached the trachytic magma reservoir of Fogo triggering an explosive trachytic eruption at the summit, whereas some of the mafic magma injected along a fault line by-passed the reservoir and fed a basaltic eruption on the lower flank of the volcano. In the case of the 1761 contemporaneous eruptions of Terceira, Pimentel et al. (2016) argue that from investigation of the petrography, mineralogy and whole-rock geochemistry, there is no evidence of interaction between the co-erupted trachybasalt and comenditic trachyte magmas and no indications of hybrid composition or mafic enclaves within the trachyte were found. Pimentel et al. (2016) propose that basaltic magma rising from depth towards the surface caused a decrease in the fumarolic activity of the Furnas do Enxofre in 1760. The dyke of basaltic magma was intruded along a WSW-ESE fracture. The injection of the dyke in a westerly direction reactivated a pre-existing WNW-ESE fracture within the Santa Bárbara volcano allowing partly degassed comenditic trachyte magma within the high-level plumbing system to erupt without noticeable explosive activity to form the Mistérios Negros Domes. Subsequently – four days later – the trachybasaltic dyke breached the surface producing the basaltic eruption.

Terceira: offshore eruptions

The basaltic fissure zone extends WNW off the west coast of Terceira at Serreta as a submarine ridge (Figure 2.4). Two submarine eruptions have occurred on this ridge in historic times in 1867 and 1998–2001.

The 1867 Eruption The 1867 eruption began on 1 June and notable activity continued until 8 June, with activity occurring about 5 km offshore (Gaspar et al. 2015) at a depth of about 180 m (Weston 1964). Details of the eruption are described by Louis Figuier in *L'Année Scientifique et Industrielle 1867* which was later translated by Ernesto do Canto (Canto 1883c, 1884b, pp. 499–501). The eruption was preceded for five months by earthquakes which damaged many houses and caused landslides in the Serreta and Raminho districts. The eruption began on the night of 1 June with detonations similar to artillery barrages and at daylight great fountains of water and steam were observed. Though scoria was ejected, no incandescent activity was recorded. Several fountains (six or seven) were active at the same time and Gaspar et al. (2015) consider that a fissure zone some 2 km in length on the seafloor was active. The eruption was accompanied by a strong smell of hydrogen sulphide which was particularly pungent near to the coast. Although vigorous activity ceased on 8 June, on calm days gas bubbles could be seen rising for several months after this.

The 1998–2001 Eruption Precursory seismic activity beneath the Serreta submarine ridge was first detected by the Azores Seismological Surveillance network on 23 November 1998 (Gaspar et al. 2003). Around 400 micro-earthquakes were recorded on 29 November, after which seismic activity reverted to normal until enhanced seismic activity returned in the middle of December. No felt earthquakes occurred. A group of fisherman witnessed the first signs of eruption on the sea surface at dawn on 18 December, with vents being located about 10 km offshore from the Serreta lighthouse (Gaspar et al. 2003). Early accounts of the eruption described the presence of 'hot steaming stones' and fishermen on a boat from Graciosa reported 'fire coming out from the seawater spreading on the air like sparks of fireworks' (Gaspar et al. 2003, p. 206). Scientific observation of this phenomenon was made on 8 January 1999 when a large area of floating lava balloons releasing columns of steam was observed during a helicopter flight (Gaspar et al. 2003). Subsequently, lava balloons were observed from small sea craft, the balloons surfacing together with gas bubbles and with the steaming diminishing as the lava cooled. The balloons normally floated for no longer than 15 minutes. Four samples were collected intact and their dimensions ranged from major axis, 47–84 cm, medium axis, 32–38 cm and minor axis, 17–35 cm (Gaspar et al. 2003, Table 1 p. 208). In the early part of the eruption there were several persistent areas of floating lava balloons along a NE-SW alignment where depths ranged from 300 to over 1000 m (Gaspar et al. 2003). Gaspar et al. (2003, 2015) propose that the balloons formed from the extrusion of basaltic lava from submarine vents along a fissure system. As a bleb of lava extrudes, a gas bubble accumulates and detaches as a lava balloon which breaks free and rises towards the surface. As the balloon rises Gaspar et al. argue that there must be still molten lava present to allow the balloon to expand as it ascends. They point to the presence of striations on the crust of the balloons as evidence in support of this process of decompression. As the hot lava skin quenches and cools against the sea water, steam is given off. The basaltic crustal skin of floating balloons will cool, crack and then fill with water and sink. Sometimes ingress of cold sea water into the hot interior causes a steam explosion shattering the balloon (Gaspar et al. 2003). Kueppers et al. (2012) in further work on the Serreta lava balloons, suggest that the balloons may have formed as bubbles in magma generated from a submarine lava lake or pond developed over the vent. Lava balloons as a product of submarine eruptions are now being more widely recognised in other parts of the world (White et al. 2015). Such balloons were probably first recognised during the Foerstner submarine eruption off the coast of Pantelleria in 1891 (Guest et al. 2003, p. 232). Kelly et al. (2014) undertook a submarine survey of the vent area of the Foerstner 1891 eruption and propose that in this eruption the balloons formed from strombolian style explosive activity.

The basaltic material erupted from the Serreta eruption is alkali basalt and much less evolved than the 1761 hawaiite with values of MgO in the range 8.94–10.52 wt% (Gaspar et al. 2003) compared to 4.5–4.67 wt% (Pimentel et al. 2016), for 1761. This aligns with the model of Zanon and Pimentel (2015, Figure 11) that shows magmas erupted from the Serreta Fracture Zone as rising directly to the surface from the base of the crust with minimal intermediate storage.

Historic earthquakes and associated tsunamis

Earthquakes in the Azores have caused around 6,300 deaths and widespread destruction (Caldeira et al. 2017, p. 29)[14]. Information on the characteristics of historic events is listed in Table 3.2 and is based on two sources. Existing compilations (e.g. Admiralty 1945; Machado 1973; Nunes 1986; Madeira 1998; Silveira et al. 2003; Nunes et al. 2004; Silva 2005; Malheiro 2006 and especially Gaspar et al. 2015) provide details on most, and in the case of Gaspar et al. (2015), all the historic events discussed below, but additional information is available on some individual events and is referenced as appropriate. Two earthquakes listed in Table 3.2 are not discussed further, these being the 1717 and 1868 events that impacted Graciosa. Although it is known that these events caused the collapse of houses and in 1868 some land-sliding, no further details are known from the archives and this reflects the extreme isolation of Graciosa at this time.

22 October 1522 – São Miguel

With a maximum Intensity of X, this is the most destructive earthquake to have affected the Azores since the islands were settled in the fifteenth century. With an epicentre to the north of Vila Franca do Campo (Figure 3.2), buildings were damaged and destroyed throughout the island, its effects being focused on Vila Franca, a settlement that was at the time the capital of the island and its seat of government. Here destruction was almost total because, in addition to direct earthquake losses, additional damage was caused by debris from landslides that were triggered in the hills above the town and which buried virtually all the urban area (Dias 1945). Deposits covered an area of 4.5 km^2, had a total volume of 6.75×10^6 m^3 and was probably the product of two landslides (Wallenstein et al. 2015; Silva et al. 2020). It is estimated that this earthquake together with a plague that followed in its wake, killed around 5,000 people, some 20% of the island's total and in Vila Franca only 70 inhabitants escaped death. One fortunate survivor was the Captain Donatary (see Chapter 1) who was occupying his home in Lagoa at the time (Bento 1994). Excavating the ruins of Vila Franca to remove bodies took over a year, the site of the town was abandoned, a new settlement being built to the west. The town never regaining its former prominence, Ponta Delgada becoming the capital of São Miguel and later the whole archipelago.

Andrade et al. (2006) note that there are few reports of coastal floods in the Azores dating from the sixteenth and seventeenth centuries, but one anonymous and undated text claims that Vila Franca was inundated on 22 October 1522, suggesting that landslide(s) may have entered the ocean and generated a small tsunami. Having carefully considered this possibility, Andrade and his colleagues believe that the anonymous author may be confusing

Frutuoso's account in *Saudades da Terra* (Frutuoso 2005) which likens the landslide to 'violent earth motions only comparable to the agitation of the sea when assaulted by raging storms' and without any collaborative evidence the authors dismiss the possibility that this was a tsunami (Andrade et al. 2006, p.176). More recently Cabral (2020), using similar historical source material, suggests that the introduction of such a large quantity of debris-flow material into the Atlantic may have generated a small tsunami.

17 May 1547 – Terceira

Relatively little is known about the physical characteristics of this earthquake, including the location of its epicentre, although based on the damage caused it was probably located in the NW of the island and had an estimated Intensity of VII. This event was responsible for at least 3 deaths, and much damage to churches and housing (Agostinho 1935a).

28 June 1563 – São Miguel

With an Intensity of X this was a major earthquake and represents the most extreme event within a seismic crisis that lasted from 22 June to 4 July and which was related to the 1563 eruptions of Fogo Volcano and Pico Sapateiro on São Miguel (see above). Recorded by Frutuoso (2005) and discussed by Wallenstein et al. (2015), this event caused total or partial collapse of buildings across the centre of the island.

26 July 1591 – São Miguel

This was the high intensity event (i.e. VIII–IX) that occurred during a period of seismic activity that affected São Miguel in July and August 1591. This earthquake was not recorded by Gaspar Frutuoso leading some commentators to doubt its veracity. Frutuoso died on 24 August 1591, less than a month after the earthquake probably allowing him insufficient time to record the event (Silveira 2002). In common with the disaster of 1522, this earthquake was associated with the generation of landslides producing destruction in Vila Franca and Agua de Pau. Casualty figures are vague, but are described as 'numerous' by the trader and historian Jan Huyghen van Linschoten[15] (Linschoten 1998). A small tsunami (maximum run-up 1 m) affected São Miguel and other islands, but this has recently been disputed by Cabral (2020).

24 May 1614 – Terceira

The epicentre of this earthquake was probably located in the NE of the island or just offshore (Figure 3.2), its estimated Intensity was X and the whole of the eastern half of Terceira was badly impacted. In particular, the north eastern settlements of Vila Nova, Lajes and Praia da Vitória were badly affected and altogether some 1600 dwellings were destroyed and an estimated 200 people were killed. The crisis began on 9 April with the destruction of the village of Fontinhas and there were aftershocks until November (Drummond 1856–1859). As in 1522 so in 1614 landsliding was a feature of this earthquake and one slide modified the coastal geomorphology and generated a tsunami with an estimated maximum run-up height of 3 m (see Table 3.3 and Andrade et al. 2006, p. 177; Cabral 2020).

2 September 1630 – São Miguel

This earthquake was associated with the 1630 Furnas sub-Plinian eruption (see above) and had a maximum estimated Intensity of VIII. Although no epicentre has been located, this was probably near to the centre of eruption at Lagoa Seca and extensive damage was caused to houses and churches in villages located in the southern part of island. Povoação was virtually destroyed and a combination of volcanic and seismic activity killed between 80 and 115 people in Ponta Garça (Dias 1936; Weston 1964; Guest et al. 1999). In Povoação only one house survived and in Ponta Garça around 100 were destroyed, with 80 survivors leaving the village for the safety of Vila Franca do Campo.

3 July 1638 – São Miguel

Seismic activity was caused by submarine volcanic activity off the west coast of São Miguel, many houses were reported to have been destroyed on the western flank of Sete Cidades volcano (Queiroz 1997), and residents of the village of Várzea left their homes fearing that they might collapse.

10 October 1652 – São Miguel

Associated with the Pico do Fogo volcanic eruption, this Intensity VIII earthquake totally or partially destroyed villages in the southern part of the island especially between Lagoa and Ponta Delgada (Weston 1964). No epicentre has been located and there are no published casualty figures. Seismic activity continued until 19 October.

14 November 1713 – São Miguel

This was a high Intensity (IX–X) event that caused total or partial destruction of buildings in the western part of São Miguel, with the villages of Várzea, Ginetes, Mosteiros and Candelária being particularly badly effected and many buildings were destroyed (Dias 1936). There were also many mass movements, a major cliff failure and seismic activity continued until 8 December, being probably associated with a submarine eruption located offshore to the west of Sete Cidades Volcano.

13 June 1730 – Graciosa

This Intensity VII earthquake caused severe damage to the southern part of Graciosa, with landsliding occurring on the inner caldera walls. With an epicentre probably positioned offshore and to the SE of the island (Figure 3.2), destruction was concentrated in the *freguesias* (i.e. parishes, see Chapter 1) of Luz and Praia, with many homes and churches being damaged (Hipólito et al. 2013).

9 June 1757 – São Jorge, Pico and Terceira

This was the largest historic earthquake to have affected the Azores. With an epicentre close to the north coast of São Jorge, the earthquake had an Intensity of X and an estimated magnitude of 7.4 (Machado 1949). Physical effects included landslides, cracks in the ground and topographic changes, while tsunamis with a run up height up to 1 m affected many coastal settlements in Terceira, Graciosa and Faial. Damage was most severe on São Jorge with large-scale building destruction, many buildings in Pico were also severely impacted and there was damage across Terceira. Estimated deaths were over 1,000 on São Jorge which was one-fifth of the population at the time, 11 on Pico and possibly a single death on Terceira.

9 July 1800 – Terceira

This event was probably associated with a submarine eruption that occurred off the south coast of the island. Recorded by Araújo (1801) this earthquake exemplifies the damage that may be caused by volcano-related seismic activity, with many buildings being destroyed and damaged across the eastern part of the island especially between Vila Nova and São Sebastião by seismicity with a maximum Intensity of VII-VII (Table 3.2). One death was reported in the village of São Sebastião.

26 January 1801 – Terceira

With an Intensity of VII-VIII and probably associated with offshore volcanic activity (Figure 3.2), seismic activity caused two deaths in the eastern half of Terceira severely damaging many buildings. The seismic crisis lasted from June 1800 to February 1801, with peak acceleration occurring on 26 January.

24 June 1810 – São Miguel

Probably associated with submarine volcanic activity that occurred in the first half of 1811, pre-eruptive seismic activity with a maximum Intensity of VII occurred between 17 and 24 June 1810, causing some houses to collapse and severely damaging many more across the western part of the island.

13 June 1811 – São Miguel

This was related to the submarine Sabrina volcanic eruption (Table 3.1) that occurred to the west of São Miguel and severe damage was reported from parishes in the western part of the island. The most severely impacted villages were Mosteiros, Várzea, Candelária, Ginetes and other smaller settlements across the flanks of Sete Cidades volcano (Silveira 2007).

21 January 1837 – Graciosa

With a maximum Intensity of VII (EMS 98) and possibly X on the Modified Mercalli scale, this large earthquake causing three deaths and damage across much of the island but with a concentration in its central area (Hipólito et al. 2013). Despite its size, little is known about this earthquake reflecting the isolation of Graciosa at the time.

15 June 1841 – Terceira

With a maximum Intensity of IX, this earthquake caused the collapse of an estimated 2,100 houses across the eastern part of the island and cracking of the ground surface in a seismic crisis that began on 12 June. Major damage was caused to the town of Praia da Vitória the second largest settlement on Terceira. Praia did not fully recover until the end of the nineteenth century (Admiralty 1945).

9 July 1847 – Flores and to a lesser extent Corvo

Despite being the deadliest tsunami to impact the Azores, few details are known. It was triggered by a large sea-cliff landslide some 350 m high, and its major impact was felt in the vicinity of Quebrada Nova on the NW coast of Flores (Cabral et al. 2010). It caused the deaths of 10 people and injured more than 100 (Gaspar et al. 2011).

6 April 1852 – São Miguel

This was a very destructive event, with a maximum Intensity of VIII and caused nine deaths, extensive injuries over the whole island and especially in the west, and in and around the town of Ribeira Grande. There were major losses to farms, vineyards, buildings and churches with the major impacts focused on the settlements of Ribeira Grande and Capelas, with Ponta Delgada and Lagoa being less badly affected. In Ribeira Grande *concelhos* (i.e. county, see Chapter 1) the *frequesia* of Santa Bárbara was almost completely destroyed with part of the church tower collapsing and the village of Rabo de Peixe was also badly impacted. On the basis of his research, Silveira (2002, p. 43) was, not only able to place the epicentre in the ocean to the north of São Miguel, but also produced an isoseismal map.

31 August 1926 – Faial

A seismic crisis began in April, lasted until September and at the end of August an extremely damaging earthquake with a maximum Intensity of IX was generated. With an estimated Magnitude of 5.6 to 5.9 and rapid attenuation SE to NW across Faial, this event caused nine deaths, over 200 injuries and produced severe damage in and around the island's capital, Horta, other settlements in the east and also affected the western half of the adjacent island of Pico. On Faial: the *freguesia* of Flamengos lost over 80% of its housing; Praia do Amoxarife, 54%; the three *freguesias* which comprise Horta, between 20 and 40%; Feteira 36%; Pedro Miguel 34%; Ribeirinha 18%; but the central parishes of Salão and Castelo Branco lost only 4% and 1%, respectively, with the western *freguesias* being virtually unaffected (Anon 1926; Fontiela 2009). Many landslides were generated and a tsunami.

5 August 1932 – São Miguel

With a maximum Intensity of VIII and an epicentre in the ocean to the SE of São Miguel, this earthquake produced extensive damage across the south and SE of the island, causing no recorded deaths but injuring several people, rendering more than 3,000 people homeless and generating many damaging landslides. Of 1811 houses surveyed in and adjacent to the town of Povoação: 18% collapsed; c.33% were ruined; c.51% were affected to some degree and c.35% of their inhabitants were rendered homeless (Silveira 2002).

27 April 1935 – São Miguel

This Intensity IX earthquake caused one death, several injuries, some building damage in Povoação *concelho* and across the central-eastern part of the island. Also associated with landslides, the most severely affected villages were Ribeira Quente and Lomba do Cavaleiro (0.5 km west of Povoação) with some damage also being noted in Ponta Garça and Ribeira Seca (Silveira et al. 2003), for location of villages see Figure 2.2.

21 November 1937 and 8 May 1939 Santa Maria

With epicentres in the ocean to the NE of Santa Maria (Figure 3.2) and probably associated with activity on the Gloria Fault (Chapter 2), these two events occurred close in time and impacted the same areas of Santa Maria damage being focused in the *freguesias* of Santo Espírito and São Pedro, respectively, in the east and west of the island. Masonry buildings showed damage and this reached Grade 4 on the EMS-98 scale (Carmo et al. 2016; Grünthal 1998, p. 15, see endnote 5). Both events had Intensities of VII, but the 1939 earthquake, with a magnitude of 6.5 to 7.1, was additionally associated with a tsunami having a maximum run-up height 0.5 m. According to a study by Carmo et al. (2016), isoseismal maps imply higher intensities in a zone extending across the middle of the island in a NE-SE direction, which is not compatible with epicentral locations to the NE of the island (Figure 3.2) and this inconsistency remains unresolved.

29 December 1950 – Terceira

This maximum Intensity VII earthquake caused damage to the northern part of the island especially in the *freguesia* of Agualva, aftershocks continued until January 1951.

26 June 1952 – São Miguel

This earthquake comprised four shocks and the first caused the most damage. Although there was only one injury, severe damage occurred across the SE coast from Povoação in the east to Vila Franca do Campo in the west. A maximum Intensity (VIII) occurred near to the coastal settlement of Ribeira Quente and activity was associated with extensive land-sliding (Silveira 2007). At least 330 buildings were destroyed in Ponta Garça, with the majority being damaged (Silveira et al. 2003).

13 May 1958 – Faial

This was a seismic crisis – comprising a swarm some 400 events – related to the 1957/58 eruption and on 13 May generated a maximum Intensity of VIII earthquake with associated landslides. Severe damage was caused to the NW part of Faial, the villages of Praia do Norte and Ribeira do Cabo were destroyed and other small settlements, Areeiro, Cruzeiro (near Capelo) and Espalhafatos, were badly affected (Figure 3.19). Some 1,037 homes were lost or severely damaged and 3,023 people were displaced (Coutinho et al. 2010).

21 February 1964 – São Jorge

Known as the Rosais earthquakes and with a maximum Intensity of VIII and a Magnitude of 4.8, this event caused severe damage in the north-western part of the island, the settlements of Rosais and Velas being badly affected. More than 900 houses were damaged, an estimated 400 destroyed and around 5,000 people had to be evacuated to Terceira, Pico and Faial (Anon 1964). This seismic crisis was related to a submarine offshore eruption.

23 November 1973 – Pico and Faial

Although only one person was injured, severe damage was caused to the northern part of Pico. Some 600 houses were damaged on Faial, 2,000 on Pico

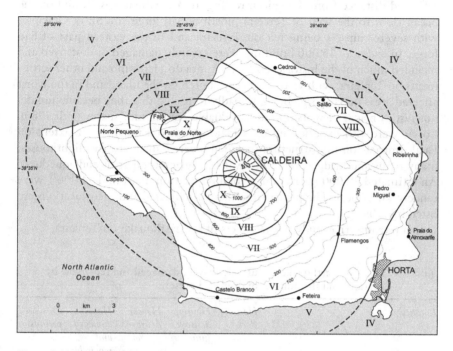

Figure 3.19 Isoseismal lines showing the maximum earthquake intensity during the
night of 13 to 14 May 1958, Faial. From Coutinho et al. 2010, Figure
3, p. 270. Copyright permission Elsevier, with additional information
from Machado 1958a and 1958b. Intensity values are mapped according
to the Modified Mercalli Scale (i.e. MM-56), which is closely related to
both the Medvedev-Sponheuer-Kárnik (i.e. MSK-64) scale and the more
recent European Macroseismsic Scale 1998 (i.e. EMS-98). A maximum
Intensity of X roughly equates to a Magnitude of *c.*7–7.6.

and 5,000 people were rendered homeless. With a maximum Intensity of VII
(EMS-98) (VIII – MM 1931) and a Magnitude 5.5, this event was related to a
seismic crisis that started in October 1973 and lasted until early 1974.

1 January 1980 – Terceira with serious effects on São Jorge, Graciosa, Pico and Faial

This Magnitude 7.2 earthquake (maximum Intensity VII EMS-98: IX MM-
1956), had a shallow depth of focus (*c.*10 km – Hirn et al. 1980; Degg &
Doornkamp 1994), generated a small tsunami (max run-up height 0.5 m on
Terceira) and was very destructive. It caused 59 deaths and more than 400
injuries. Mortality and morbidity would have been greater, but for the fact
that the earthquake occurred at 15:42 (local time) when many people were
outside on a sunny day enjoying the New Year holiday. Broadly damage

reflected distance from the epicentre (Figure 3.2) which was located on the ocean floor to the west of Terceira, north of São Jorge and SE of Graciosa, with severe impacts being felt on Terceira and in the eastern part of São Jorge. In excess of 15,000 buildings were severely damaged or destroyed and included much of the historic 'cores' of Angra do Heroísmo and other settlements on Terceira, with virtually every church on the island and in the worst affected areas of São Jorge being so badly impacted that they became unsafe, not just as places of worship but as valuable community spaces (Pell 1980). Losses were also experienced on Graciosa and to a much lesser extent on Faial. Impacts on infrastructure included severe damage to five lighthouses; many roads and bridges; retaining walls in agricultural areas and aqueducts. An estimated 20,000 (*c*.21%) out a total population of around 94,000[16] were rendered homeless. The principal settlements affected and their locations are listed in Table 3.4.

In a retrospective study of the effects of the earthquake on Terceira,

Table 3.4 The 1980 earthquake: Houses destroyed, damaged and repairable by location, and the number of displaced persons

Locality	Housing stock	Houses destroyed (number)	Houses destroyed (ca %)	Houses damaged and repairable	Displaced persons
Terceira Island	19,075	4,727	25	2,447	18,741
Angra do Heroísmo (southern) town[1]	11,220	3,962	35	2,203	15,978
Rural *freguesias* (i.e. parishes) in Angra *concelho*[2]	6,613	2,148	32	1,399	8,312
Praia da Vitória (eastern) *concelho*[3]	7,855	765	10	244	2,863
São Jorge Island Calheta (eastern) *concelho*[4]	2,241	450	20	1,424	2,064
Graciosa Island (most damage in the south and south west)	2,991	120	4	478	489

1 The Angra do Heroísmo *town* includes the urban *freguesias* of Nossa Senhora de Conceição, Santa Luzia, São Bento, São Pedro and Sé.
2 Rank of rural *freguesias* by percentage of houses destroyed and their location: Doze Ribeiras (99% – west of island); Santa Bárbara (59% – south west of island); Cinco Ribeiras (50% – south west of island); São Bartolomeu de Regatos (45% – south of island); Serreta (41% – west of the island); Terra Chã (36% – north of Angra); Ribeirinha (30% – south of the island); Altares (25% north west of the island); São Sebatião (20% – south east of the island); Raminho (20% north west of the island); São Mateus da Calheta (18% – south of the island) and Porto Judeu (1% – north of Angra).
3 Rank of *frequesias* by percentage of houses destroyed and their location: Quatro Ribeiras (57% – north of the island); Biscoitos (51% – north of the island); Vila Nova (15% – north of the island); São Brás (4% – east of the island); Fontainhas (5% – east of the island); Lajes (1% – east of the island); Cabo da Praia (1% – east of the island) and Santa Cruz (1% – east of the island).
4 Rank of *frequesias* by percentage of houses destroyed and their location: Topo (50% – east of the island); Santo Antão (32% – east of the island); Norte Pequeno (16% – north of the island); Ribeira Seca (12% – central south of the island) and Calheta (4% – central south of the island).
Data from Pell (1980).

Teves-Costa et al. (2007) and Teves-Costa and Veludo (2013) demonstrated that, whereas there was a broad relationship between impact and intensity, which varied spatially from a maximum of IX (MM) in the west to V (MM) in the extreme NE, damage also reflected other factors. For instance, in the island's capital, Angra do Heroísmo, Paula Teves-Costa and her colleagues tested a number of possible factors that could have influenced the spatial pattern of losses (e.g. surface geology type, soil peak frequency, building type and the number of storeys within an edifice), classifying damage from: Level 1 – slight or no damage; to Level 4 – collapse. Pre-1960 buildings, which equate to Classes A and B of EMS-98 (Neves et al. 2012, see above), show a distribution across all damage levels. In contrast, those constructed after 1960, when construction incorporated reinforced concrete beams and columns with links between interior walls and facades, showed damage heavily skewed towards the minimum (i.e. Level 1) category. Surface geology could not on its own explain the observed distribution of losses, though buildings on 'harder' formations such as trachytic lava showed a better performance than those with foundations on lapilli tuff, pumice and ignimbrite. Buildings with a single storey displayed more Level 1 damage than edifices with two or three storeys which were more severely impacted. The most important finding was that the soil factor has greatest explanatory power. Although correlated broadly with surface geology, Paula Teves-Costa and her colleagues conclude that care must be taken because geological maps and memoirs do not provide sufficient detail on the seismic behaviour of soils, which is related to small-scale variations in lithology and thickness especially in shallow layers (Teves-Costa et al. 2007, p. 998). The implication of this research is that in earthquake zones such as the Azores geotechnical studies of soils are required in the pre-planning of housing and other building development (see Chapter 5).

22 May 1980 – Flores

In common with the 1847 tsunami, this event was triggered by a large sea-cliff landslide and had its greatest effect on the south coast of Flores below the cliffs of Rocha Alta. A small tsunami was generated (Cabral 2020). Despite its recent date, details of this event are sparse (Cabral et al. 2010), but a small tsunami was generated.

9 July 1998 – Faial also damage in Pico and São Jorge

This maximum Intensity VIII (M 5.8) earthquake had its epicentre in the ocean to the NE of Faial and NW of Pico (Figure 3.2), with attenuation occurring in an east to west direction across Faial, and west to east across São Jorge. It caused nine deaths, injured over 100 people and produced severe damage on Faial and some parishes on Pico. Fifteen hundred buildings were totally or partially damaged, and 1,457 residents were made homeless (Ferreira et al. 2016), landslides were generated, roads and retaining walls collapsed together with many bridges and aqueducts. Significant seismic wave amplification was noted in association with 'soft' substrates and on deep soils in the villages of Castelo Branco and Flamengos located, respectively, in southern and eastern Faial, with Flemengos and Ribeirinha (in the NE) being the most severely affected settlements. The earthquake also triggered landslides: along the coasts of Faial and São Jorge; across fault scarps; along the NW slopes of the central volcano of Faial including the walls of the caldera and in areas of unconsolidated slope material. One notable landslide occurred on the outer caldera slopes near to Ribeira Funda (NW Faial) and was 200 m wide, 1,400 m long, had a height different between its top and bottom of 320 m and a volume of 600,000 m^3 (Malheiro 2006).

On Faial out of a total of 3,950 houses, 54% experienced no impact, 10% totally collapsed and 36% suffered some damage. Following the 1926 earthquake, some buildings in Horta had been strengthened using tie rods and screw bolts and these buildings fared better than those retrofitted after the 1973 event when an alternative 'collar beam' technique was used. Buildings in Horta remained vulnerable to seismic activity and suffered major losses (Spencer et al. 2000), the vast majority being old and falling into EMS-98 vulnerability Classes A and B (Neves et al. 2012, see above).[17]

Other tsunamis

Earthquakes such as those of 1522, 1614, 1757, 1939 and 1980 (January), not only produced damage on land, but were also associated with tsunamis. This may also have been the case with 1591 and 1926 earthquakes though the evidence is far more equivocal (Cabral 2020). As Table 3.3 shows, there were other local and distal marine earthquakes that, whilst having had no direct impact on-shore, caused significant damage to coastal settlements and infrastructure because of wave impact and inundation. Little is known in detail about the event of 1571, but in December 1641 Andrade et al. (2006) found evidence of damage along the south coast of São Jorge, with three waves destroying a Franciscan convent and causing flooding in a nearby settlement. A run-up height of *c.*9 m is implied by the local topography, the town was probably Velas and there may have been 50 victims (Agostinho 1935b). The veracity of both these events, however, is disputed by one recent author (Cabral 2020). Although an event of 1653 could have been a tsunami, Andrade et al. (2006, p. 177) also suggest the possibility of a large storm. There is little certainty about the putative tsunami of 1668. Boulder entrainment prevented navigation into the port of Calheta (south coast of São Jorge), with the sea rushing into low lying areas, destroying housing and much of the sea wall, but there is a possibility that these effects were caused by a storm (Table 3.3). Relatively little is known about possible tsunamis in 1676 and 1691, but according to Drummond (1856–1859) the former caused much damage along the south coast of Terceira. In contrast, the 1755

Table 3.5 The impact of tsunamis generated by the 1755 'Lisbon' earthquake in the Azores

Island	Impact
Terceira	Sea-water inundated the southern coast, destroyed many houses, spread marine sediments across pastures and cropped land and demolished fortifications. At Porto Judeu there was exceptionally ebb and flow of the tide and in Angra do Heroísmo waters reached the main square (*Praça Velha*), some 250 m inland and located at an elevation of 11–15 m.
	At Vila da Praia (now Praia da Vitória) on the east coast of the island, unprecedented tides were noted and drowned a wetland some 750 m inland. Along the coast to the south of Praia, 15 houses were destroyed and land covered by sediment took many years before it produced crops albeit with reduced yields.
São Miguel	In Ponta Delgada many buildings were ruined, with the pattern of losses implying a run-up height of 4–5 m
Corvo and Flores	Exceptional ebb and flow of the tide, but the damage caused if any is uncertain.
Faial	High waves destroyed mills in the Conceição *freguesia* (parish) within Horta, and several boats and a ship were swept offshore.

Based on information in: Drummond (1856–59) and Andrade et al. (2006).

(M 8.5–8.8) Lisbon earthquake caused well recorded damage throughout the islands. Conventionally it has been assumed that this huge earthquake had its epicentre on the Gorringe Bank, located some 210 km west of Cabo de São Vicente (Cape S. Vincent) in the Algarve, though more recently alternative epicentral locations have been suggested (Chester & Chester 2010; ten Brink et al. 2014), but there is no doubt about the scale of the damage in the Azores (Table 3.5). The 1761 tsunami caused destruction across the North Atlantic littoral with effects being felt in Great Britain and the West Indies as well as in the Azores, although accounts are sparse. Later in 1792 waves – possibly with a run-up height of 8 m (Table 3.3) – occurred at São Jorge damaging housing and sea defences, but whether caused by a tsunami or storm waves is uncertain (Cabral 2020).

In the nineteenth century, only three, seismically induced tsunamis are recorded in 1855, 1856 and 1899, however the reliability of the records upon which they are recognised are disputed (Table 3.3). In 1855 Terceira was the principal island affected with damage being concentrated along its southern and eastern coasts, while in 1856 São Miguel was the focus of impact with houses and sea-walls being affected. Andrade et al. (2006, p. 179) admit that this event may have been a severe storm rather than a tsunami. It is known that there were large storms in February 1899, but at Velas (São Jorge) recorded inundation exceeded all known storm-surges in terms of damage and a tsunami is strongly implied (Avellar 1902).

In view of the fact that it occurred less than 90 years ago, little is known about the 1931 tsunami except that it impacted the coastal settlements of Horta and Feteiras on Faial Island (Heek 1944), but the veracity of this event and that of 1935 are questioned. It is possible that the damage was caused by storms (Cabral 2020). Events in 1939 (see above) and 1941 were small scale and left no recorded impact on the Azores coast (Andrade et al. 2006) and, even events in 1969 and 1975, which had a maximum run-up heights of 1 m, failed to produce any recorded damage across the archipelago.

Concluding Remarks

Information contained in this chapter will be used to assess the exposure of the Azores to future eruptions, earthquakes and tsunamis (Chapter 4); to discuss the ways in which the people have coped with historic disasters (Chapter 5) and to inform policies of disaster risk reduction (Chapter 6). A feature of historic catalogues is that they represent only a small fraction of the geological age of the archipelago (Chapter 2). This means that for volcanic eruptions and possibly also for earthquakes, the population, their settlements, infrastructure, economy and culture may have to face far more extreme disaster scenarios than have occurred historically. In the Holocene there have been many examples of Plinian scale volcanic activity and, for example, if a Fogo A (4.6 ka BP) or Furnas C (*c*.1.9 ka BP) style eruption (Guest et al. 2015; Wallenstein et al. 2015), were to recur then it would have a devastating effect on São Miguel and could affect most or all of the islands.

Notes

1 The size of an earthquake may be measured in a number of ways and Intensity and Magnitude are the most frequently used metrics.

   Intensity is a descriptive measure of the size of an earthquake, which is based on the impact on the ground surface, buildings and people. Isoseismal lines are used to join places of equal intensity. There are several scales, of which the Modified Mercalli (MM) and the Medvedev-Sponheuer-Kárnik (MSK) scales, the latter widely used by European seismologists, are the most well known. A new scale was developed in 1998 and is particularly suitable for earthquakes in Europe. Known as the EMS-98 and is very similar to the MSK scale. All intensity scales run from I (minimum) to XII (maximum). It is because they are based on descriptions of recorded damage, that Intensity may be calculated for both historic and present-day earthquakes. In this volume, the maximum Intensity is quoted without reference to a particular scale.

   Magnitude is a quantitative measure of earthquake size. For much of the twentieth century a scale devised by Charles Richter in California in 1931 was widely used and is still frequently (and often erroneously) quoted in popular accounts. Known as the Richter (or local) magnitude (M_L), for each unit increase in magnitude, the amplitude of seismic waves increases 10 times and the energy released over 30 times greater. Other magnitude scales are the surface wave magnitude (M_S); body-wave magnitude (m_b) and the moment magnitude (M_W) (Bolt 2004, p. 163–70). It is the current convention to quote the magnitude as (M) using the most appropriate measure and this convention is followed in the present volume.

2 *Archivo dos Açores* – available from the University of the Azorses - https://biblioteca.uac.pt/pt-pt/recursos-arquivo-acores - Accessed 7 August 2919

3 The *Serviços Geológicos de Portugal* has had a complex recent administrative history. After 1993 it became part of the *Instituto Geológio Mineiro* (Institute for Geology and Mining), between 2003 and 2006 it was incorporated into the *Instituto da Engenharia Tecnologia e Inovação* (Institute of Engineering Technology and Innovation) and finally became integrated into the *Laboratório Nacional de Energia e Geologia* (National Laboratory of Geology and Energy).

4 This means that the islands of Flores and Corvo have not been affected by earthquakes.

5 Under the EMS-98 classification buildings most at risk (i.e. Classes A and B) are constructed from *rubble stone* and *simple-stone*. *Rubble-stone* is a construction technique 'in which undressed stones are used as the basic building material, usually with poor quality mortar, leading to buildings which are heavy and have little resistance to lateral stress.' Floors are usually wooden and provide little horizontal rigidity. *Simple stone* differs from *rubble-stone* construction 'in that the building stones have undergone some dressing prior to use. These hewn stones are arranged ... to improve the strength of the structure, e.g. using larger stones to tie in the walls at the corners. In the normal case, such buildings are treated as vulnerability class B, and only as class A when in poor condition or put together with particularly poor workmanship' (Grünthal 1998, p. 34-5).

6 The true cost of losses is usually much greater following an earthquake. Losses may be divided into direct (or immediate) and indirect (or delayed). The former may themselves be sub-divided into quantifiable (e.g. property damage, evacuation costs, insured losses, fatalities, injuries, homelessness, crop damage and disruption of infrastructure) and non-quantifiable (e.g. damage to the cultural heritage, habitat loss, mental distress). For indirect costs, again subdivision is possible into quantifiable (e.g. declines in economic activity, supply chain disruption, business interruption and lost income) and non-quantifiable (e.g. electric power failure, lost education, fewer chances to develop business opportunities and family dislocation including orphaning). The true costs of a future earthquake similar in size to those occurring in 1980 and 1998 would, therefore, greatly exceed a calculation based solely on direct costs (Gall et al. 2014).

7 A friar is a member of a Catholic religious order who is devoted to the religious life but works in the community rather than a monastery. He may or may not be a priest.

8 English – burnt peak.

9 This is a very imprecise measurement and indeed until the modern era Leagues varied in length in different European countries. Today 1 League = 4.83 km (or 3 statute miles) and in 16th century Portugal 1 League = *c*.6.17 km. Both conversions are clearly incorrect because measurement from the fissure to the lava delta is about 4 km. Therefore a league in Frutuoso's account represents *c*.2.7 km (see Wallenstein et al. 2005, p. 132).

10 The term *mistério* (mystery) is used in the Azores to describe a black lava flow-field, with rocky and infertile soil formed by an historic eruption that was witnessed by the population.

11 Since the eruption of Mt. Pelée in 1902, it has been widely accepted in the volcanological literature that the term *nuées ardentes* was introduced by Lacroix (1904), who described the pyroclastic flows that he had witnessed. Lacroix acknowledges that the Portuguese term *nuvem ardente* was first translated as *nuée ardente* and introduced into the scientific by the French geologist Ferdinand Fouqué (1828-1904). Indeed Lacroix (1904) provides an account of the 1580 and

1808 eruptions of São Jorge in his work La Montagne Pelée et ses Éruptions. The seminal work of Frank Alvord Perret (1935) on the 1929-32 eruptions of Mt. Pelée and his description of *nuées ardentes* reinforced the association between the term and Mt. Pelée (Wallenstein et al. 2018).

12　The role of the Dabney family as Consuls of the United States of America is discussed in more detail in Chapter 5.

13　Reference is also made to William Hamilton describing similar circumstances at Vesuvius in Italy.

14　The historical record may not include examples of some of the largest tsunamis to have affected the Azores. For instance, Omira et al. (2016) have reconstructed the possible effects of a putative flank collapse off the NE slope of Pico at around 70 ka. With an estimated volume of 8 km^3, this would have caused a mega-tsunami which they estimate would have impacted the coasts of Pico and São Jorge with run-up heights of up to 50m.

15　Jan Huyghen van Linschoten (1563-1611) was a Dutch trader and merchant who spent the years 1590-1592 living in Terceira.

16　Based on 1970 census figures for: Faial; Pico; São Jorge; Graciosa and Terceira. These are known as the Central Islands of the archipelago.

17　Collars are horizontal timber or concrete beams connecting rafters together to form a strong 'A' shaped roof truss.

References

Admiralty 1945. *Spain and Portugal Volume IV: The Atlantic Islands*. Naval Intelligence Division, Geographical Handbook Series BR 502c, London.

Agostinho, J. 1931. The volcanoes of the Azores. *Bulletin Volcanologique* **14**, 123–138.

Agostinho, J. 1935a. *Tectónica, sismicidade e vulcanismo das ilhas dos Açores*. Separata de Açores de Açoreana - Boletim da Sociedade Afonso Chaves, Angra do Herorísmo.

Agostinho, J. 1935b. Notices sur quelques raz de mareé aux Açores. Annales de la Commission pour l'etude des raz de mareé. *International Union Geodesy and Geophysics* **4**, 21–24.

Agostinho, J. 1936. Tectónica, sismicidade e vulcanismo dos ilhas dos Açores. *Açoreana* **I9II**, 87–98.

Andrade, C., Borges, P. and Freitas, M.C. 2006. Historical tsunami in the Azores archipelago (Portugal). *Journal of Volcanology and Geothermal Research* **156**, 172–185.

Anon 1926. The earthquakes in the Azores. Casualties over two hundred - a town wrecked. *The Times* (London), September 2, page 10.

Anon 1964. Azores earthquake rescue starts. *The Time (London)*, February 20, p. 10.

Anon 1983. *Problemática da Reconstrução. Sismo de 1 de Janeiro de 1980*. Instituto Açoriano de Cultura, Angra do Heroismo (2 vols.).

Anon 2014. Fieldwork in the Azores. *Archaeology of Mediterranean Earthquake in Europe 1000–1550 AD*. Department of Archaeology, University of Durham. http://armedea. wordpress.com/2014/12/10/fieldwork-in-the-azores/. Accessed 7 August 2919.

Araújo, L.A. 1801. *Memoria chronológica dos tremores mais notaveis e irrupções de fogo acontecidos nas ilhas dos Açores: com a relação dos tremores que houverao nesta ilha Terceira. Acrescenta-se a noticia de hum fenomeno observado no dia 25 de Junho*. Typgraphia Chalcographica e Litteraria do Arco do Cego, Lisboa.

Avellar, J.C.S. 1902. *Ilha de São Jorge (Açores): apontamentos para a sua história*. Typographia Minerva Insulana, Horta.

Baptista, M.A., Miranda, J.M., Batlló, J. Lisboa, F., Luis, J. and Marciá, R. 2016. New study of the 1941 Gloria Fault earthquake and tsunami. *Natural Hazards and Earth System Sciences* **16**, 1967–1977.

Bebo C. 1992. Política financeira da reconstrução. Fontes de financiamento e situação actual. *In*: Oliveira, C., Lucas, A. and Guedes, J.H.C. (eds) *Monografia 10 anos após o sismo dos Açores de Janeiro de 1980*. Secetaria Regional de Habitação e Obras Públicas - delagação de ilha Terceira. Laboratório Nacional de Engenharia Civil, Lisboa 1, 797–806.

Behncke, B., Calvari, S., Giammanco, S., Neri, M. and Pinkerton, H. 2008. Pyroclastic density currents resulting from interaction of basaltic magma with hydrothermally altered rock: An example from the 2006 summit eruptions of Mount Etna (Sicily). *Bulletin of Volcanology* **70**, 1249–1268.

Bento, C. M. 1989. *Escavações Arqueológicas em Vila Franca do Campo 1967–1982*. Associação Arqueológica do Arquipélago dos Açores, São Miguel.

Bento, C.M. 1994. *History of the Azores*. Empress Gráfica Açoreana, Ponta Delgada (Translated By Pilar Vaz Rego Pereira).

Bolt, B.A. 2004. *Earthquakes*. Freeman, New York, 5th Edition.

Booth, B., Croasdale, R. and Walker, G.P.L. 1978. A quantitative study of five thousand years of volcanism on São Miguel, Azores. *Philosophical Transactions of the Royal Society of London* **228A**, 271–319.

Booth, B., Croasdale, R. and Walker, G.P.L. 1983. Volcanic hazard on São Miguel, Azores, *In*: Tazieff, H. and Sabroux, J-C. (eds) *Forecasting Volcanic Events*, Elsevier, Amsterdam, 99–109.

Branca, S., De Beni, E., Chester, D., Duncan, A. and Lotteri, A. 2017. The 1928 eruption of Mount Etna (Italy): Reconstructing lava flow evolution and the destruction and recovery of the town of Mascali. *Journal of Volcanology and Geothermal Research* **335**, 54–70.

Brown, R.J. and Andrews, G.D.M. 2015. Deposits of pyroclastic density currents. *In*: Sigurdsson, H. (ed) *The Encyclopedia of Volcanoes*. Academic Press, London, 631–648, 2nd Edition.

Cabral, N. 2020. Revision of the Azorean catalogue of tsunamis. *In*: Dilek, Y., Ogawa, Y. and Okubo, Y. (eds) *Characterization of Modern and Historic Seismic-Tsunamic Events and Their Global-Societal Impacts*. Geological Society of London, Special Publications **501** http://doi.org/10.1144/SP501-2019-107

Cabral, N., Ferreira, T. and Queiroz, M.G. 2010. Tsunami hazard assessment for the Azores archipelago: A historical review. European Geosciences Union General Assembly 2010, Vienna, Austria, 02–07 May (Poster).

Calado, H., Borges, P., Ng, K. and Alves, F. 2011. The Azores archipelago, Portugal: Improved understanding of small island coastal hazards and mitigation measures. *Natural Hazards* **58**, 427–444.

Caldeira, B., Fontiela, J., Borges, J.F. and Bezzeghoud, M. 2017. Grandes terremotos en Azores. *Fisica de la Terra* **29**, 29–45.

Callender, J.M. and Henshall, J.D. 1968. The land use of Faial in the Azores. *In*: Cook, A.N. (ed) *Four island studies. The World Land Use Survey. Monograph 5*, Geographical Publications, Bude, Chapter 11, 367–395.

Calvert, A.T., Moore, R.B., McGeehin, J.P. and Rodrigues de Silva, A.M. 2006. Volcanic history and $^{40}Ar/^{39}Ar$ and ^{14}C geochronology of Terceira Island, Azores, Portugal. *Journal of Volcanology and Geothermal Research* **156**, 103–115.

Camus, G., Boivin, P., De Goër de Herve, A., Kieffer, G., Mergoil, J. and Vincent, P.M. 1981. Le Capelinhos (Faial, Açores) vingt ans après son eruption: le modèle éruptif "surtseyen" et les anneaux de hyaloclastiques. *Bulletin Volcanologique* 44 (1), 31–42.

Canto, E.P. 1879a. Erupção na Ilha de S. Miguel. Anno de 1563. *Archivo dos Açores* 1 (5), 453–466 & 1 (6), 536–541.

Canto, E.P. 1879b. Erupção na Ilha do Pico. Anno do 1562. *Archivo dos Açores* 1 (4), 360–367.

Canto, E.P. 1880a. Erupção na Ilha de S. Miguel. Anno de 1564. *Archivo dos Açores* 2 (8), 186.

Canto, E.P. 1880b. Erupção no Ilha de S. Jorge. Anno de 1580. *Archivo dos Açores* 2, 188–193.

Canto, E.P. 1881a. Vulcanismo nos Açores. Anno de 1630: erupção na valle das Furnas. *Archivo dos Açores* 2 (12), 527–547.

Canto, E.P. 1881b. Erupção na Ilha do Fayal. Anno de 1672. *Archivo dos Açores* 3 (16), 344–351.

Canto, E.P. 1882. Erupção na Ilha do Pico. Anno de 1718. *Archivo dos Açores* 3 (18), 497–506.

Canto, E.P. 1883a. Erupção na Ilha do Pico. Anno de 1720. *Archivo dos Açores* 4 (22), 343–345.

Canto, E.P. 1883b. Terremoto e erupção na Ilha Terceira. Anno de 1760–1761. *Archivo dos Açores* 4 (26), 362–365.

Canto, E.P. 1883c. Erupção submarina junto a Ilha Terceira. Anno de 1867. *Archivo dos Açores* 5 (3), 499–503.

Canto, E.P. 1884a. Erupção na Ilha de S.Jorge. Anno de 1808. *Archivo dos Açores* 5 (29), 437–447.

Canto, E.P. 1884b. Erupção submarina junto a Ilha Terceira. Anno de 1867. *Archivo dos Açores* 5 (30), 499–501

Canto, E.P. 1888. Erupção no Capello, Ilha do Fayal. Anno de 1672. *Archivo dos Açores* 9 (53), 425–432.

Cappello, A., Zanon, V., Del Negro, C., Ferreira, T.J.L. and Queiroz, M.G.P.S. 2015. Exploring lava-flow hazards at Pico Island, Azores archipelago (Portugal). *Terra Nova* 27, 156–161.

Carmo, R., Ferreira, T. and Marques, R. 2016. Macroseismic evaluation of Santa Maria Island (Azores) 1937 and 1939 earthquakes. 35th General Assembly of the European Seismological Commission, Trieste, *ESC2016- Preview*, no page numbers.

Carneiro, A., Simões, A., Diogo, M.P. and Mota, T.S. 2013. Geology and religion in Portugal. *Notes and Records of the Royal Society* 67, 331–354.

Carvalho, J., Torres, L., Castro, R., Dias, R. and Mendes-Victor, L. 2009. Seismic velocities and geotechnical data applied to the soil microzoning of western Algarve, Portugal *Journal of Applied Geophysics* 68 (2) 249–258

Cas, R.A.F. and Wright, J.V. 1987. *Volcanic Successions Modern and Ancient*. Allen and Unwin, London.

Castello Branco, A., Zbyszewski, G., Moitnho de Almeida, F. and da Veiga Ferreira, O. 1959. Rapport de la premiere mission Geologique. *Serviços Geológicos de Portugal Memória* 4, 9–27.

Chaves, F.A. 1960. Erupções submarinas nos Açores. Informações que os navegantes podem prestar sobre tal assunto. *Açoreana* 5, 50.

Chester, D.K. and Chester, O.K. 2010. The impact of eighteenth century earthquakes on the Algarve Region. Southern Portugal. *Geographical Journal* **176** (4), 350–370.

Chester, D., Duncan, A., Coutinho, R., Wallenstein, N. and Branca, S. 2017. Communicating information on eruptions and their impacts from the earliest times until the 1980s. *In:* Bird, D., Fearnley, C., Haynes, K., Jolly, G. and McGuire, B. (eds) *Observing the Volcano World; Volcano Crisis Communication.* Springer, Berlin, 1–25.

Chester, D.K., Duncan, A.M., Coutinho, R. and Wallenstein, N. 2019. The role of religion in shaping responses to earthquake and volcanic eruptions: A comparison between Southern Italy and the Azores, Portugal. *Philosophy, Theology and the Sciences* **6**, 33–45.

Cole, P.D., Queiroz, G., Wallenstein, N., Gaspar, J., Duncan, A.M. and Guest, J.E. 1995. An historic subplinian/phreatomagmatic eruption: The 1630 AD eruption of Furnas Volcano, São Miguel, Azores. *Journal of Volcanology and Geothermal Research* **69**, 117–135.

Cole, P.D., Guest, J.E. and Duncan, A.M. 1996. Capelinhos: The disappearing volcano. *Geology Today* **12**, 68–72.

Cole, P.D., Guest, J.E., Quieroz, G., Wallenstein, N., Pacheco, J.M., Gaspar, J.L., Ferreira, T. and Duncan, A.M. 1999. Styles of volcanism and volcanic hazards on Furnas volcano, São Miguel, Azores. *Journal of Volcanology and Geothermal Research* **92**, 39–53.

Cole, P.D., Guest, J.E., Duncan, A.M. and Pacheco, J.-M. 2001. Capelinhos 1957–1958, Faial, Azores: Deposits formed by an emergent surtseyan eruption. *Bulletin of Volcanology* **63**, 204–220.

Cole, P.D., Neri, A. and Baxter, P.J. 2015. Hazards from pyroclastic density currents. *In:* Sigurdsson, H. (ed) *The Encyclopedia of Volcanoes.* Academic Press, London, 943–956, 2nd Edition.

Conte, E., Widom, E., Kuentz, D. and França, Z. 2019. [14]C and U–series disequilibria age constraints from recent eruptions at Sete Cidades volcano, Azores. *Journal of Volcanology and Geothermal Research* **373**, 167–178.

Corrêa, J. 1924. *Leituras sobre a História do Valle Valle das Furnas.* Oficina de Artes Graficas, S. Miguel, Açores.

Coutinho, R., Chester, D.K., Wallenstein, N. and Duncan, A.M. 2010. Responses to, and the short and long-term impacts of the 1957/1958 Capelinhos volcanic eruption and associated earthquake activity on Faial, Azores. *Journal of Volcanology and Geothermal Research* **196**, 265–280.

Cruz, J.V., Nunes, J.C., França, Z., Carvalho, M.R. and Forjaz, V.H. 1995. Estudo vulcanológico das erupções históricas da Ilha do Pico – Açores. IV Congresso Nacional de Geologia. Porto. *Memórias Museu Laboratorio Mineral Faculdade de Ciências Porto* **4**, 985–987.

Degg, M. 1992. Some implications of the 1985 Mexican earthquake for hazard assessment. *In:* Mc Call, G.J.H., Laming, D.J.C. and Scott, S.C. (eds) *Geohazards: Natural and Man-made.* Chapman and Hall, London, 105–114

Degg, M. 1995. Earthquakes, volcanoes and tsunamis: Tectonic hazards and the built environment. *Built Environment* **21** (1) 94–113.

Degg, M. and Doornkamp, J. 1994. *Earthquake Hazard Atlas. 7 Iberia including the Azores, Balearics, Canary and Madeira Islands.* London Insurance and Reinsurance Market Association, London.

Dias, U.M. 1936. *História dio Valle das Furnas.* Empressa Tip Lda, Vila Franca do Campo.

Dias, A.A.M. 1945. *O sismo de 1522 em S. Miguel*. Insulana, Ponta Delgada, 1, 5–15.

Di Chiara, A., Tauxe, L. and Speranza F. 2014. Paleointensity determination from São Miguel (Azores Archipelago) over the last 3 ka. *Physics of the Earth and Planetary Interiors* **234**, 1–13.

Drummond, F.F. [1856–59] 1981. *Annaes da Ilhas Terceira, I, III, IV. Reimpressão fac-similada da edição de 1859- Governo da Região Autónoma dos Açores*. Secretaria Regional da Educação et Cultura, Angra do Heroísmo.

Duncan, A. M., Guest, J. E., Stofan, E. R., Anderson, S. W., Pinkerton, H. and Calvari, S. 2004. Development of tumuli in the medial portion of the 1983 aa flow-field, Mount Etna, Sicily. *Journal of Volcanology & Geothermal Research* **132**, 173–187.

Feio, A., Bologna, R., Monteiro, D. and Félix, D. 2015. The Earthquakes in Lisbon (1755) and Angra do Heroísmo (1980). *In*: Filion, P., Sands, G. and Skidmore, M. (eds) *Cities at Risk: Planning for and Recovering from Natural Disasters*. Routledge, Abingdon, 81–95.

Ferlito, C., Viccaro, M., Nicotra, E. and Cristofolini, R. 2010. Relationship between the flank sliding of the South East Crater (Mt. Etna, Italy) and paroxysmal event of November 16, 2006. *Bulletin of Volcanology* **72**, 1179–1190.

Ferreira, T., Gomes, A., Gaspar, J.L. and Guest, J. 2015. Distribution and significance of basaltic centres: São Miguel, Azores. *In*: Gaspar, J.L., Guest, J.E., Duncan, A.M., Chester, D.K. and Barriga, F. (eds) *Volcanic Geology of S. Miguel Island (Azores, Archipelago)*. Memoir Geological Society of London **44**, 135–146.

Ferreira, T.M., Maio, R., Vicente, R. and Costa, A. 2016. Earthquake risk mitigation: The impact of seismic retrofitting strategies on urban resilience. *International Journal of Strategic Property Management* **20** (3), 291–304.

Fontiela, J. 2009. Human losses and damage expected in future earthquakes in Faial Island - Azores: A contribution to risk mitigation. M.Sc. Thesis, Universidade dos Açores.

Fontiela, J., Oliveira, C.S. and Rosset, P. 2018. Characterisation of Seismicity of the Azores Archipelago: An overview of historical events and a detailed analysis for the period 2000–2012. *In*: Kueppers and Beier, C. (eds) *Volcanoes of the Azores, Active Volcanoes of World*, Springer-Verlag, Germany, 127–153.

Forjaz, V.H. (ed) 2007. *Vulcão dos Capelinhos Memórias 1957–2007*. OVGA – Observatório Vulcanológico e Geotérmico dos Açores, Ponta Delgada.

Fouqué, F. 1867. Sur les phénomènes volcanique observés aux Açores. *Académie des Sciences à Paris, Compte Rendus* **65**, 1050–1053.

Fouqué, F, 1873. Voyages géologiques aux Açores. *Revue des deux mondes (2 ème période)* **103**, 40–65.

França, Z. 2002. *Origem e evolução petrológica e geoquimica do vulcanismo da Ilha do Pico, Açores*. Câmara São Roque do Pico, Azores.

França, Z., Tassinari, C., Cruz, J., Aparicio, A., Araña, V. and Rodriques, B. 2006. Petrology, geochemistry and Sr-Nd-Pb isotopes of the volcanic from Pico – Azores (Portugal). *Journal of Volcanolology and Geothermal Research* **156**, 71–89.

França, Z., Forjaz, V.H., Ribeiro, L.P. França, I.E. and Neves, G. 2014. *Ilha do Pico erupção de 1562–64 450 anos após*. Câmera Municipal de São Roque do Pico, Azores.

Frutuoso, G. 2005. *Saudades da Terra*. Instituto, Cultural de Ponta Delgada, Pont Delgada, Açores (6 volumes) - original manuscript 1586–1590.

Gall, M. Emrich, C.T. and Cutter, S.L. 2014. Who needs loss data? Background paper prepared for the 2015 Global Assessment Report on Disaster Risk Reduction. Global Assessment Report On Disaster Risk Reduction. United Nations International Strategy for Disaster Risk Reduction, United Nations, Geneva.

Garrard, C., Forlin, P., Froude, M., Petley, D., Gutiérrez, A., Treasure, E., Milek, K. and Oliveira, N. 2021. The Archaeology of a Landslide Disaster of 1522 and its Consequences. *European Journal of Archaeology* 24 pages – pre-publication.

Gaspar, J.L., Queiroz, G., Pacheco, J.M., Ferreira, T., Wallenstein, N., Almeida, M.H. and Coutinho, R. 2003. Basaltic lava balloons produced during the 1998–2001 Serreta Submarine Ridge eruption (Azores). *In*: White, J.D.L., Smellie, J.L. and Clague, D.A. (eds) *Explosive Subaqueous Volcanism*, Geophysical Monograph **140**, American Geophysical Union, Washington DC.

Gaspar, J.L., Queiroz, G., Ferreira, T., Amaral, P., Viveiros, F., Marques, R., Silva, C and Wallenstein, N. 2011. Geological Hazards and Monitoring at the Azores (Portugal). *Earthzine*, http://www.earthzine.org/2011/04/12/geological-hazards-and-monitoring-at-the-azores-portugal/.

Gaspar, J.L., Queiroz, G., Ferreira, T., Medeiros, A.R., Goulart, C. and Medeiros, J. 2015. Earthquakes and volcanic eruptions in the Azores region: Geodynamic implications from major historical events and instrumental seismicity. *In*: Gaspar. J.L., Guest, J.E., Duncan, A.M., Barriga, F.J.A.S. and Chester, D.K. (eds) *Volcanic Geology of São Miguel Island (Azores Archipelago)*, Geological Society, London Memoir **44**, 33–49.

Gomes, A., Gaspar, J.L. and Queiroz, G. 2006. Seismic vulnerability of dwellings at Sete Cidades Volcano (S. Miguel Island, Azores). *Natural Hazards and Earth System Sciences* **6**, 41–48.

Grünthal, G. 1998. *European macroseismic scale 1998*. Conseil de L'Europe Cahiers du Centre Européen de Géodynamique et de Séismologie vol. 15, Luxembourg.

Guest, J.E., Kilburn, C.R.J., Pinkerton, H. and Duncan, A.M. 1987. The evolution of lava flow-fields: Observations of the 1981 and 1883 eruptions of Mount Etna, Sicily. *Bulletin of Volcanology* **49**, 527–540.

Guest, J.E., Gaspar, J.L., Cole, P.D., Queiroz, G., Duncan, A.M., Wallenstein, N. and Ferreira, T. 1994. Preliminary report on the volcanic geology of Furnas Volcano, São Miguel, the Azores. *CEC Environment: ESF Laboratory Volcano. Eruptive History and Hazard Open File Report: Number 1.* January 1994, 24pp.

Guest, J.E., Gaspar, J.L., Cole, P.D., Queiroz, G., Duncan, A.M., Wallenstein, N., Ferreira, T. and Pacheco, J-M. 1999. Volcanic geology of Furnas Volcano, São Miguel, Azores. *Journal of Volcanology and Geothermal Research* **92**, 1–29.

Guest, J., Cole, P., Duncan, A. and Chester, D. 2003. *Volcanoes of Southern Italy*. Geological Society of London, Bath.

Guest, J.E., Duncan, A.M., Stofan, E.R. and Anderson, S.W. 2012. Effect of slope on development of pahoehoe flow fields: Evidence from Mount Etna. *Journal of Volcanology & Geothermal Research* **219–220**, 52–62.

Guest, J.E., Pacheco, J.M., Cole, P.D., Duncan, A.M., Wallenstein, N., Queiroz, G., Gaspar, J.L. and Ferreira, T. 2015. The volcanic history of Furnas Volcano, São Miguel, Azores. *In*: Gaspar. J.L., Guest, J.E., Duncan, A.M., Barriga, F.J.A.S. and Chester, D.K. (eds) *Volcanic Geology of São Miguel Island (Azores Archipelago)*, Geological Society, London Memoir 44, 125–134.

Heek, N. 1944. List of Seismic Sea Waves. *Seismological Society of America Bulletin* **37** (4), 269–284.

Hipólito, A., Madeira, J., Carmo, R. and Gaspar, J.L. 2013. Neotectonics of Graciosa island (Azores): A contribution to seismic hazard assessment of a volcanic area in a complext geodynamic setting. *Annals of Geophysics* **56** (6) 2013, SO677. doi: 10.4401/ag-6222.

Hirn, A., Haessler, H., Hoang Trong, P., Wittlinger, G. and Mendes Victor, L.A. 1980. Aftershock sequence of the January 1st, 1980, earthquake and present-day tectonics in the Azores. *Geophysical Research Letters* **7** (7), 501–504.

Jerónimo, G.M. 1989. Nós, açorianos e os sismos. Correio dos Açores, Ponta Delgada.

Kauahikaua, J., Sherrod, R., Cashman, K., Heliker, C., Hon, H., Mattox, T.N. and Johnson, J.A. 2003. Hawaiian lava-flow dynamics during the Pu'u 'Ō'ō-Kūpaianaha eruption: A tale of two decades. *U.S. Geological Survey Professional Paper* **1676**, 63–87.

Kelly, J.T., Carey, S., Pistolesti, M., Rosi, M., Croff-Bell, K.L., Roman, C. and Marani, M. 2014. Exploration of the 1891 Foerstner submarine vent site (Pantelleria, Italy): Insights into the formation of basaltic balloons. *Bulletin of Volcanology* **76**, 844.

Kueppers, U., Nichols, A.R.I., Zanon, V., Potuzak, M. and Pacheco, J.M.R. 2012. Lava baloons – peculiar products of basaltic submarine eruptions. *Bulletin of Volcanology* **74**, 1379–1393.

Le Bas, M.J., Le Maitre, R.W., Streckeisen, A. and Zanettin, B. 1986. A chemical classification of volcanic rocks based on the total alkali-silica diagram. *Journal of Petrology* **27**, 745–750.

Lacroix, A. 1904. *La Montagne Pelée et ses éruptions.* Masson, Paris.

Linschoten, J.H. 1596[1998]. *Itinerário, viagem ou navegação para as Indias Ocidentais ou Portuguesas.* 2nd Edition. Comissã Nacional para Comemoração dos Descombrimentos Portuguese, Lisboa (Original 1596).

Luz, J. B. 1996. O homem e a história em Gaspar Frutuoso. Homenagem ao Prof. Doutor Lúcio Craveiro da Silva. *Revista Portuguesa de Filosofia* **52**, 475–486.

Macedo, A.L.S. 1871. *História das quarto ihlas que forman o Distrito do Horta.* Typographia de L. P. Silva Correa, Horta, 3 vols. (reissued in facsilile by Secretaria Regional da Educação Região Autónoma dos Açores, 1981).

Machado, F. 1949. O terramoto de S. Jorge em 1757. *Açoreana* **4** (4), 311–324.

Machado, F. 1958a. Actividade Vulcânica de Ilha do Faial (1957–1958). Notícia preliminar relativa aos meses de Setembro a Dezembro de 1957. *Atlantida* **vII**: 225–236.

Machado, F. 1958b. Actividade Vulcânica de Ilha do Faial (1957–1958). *Atlantida* **vII**: 305–315.

Machado, F. 1959. A erupçãodo Faial em 1672. *Serv. Geol. de Portugal* **4**, 89–99.

Machado, F. 1962. Erupções históricas do sistema volcânico Faial – Pico – S. Jorge. *Atlântida* **6**, 84–91.

Machado, F. 1973. *Periodicidade sísmica nos Açores.* Ministério da Economia. Secretaria de Estado da Indústria. Direcção - Geral de Minas e Serviços Geológicos Lisboa, 476–484.

Machado, F., Parsons, W.H., Richards, A.F. and Mulford, J.W. 1962. Capelinhos eruption of Fayal Volcano, Azores, 1957–1958. *Journal of Geophysical Research* **67**, 3519–3529.

Madeira, J.E. 1998. *Estudos de Neotectónica nas ilhas do Faial, Pico e S. Jorge: uma contribuição para o conhecimento Geodinâmico da junção tripla dos Açores.* Dissertação apresentada à Universidade de Lisboa para a obtenção do grau de doutor em geologia, na especialidade de Geodinâmic`a Interna. Faculdade de Ciências da Universidade de Lisboa, Departamento de Geologia.

Madeira, J. 2005. *The Volcanoes of the Azores Islands: World Class Heritage (Examples from Terceira, Pico).* Field Trip Guide Book, IV International Symposium PROGEO on the Conservation of Geological Heritage. Univ. Lisbon, Portugal

Madeira, J. and Brum da Silveira, A. 2003. Active tectonics and first paleoseismological results in Faial, Pico and S. Jorge islands (Azores, Portugal). *Annals of Geophysics* **46** (5), 733–761.

Madeira, J.E, Brum da Silveira, A., Serralheiro, A., Monge Soares, A. and Rodrigues, C.F. 1998. Radiocarbon ages of recent volcanic events from the Island of S. Jorge (Azores). *Comunicações I.G.M.* **84** (1), A189–192.

Malheiro, A. 2006. Geological hazard in the Azores archipelago: Volcanic terrain instability and human vulnerability. *Journal of Volcanology and Geothermal Research* **156**, 158–171.

Marques, R., Zêzere, J.L. and Amaral, P. 2009. Reconstituição e modelação probabilística da escoada detrítica de Vila Franca do Campo desencadeada pelo sismo de 22 de Outubro de 1522 (S. Miguel, Açores). *Associação Portuguesa de Geomorfologia,* **VI**, 175–182.

Mendes-Victor, L.A, Oliveira C.S, Pais, I. and Teves-Costa, P. 1994. Earthquake damage scenarios in Lisbon for disaster preparedness. *In:* Tucker, B. E., Erkik, M. and Hwang, C. N. (eds) *Issues in Urban Earthquake Risk,* NATO ASI series e Applied Sciences vol. 271, Kluwer Academic Press, Dordrecht, 265–285.

Mendonça, J.J.M. 1758. *História Universal dos Terramotos que tem havido no Mundo, de que há notícia, desde a sua Criação até o Século presente, com huma Narraçam Individual do Terramoto do primeiro de Novembro de 1755, e notícia Verdadeira dos seus effeitos em Lisboa, todo Portugal, Algarves, e mais partes da Europa, África, e América, aonde se estender.* Oficina de António Vicente da Silva, Lisboa.

Middlemost, E.A.K. 1997. *Magmas, Rocks and Planetary Development.* Longman, Harlow.

Mitchell, N.C., Beier, C., Rosin, P., Quartau, R. and Rosin, P. 2013. Lava penetrating water: Submarine lava flows around the coasts of Pico Island, Azores. *Geochemistry Geophysics Geosystems* **9**, Q03024.

Monte Alverne, Frei A. [1695] 1994. *Cronicas da Provincia de S. Joao Evangelista da Ilhas dos Acores.* II Edicao do Instituto Cultural de Ponte Delgada, 2nd Edition.

Moore, R. 1991a. *Geological Map of São Miguel, Azores.* USGS Miscellaneous Investigations Series Map 1–2007.

Moore, R. 1991b. Geology of three late Quaternary stratovolcanoes on São Miguel, Azores. *United States Geological Survey Bulletin* **1900**, 1–26.

Moore, J.G., Phillips, R.L., Grigg, R.W., Peterson, D.W. and Swanson, D.A. 1973. Flow of lava into the sea, 1969–1971, Kilauea volcano, Hawaii. *Geological Society of America Bulletin* **84**, 537–546.

Mota, T. S. and Caneiro, A. 2013. 'A time for engineers and a time for geologists': Scientific lives and different pathways in the history of Portuguese geology. *Earth Science History* **32** (1), 23–38.

Neumann van Padang M, Richards A F, Machado F, Bravo T, Baker, E. and Le Maitre W. 1967. *Atlantic Ocean: Catalogue of the Active Volcanoes of the World.* International Association of Volcanology, Rome, Italy, Part XXI.

Neves, F., Costa, A., Vicente, R., Oliveira, C.S. and Varum, H. 2012. Seismic vulnerability assessment and characterisation of the buildings on Faial Island, Azores. *Bulletin of Earthquake Engineering* **10** (1), 27–44.

Nunes, J.C. 1986. *Sismicidade Histórica e Instrumental do Arquipélago dos Açores. Catálogo Preliminar, 1444–1980*. Relatório INMG/LNEC, Lisboa.

Nunes, J.C. 1999. *A actividade vulcânica na Ilha do Pico do plistocénica ao holocénico: mecanismo eruptive e hazard vulcânica*. PhD thesis, Dept. Geociências, Universidade dos Açores.

Nunes, J.C., Forjaz, V.H. and Oliveira, C.S. 2004. *Catálogo Sísmico Região dos Açores Versão 1.0 (1850–1998)*. SÍSMICA 2004–6th Congresso Nacional de Sismologia e Engenharia Sísmica, Guimarães, 14, 15 e de Abril 2004.

Oliveira, C. and Ferreira, M.A. 2008. Impacto do sismo de 1998 no território dos Açores. Principais consequêcias e indicatores. *In*: Oliveira, C., Costa, A. and Nunes, J. (eds) *Sismo 1998: uma década depois*. Governo dos Açores/SPRI, Ponta Delgada, 717–726.

Oliveira, C.S., Lucas, A.R.A. and Correia Guedes, J.H. 1992. *10 Anos Após o Sismo dos Açores de 1 de Janeiro de 1980*. Secretarti Regional de Habitação e Obras Públicas, Acores.

(eds) Omira, R., Quartau, R., Ramalho, I., Baptista, M.A. and Mitchell, N.C. 2016. The tsunami effects of a collapse of a volcanic island on a semi-enclosed basin: The Pico-São Jorge Channel in the Azores Archipelago. *In*: Duarte, J.C. and Schellart, W.B. (eds) *Plate Boundaries and Natural Hazards*. American Geophysical Union, Geophysical Monograph **219**, 271–287.

Pell, C. 1980. *Earthquake in the Azores, a tragedy and the American Response: A Report of the Committee on Foreign Relations, United States Senate*. US Government Printing Office, Washington, DC.

Perret, F.A. 1935. *The Eruption of Mount Pelée 1929–1932*. Carnegie Institution of Washington, Publication **458**, Washington DC.

Pimentel, A, Zanon, V., de Groot, L.V., Hipólito, A., Di Chiara, A. and Self, S. 2016. Stress-induced comenditic trachyte effusion triggered by trachybasalt intrusion: Multidisciplinary study of the AD 1761 eruption at Terceira Island (Azores). *Bulletin of Volcanology* **78**, 22.

Poland, M.P. and Orr, T.R. 2014. Identifying hazard associated with lava deltas. *Bulletin of Volcanology* **76**, 880.

Queiroz, G. 1997. *Vulcão das Sete Cidades (S. Miguel, Açores): história eruptiva e avaliação do hazard*. Tese de doutoramento no ramo de Geologia, especialidade de Vulcanologia, Departamento de Geociências, Universidade dos Açores.

Queiroz, G., Gaspar, J.L., Cole, P.D., Guest, J.E., Wallenstein, N., Duncan, A.M. and Pacheco, J. 1995. Erupções volcanicas no valle das Furnas (Ilha de S. Miguel, Açores) na primeira metade do seculo XV. *Açoreana* **8**, 159–165.

Queiroz, G., Gaspar, J.L., Guest, J.E., Gomes, A. and Almeida, M.H. 2015. Eruptive history and evolution of Sete Cidades Volcano, São Miguel island, Azores. In: Gaspar, J.L., Guest, J.E., Duncan, A.M., Barriga, F. and Chester, D.K. (eds) *Volcanic Geology of S. Miguel Island (Azores, Archipelago)*. Memoir Geological Society of London **44**, 87–104.

Rangel, G., Ponte, C. and Franco, A. 2015. Use of the geothermal resources in the Azores islands: A contribution to the energy self-sufficiency of a remote and isolated region. In: Gaspar, J.L., Guest, J.E., Duncan, A.M., Barriga, F. and Chester,

D.K. (eds) *Volcanic Geology of S. Miguel Island (Azores, Archipelago)*. Memoir Geological Society of London **44**, 297–301.

Richter, C. F. 1958. *Elementary Seismology.* W.H. Freeman, San Francisco.

Roca, A., Oliveira, C.S., Ansal, A. and Figueras, S. 2008. Local site effects and microzonation. *In*: Gaspar, J.L., Roca, A. and Goula, X. (eds) *Assessing and Managing Earthquake Risk,* Springer, Dordrecht, 67–89.

Scarth, A. and Tanguy, J-C. 2001. *Volcanoes of Europe.* Terra Publishing, Harpenden.

Self, S. 1976. The recent volcanology of Terceira, Azores. *Journal of the Geological Society of London* **132**, 228–240.

Senos, M.L., Teves-Costa, P. and Nunes, J.C. 2000. Seismicity studies of the Azores Islands - An Application to the July 9, 1998 Earthquake. 12 WCC 2000: 12 World Conference on Earthquake Engineering, Auckland, New Zealand, Abstract 1439.

Serralheiro, A., Fojaz, V.H., Alves, C.A.M. and Rodrigues, E.B. 1989. *Carta Vulcanológica dos Açores, Ilha do Faial.* Centro de Vulanologia, Servizio Regional de Protecção Civile, University of the Azores.

Silva, M. 2005. *Caracterização da sismicidade histórica dos Açores com base na reinterpretação de dados de macrossísmica: contribuição para a avaliação do risco sísmico nas ilhas do Grupo Central.* Dissertação de Mestrado em Vulcanologia e Riscos Geológicos, Departamento Geociências, Universidade dos Açores.

Silva, R., Carmo, R. and Marques, R. 2020. Characterization of the tectonic origins of historical and modern seismic events and their societal impact on the Azores Archipelago, Portugal. *Geological Society London Special Publication* 301 http://doi.org/10.1144/SP501-209-106 (accepted manuscript).

Silveira, D. 2002. *Caracterização da sismicidade histórica da ilha de S. Miguel com base na reinterpretação de dados de macrossísmica: contribuição para a avaliação do risco sísmico.* Tese realizada no âmbito do mestrado em vulcanologia e riscos geológicos. Departamento Geociências, Universidade dos Açores.

Silveira, D. 2007. Caracterização da sismicidade histórica da Ilha de São Miguel. *Boletim do Núcleo Cultural da Horta* **16**, 81–102

Silveira, D., Gaspar, J.L., Ferreira, T. and Queiroz, G. 2003. Reassessment of the historical seismic activity of S. Miguel Island (Azores). *Natural Hazards and Earth System Science* **3**, 615–623.

Spencer, R.J.S., Oliveira, C.S., D' Ayala, D.F., Papa, F. and Zuccaro, G. 2000. *The performance of strengthened masonry buildings in recent European Earthquakes.* 12th World Conference on Earthquake Engineering (12WCEE), 30 January - Friday 4 February 2000, paper **1366**, 8pp.

ten Brink, U.S., Chaytor, J.D., Geist, E.L., Brothers, D.S. and Andrews, B.D. 2014. Assessment of tsunami hazard to the U.S. Atlantic margin. *Marine Geology* **353**, 31–54.

Teves-Costa, P., Oliveira, C.S. and Senos, M.L. 2007. Effects of local site and building parameters on damage distribution in Angra do Heroísmo - Azores. *Soil Dynamics and Earthquake Engineering* **27**, 986–999.

Teves-Costa, P. and Veludo, I. 2013. Soil characterization for seismic damage scenarios purposes: Application to Angra do Heroísmo (Azores). *Bulletin of Earthquake Engineering* **11**, 401–421.

Tillard, S. 1812. A narrative of the eruption of a volcano in the sea off the island of St. Michael. *Philosophical Transactions of the Royal Society of London* **102**, 152–158.

Walker, G.P.L. and Croasdale, R. 1971. Two Plinian-type eruptions in the Azores. *Journal of the Geological Society of London* **127**, 17–35.

Wallenstein. N. 1999. Estuda da história recente do comportamento eruptivo do Vulcão do Fogo (S. Miguel, Açores). Avaliação preliminar do hazard. PhD thesis, University of the Azores.

Wallenstein, N., Duncan, A.M., Almeida, H. and Pacheco, J. 1998. A erupção de 1563 do Sapateiro, São Miguel (Açores). *Proceedings of the 1ª Assembleia Luso Espanhola de Geodesia e Geofísica (electronic format)*. Espanha, Almeria.

Wallenstein, N., Chester, D. and Duncan, A.M. 2005. Methodological implications of volcanic hazard evaluation and risk assessment: Fogo Volcano, Sao Miguel, Azores. Zeitschrift für *Geomorphologie* **140**, 129–149.

Wallenstein, N., Duncan, A.M., Chester, D.K. and Marques, R. 2007. Fogo Volcano (São Miguel, Azores): A hazardous landform. *Géomophologie: Relief, Processus, Environment* **3**, 259–270.

Wallenstein, N., Duncan, A., and Chester, D.K. 2015. Eruptive history of Fogo Volcano, São Miguel, Azores. *In*: Gaspar, J.L., Guest, J.E., Duncan, A.M., Barriga, F.J.A.S. and Chester, D.K. (eds) *Volcanic Geology of São Miguel Island (Azores Archipelago)*, Geological Society, London Memoir **44**, 105–123.

Wallenstein, N., Duncan, A., Coutinho, R. and Chester, D. 2018. Origin of the term *nuées ardentes* and the 1580 and 1808 eruptions on São Jorge Island, Azores. *Journal of Volcanology and Geothermal Research* **358**, 165–170.

Waters, A.C. and Fisher, R.V. 1971. Base surges and their deposits: Capelinhos and Taal volcanoes. *Journal of Geophysical Research* **76**, 5596–5614.

Webster, J.W. 1821. *A description of the Island of St Michael. Comprising an account of its geological structure, with remarks on the other Azores or western islands*. R.P. & C.William, Boston.

Weston, F. 1964. List of recorded volcanic eruptions in the Azores with brief reports. *Boletim do Museu e Laboratório Mineralógico da Faculdade de Ciências de Lisboa* **10** (1), 3–18.

White, J.D.L., Schipper, C. I. and Kano, K. 2015. Explosive submarine volcanism. *In*: Sigurdsson, H. (ed) *The Encyclopedia of Volcanoes*, 2nd Edition. Academic Press, London, 553–569.

Woodhall, D. 1974. Geology and volcanic history of Pico Island volcano Azores. *Nature* **248**, 663–665.

Zanon, V. and Frezzotti, M.L. 2013. Magma storage and ascent conditions beneath Pico and Faial islands (Azores archipelago): A study on fluid inclusions. *Geochemistry, Geophysics, Geosystems* **14** (9), 3494–3514.

Zanon, V. and Pimentel, A. 2015. Spatio-temporal constraints on magma storage and ascending conditions in a transtensional tectonic setting: The case of the Terceira Island (Azores). *American Mineralogist* **100**, 795–805.

Zanon, V. and Viveiros, F. 2019. A multi-methodological re-evaluation of the volcanic events during the 1580 CE and 1808 eruptions at São Jorge Island (Azores Archipelago, Portugal). *Journal of Volcanology and Geothermal Research* **373**, 51–67.

Zanon, V., Pimentel, A., Auxerre, M., Marchini, G. and Stuart, F.M. 2020. Unravelling the magma feeding system of a young basaltic oceanic volcano. *Lithos* **352–353**, 105325.

Zbyszewski, G. 1960. L'eruption du Vulcan de Capelinhos (Ilhe de Faial, Açores). *Bulletin Volcanologique* **Ser II**, XXIII, 78–100.

Zbyszewski, G. 1961. Étude geologique de l'ile de S. Miguel (Açores). *Comunicações Serviços Geológicos de Portugal* **45**, 5–79.

Zbyszewski, G. 1963. Les phénomènes volcaniques modernes dans l'archipel Açores. *Comunicações Serviços Geológicos de Portugal* **47**, 227p.

Zbyszewski, G., Moitinho de Almeida, F., Veiga Ferreira O. and Torre de Assunção, C. 1958. *Notícia explicativa da Folha "B", da ilha S. Miguel (Açores) da Carta Geológica de Portugal na escala 1:50000.* Serviços Geológicos de Portugal, Lisboa.

Zbyszewski, G., Ferreira, O. V. and Assunção, C. T. 1959. *Notícia explicativa da Folha "A", da ilha S. Miguel (Açores) da Carta Geológica de Portugal na escala 1:50000.* Publ. Serviços. Geológicos de Portugal, Lisboa.

Zbyszewski, G., Medeiros, A.C., Ferreira, O. V. and Assunção, C. T. 1971. *Carta geológica de Portugal, escala de 1:50000. Notícia explicativa da Folha da ihla Terceira Island (Azores).* Publ. Serviços. Geológicos de Portugal, Lisboa.

Zhao, Z., Mitchell, N., Quartau, R., Tempera, F. and Bricheno, L. 2019. Submarine platform development by erosion of a Surtseyan cone at Capelinhos, Faial Island, Azores. 2019. *Earth Surface Processes and Landforms* **44**, 2982–3006.

4 Evaluation and prediction of hazards

The *raison d'être* for the evaluation and prediction of earthquakes, volcanoes and related hazards is twofold: first it is desirable to know the places that will be affected by extreme events over differing time scales and what the impacts will be on populations and, second, to forecast the time and character of a future earthquake or volcanic eruption (Chester 1993). In the context of volcanoes, the former approach has been termed *general prediction* and is defined as an examination of 'the past behaviour of a volcano so as to determine the frequency, magnitude and style of eruptions, and to delineate high risk areas' (Walker 1974, p. 23). More recently, this procedure has been designated *long-term forecasting* (Cashman & Sparks 2013, p. 683) and, using historical catalogues (Chapter 3), has been applied to both earthquakes and eruptions, with results being either probabilistic statements of future events and/or probabilistic hazard maps. Traditionally output has been presented cartographically, but in the past three decades Geographical Information Systems (GIS) have increasingly been adopted both in the Azores and elsewhere.[1] Alternative names for *general prediction/long-term forecasting* are hazard mapping and assessment and this approach has been adopted not just to highlight risky locations, but pro-actively: to steer new development into less hazard-exposed locations; to develop civil-protection policies, evacuation plans and zoning maps for insurance and land-use planning purposes and, retro-actively, to modify (i.e. retro-fit) existing buildings so they become more resilient (see Chapter 5).

A second set of procedures is named *specific prediction* by George Walker and more latterly *short-term forecasting* (Cashman & Sparks 2013, p. 683)[2] being based on the 'surveillance of... (a)... volcano and monitoring of changes, for example, in seismic activity,... so as to forecast the time, place and magnitude of an eruption' (Walker 1974, p. 23, see also Pallister & McNutt 2015). Similar procedures may be applied to earthquakes and entail the recognition of pre-cursory signs of activity which necessitates the observation of a volcano or seismically active region – both visibly and instrumentally – for months and in some cases years before an extreme event.

As Bruce Bolt (2004, p. 215) has noted, there is a fundamental issue raised by the *short-term prediction* of earthquakes and, by implication volcanic eruptions, because they both involve correlations between time series which may

DOI: 10.4324/9780429028007-4

imply, but not establish causal links between events and supposed causes. He writes:

> suppose that seismological measurements hint that an earthquake of a certain magnitude will occur in a certain area during a certain period of time. Presumably this area is a seismic one, or the study would not have been initiated in the first place. Therefore, it follows that by chance alone... an earthquake will occur during the period suggested. Thus, if an earthquake occurs, it cannot be taken as decisive proof that the methods used to make the prediction are correct, and they may fail on future occasions. Of course, if a firm prediction is made and nothing happens, that must be taken as proof that the method is invalid.

Notwithstanding these issues, in the 1970s much was promised by procedures which linked putative precursory phenomena to the occurrence of earthquakes, by means of what at the time were assumed to be robust causal mechanisms.[3] By the 1990s and despite what was claimed to be some limited success,[4] persistent failures successfully to predict damaging earthquakes led many scientists to question whether *short-term forecasting* was possible given the current state of theoretical knowledge and available monitoring techniques (Anon 2011; Cimellaro & Marasco 2018). It was because of such uncertainties that institutional and personal liability for incorrect forecasts became serious concerns, with the most well-known case taking place in the aftermath of the 2009 L'Aquila earthquake in Italy when seven scientists and technicians were convicted of manslaughter, a judgement that was subsequently overturned (Joffe et al. 2018).

Although similar caveats apply over correlation and cause, *short-term forecasting* of eruptions on well-monitored volcanoes has been more successful. Indeed, in the two decades up to 2015, it has been estimated that 25 successful predictions were made with examples being drawn from Colombia, Indonesia, Japan, Mexico, Papua New Guinea, the Philippines, Russia, the USA and the West Indies (Montserrat). Seismic and infrasound[5] monitoring (McNutt et al. 2015), together with studies of ground movement (Kilburn 2018), have been the most effective tools.

It was only with the establishment of the *Universidade dos Açores* in 1976 and the stimulus of, and funding for, research that followed the 1980 (Terceira) earthquake, that Azorean earth science became engaged in prediction.[6] One consequence was that the period of false optimism of the 1970s, which held that reliable *short-term forecasts* were both possible and imminent, was avoided. More negatively a late start has meant that monitoring has only taken place over the past four decades; it is only from the 1990s that surveillance has become thorough because of the deployment of a wide range of monitoring techniques (e.g. seismic, deformation, geochemical, gases and infrasound) in order to search for possible precursors and that development of reliable historic data-bases, upon which longer-term forecasts may be based, has had to wait for the publication of historic catalogues of earthquakes and eruptions. These have included research by a number of scholars including Nunes (1986, 1999), Nunes et al. (2004) and Gaspar et al. (2015a), in the case of earthquakes; and Booth et al. (1978); Queiroz et al. (1995), Madeira (1998), Gaspar et al. (2015a), with respect to volcanic eruptions. Chapter 3 of the present volume adds to this historical database.

Notwithstanding the late start, substantial progress has been made in *long-term forecasting* of earthquakes and eruptions, the results of which are reported in this chapter. In the case of *short-term* forecasts and in common with other seismically active areas of the world, research has not yet advanced to the stage where there are a set of reliable precursory signals that can be used to predict the times, dates, magnitudes and locations of future events though, as in other parts of the world, the prospects of eruption prediction are more advanced. The research frontier is characterised by the monitoring of signals, which represent a range of possible earthquake and eruption precursors and to exchange and bank these data internationally. As Cashman and Sparks (2013, p.683) have argued with respect to eruptions, the large bodies of data so created 'can be used not only for pattern recognition, but also to provide input for the development and testing of predictive models based on eruptive precursors.'

Long-term prediction

Earthquakes

As far as the Azores as a whole is concerned, Fontiela et al. (2017) have summarised historic records (see Chapter 3) and have used these to produce *Maximum Observed Intensity* (MOI) maps. The data used comprise 325 geo-referenced point records of intensity from 67 historical earthquakes. These include 33 large historical events that had Maximum Intensities (I max) greater than VII and which occurred between 1522 and 2012. Using these data, the authors interpolated values[7] on to a grid, with the calibration being confirmed by comparing the expected (i.e. calculated) and observed spatial intensities of four well-studied earthquakes: São Miguel, 1522; Terceira 1614; São Jorge, 1757 and Terceira and other islands in 1980. Maps of Maximum *Observed Intensity* were produced with I max values reported using the *Modified Mercalli Scale* (see Chapter 3). The principal findings of the research are as follows:

a   The Central Group of islands, Terceira, Graciosa, São Jorge, Pico and Faial, show the highest calculated Intensities (XI) in eastern São Jorge.

b   MOI values of X are observed along a NW/SE strip across the centre of Faial, with the highest values recorded in the vicinity of Horta, the island's capital.

c   For Graciosa, Terceira and Pico MOI values are, respectively, IX, VIII and VII.

d   Located on the North American Plate, Flores and Corvo show very low I max values.

e   In the Eastern Islands, highest Intensities of X are to be found in south east São Miguel in the vicinities of the coastal settlements of Vila Franca do Campo and Ponta Garça, with attenuation of intensity occurring inland. On Santa Maria the highest I max values are in the eastern third of the island.

The authors are aware that their data represent just under 500 years (i.e. 1522–2012) of earthquake records and may not include the largest earthquakes that have occurred in the Azores. In addition, they do not address human vulnerability as expressed, *inter alia* in: population numbers; building types and densities; economic activities and the location of the socially vulnerable (see Chapter 5), but rightly maintain that their research findings represent 'a valuable spatial instrument that stakeholders can use to establish seismic-hazard mitigation plans' (Fontiela et al. 2017, p. 1183) (see Chapters 5 and 6).

More detailed studies have been produced for individual islands. These include works by: Carvalho et al. (2001) and Silva (2005), on the Central Islands; Silveira et al. (2003), on São Miguel and Hipólito et al. (2013), on Graciosa. Using the Central Islands and São Miguel as examples, in the case of the former Carvalho et al. (2001) have produced probability maps showing *Peak Ground Acceleration* (PGA) based on events exceeding a probability of 10% in 50 years. Using historical information from Nunes et al. (2000, 2004), the University of the Azores and other sources which the authors admit are far from complete, a database of more than 15,000 items (9,833 with epicentral locations) was produced. Many of the data relate to the post-1980 period of detailed instrumentation although major earthquakes from the past, such as events in 1522, 1757, 1926, 1931, 1959, 1980 and 1998 (see Chapter 3), are included. Intensities are converted to magnitudes using regression techniques based on earthquakes for which both magnitude and intensity were known and, once events with a magnitude less than 3.7 and all foreshocks and aftershocks are eliminated, the catalogue is reduced to 451 items. Using two complementary methods, probability maps were drawn and several *concelhos* (i.e. counties, see Chapter 1) emerge as being particularly earthquake exposed. These are the *concelhos* of Horta, on Faial in the west; Praia da Vitória on Terceira in the east; with the least exposed *concelho* being Santa Cruz, on Graciosa.

This study is notable because it, not only uses the historical catalogue, but also more recent instrumental records which highlight zones of contemporary unrest. The authors call for improved seismic codes, but note that these should not be the same across the whole Central Group because the probabilities of future earthquakes of different magnitudes are spatially variable. In addition, future assessment needs to take into account site specific conditions such as possible soil amplification and the influence of local geological factors (see below).

Silveira et al. (2003) provided an earthquake hazard map for São Miguel, which with *c*.140,000 inhabitants is the most populous island in the archipelago (Chapter 1). Historical data (see Chapter 3) are again employed, and the authors adopt a methodology that concentrates on the detailed reconstruction of the spatial variation in intensity for six major earthquakes (i.e. those of 1522, 1591, 1852, 1932, 1935 and 1952) and six seismic swarms (i.e. those of 1563, 1630, 1652, 1713, 1810 and 1811). With the exception of the 1591 earthquake, for which there are insufficient data, isoseismal maps are drawn using criteria defined by the *European Macroseismic Scale* (EMS-98) (Grünthal 1998).[8] For the 1591 event and the seismic swarms local intensities are plotted. By combining isoseismal plots, a map showing maximum historical intensity was produced (see Figure 4.1). As Figure 4.1 shows, no location on São Miguel may be considered safe, with I max values ranging from VIII to X. An Intensity X zone is centred on Fogo volcano and trends NW/SE and includes important towns and villages such as: Ribeira Grande (pop 6,393) and Ribeira Seca (pop 2,950), on the north coast; Santa Bárbara (pop 1,275), in the centre of the island and on the south coast Água de Pau (pop 3,058), Ribeira Chã (pop 396), Água d'Alto (pop 1,788), Vila Franca do Campo (pop 4,085), Ribeira das Tainhas (pop 703) and Ponta Garça (pop 3,547).[9]

An important theme in the research reported by Silveira et al. (2003) is the identification of areas where historical intensity values are higher than predicted (i.e. positive anomalies) and areas where they are lower than predicted (i.e. negative anomalies). As Table 4.1 shows, the authors suggest several possible reasons for these anomalies and call for further research to test the veracity of their assertions. Silva (2005) applied, a similar methodology to that used by Silveira et al. (2003) in her study of historical earthquakes in the Central Islands.

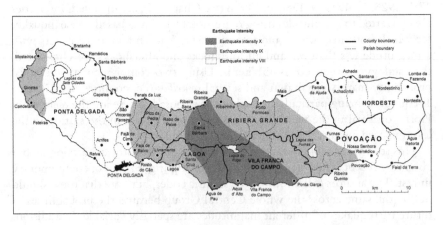

Figure 4.1 Map showing maximum historical intensity of São Miguel using the EMS-98 Intensity scale. Based on Silveira et al. 2003, Fig. 5, page 621. Copyright permission Dr Dina Silveira.

Table 4.1 São Miguel: Suggested reasons for positive and negative intensity anomalies according to Silveira et al. (2003).

	Suggested causes
Negative anomalies	
Furnas and Sete Cidades calderas	Presence of shallow magma chambers
Povoação and Ribeira Grande villages	More earthquakes resistant buildings (i.e. EMS – 98 – Class B).
Picos Volcanic System	Most of the settlement is constructed on basaltic lava flows with minimal soil or substrate amplification (see Chapter 3). The fact that damage was less than expected in these areas was noted during the 1932 earthquake.
Positive anomalies	
The villages of Santo António, Lomba do Cavaleiro and Salga	It was observed during the 1932 and 1935 earthquakes that these ridge top villages had higher than expected values for intensity in comparison to surrounding lower lying areas. It is well known from many parts of the world that the destructiveness of ground shaking during earthquakes is affected by *topographic amplification* on the crests of ridges (Paolucci 2002), and this is suggested by Silveira et al. (2003) as a possible cause of the seismic enhancement.
Lomba de Santa Bárbara and Lomba da Fazenda	The reasons are less clear but could be related to the presence of the well-defined NE-SW trending Ribeira do Guilherme fault.

Since 2010 one trend in the *long-term prediction* of earthquakes has been the publication of more detailed studies of the exposure of individual settlements. Some published research (e.g. Ferreira et al. 2016; Fagundes et al. 2017; Maio et al. 2017, 2019) has sought to link hazard evaluation to attempts to improve building resilience through policies aimed at making traditional buildings safer. Based on civil engineering techniques, this has involved what is known as *retro-fitting*[10] of existing buildings and is further discussed in Chapters 5 and 6. A second approach is to examine detailed vulnerability at the level of the individual city, town or village, and here the research of Martins et al. (2012a, 2012b) and Ferreira et al. (2017) is noteworthy.

Following the 1998 earthquake, which caused 9 deaths, injured over 100 and caused severe damage to around 1,500 buildings (see Chapter 3), Ferreira et al. (2017) undertook a vulnerability assessment of 192 traditionally constructed buildings located in the old city centre of Horta.[11] Making use of data collected on buildings damaged in the 1998 earthquake, the authors use a methodology first developed in Italy and later adapted by Vicente et al. (2011) to Portuguese building practices, to derive a *Vulnerability Index* (I_v).[12] This is normalised to a range of 0–100: the lower the value the lower the seismic vulnerability, with the index being calibrated by means of damage data from the 1998 earthquake. On the basis of a sample of 90 buildings, a mean I_v value of 32 was calculated, some 27% of buildings showing values of over 40 and 17% values of over 45 (equivalent to Class A on the EMS 98 scale, see Chapter 3).

Information was integrated into a GIS (see above) and maps of I_v were drawn for the old city centre of Horta, which allowed the identification of buildings with an $I_v > 40$ (Figure 4.2), and from this analysis it is possible to highlight that corner and row-end buildings were particularly vulnerable because of their location and lack of buttressing. Other conclusions drawn from this study are tabulated in Table 4.2 and demonstrate the degree of exposure to large earthquakes, not only within Horta's traditionally constructed city centre, but also by implication in many settlements across the archipelago. The authors conclude by arguing that local authorities need to develop comprehensive databases of traditionally constructed settlements and individual buildings within them so as to allow the estimation and forecasting of the direct and indirect consequences of earthquakes on buildings, local economies and populations (Ferreira et al. 2017, p. 2898).

As Figure 4.1 shows, most of the land area of the *concelho* (i.e. county) of Vila Franca do Campo falls within the area where the most severe historic

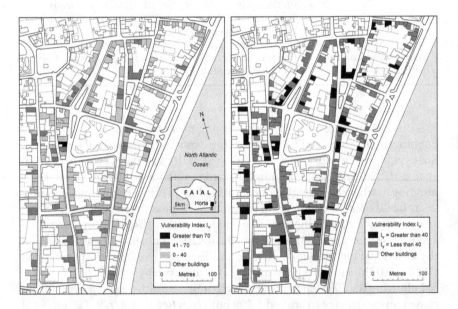

Figure 4.2 Traditionally constructed buildings in the historic centre of Horta (Faial): a. Map of the Vulnerability Index (I_v). b. Buildings with an $I_v \geq 40$. The category 'Other Buildings' includes both buildings that were not surveyed and those that are constructed using reinforced concrete. Simplified and based on Ferreira et al. (2017, Fig. 8, page 2891). Copyright permission Springer Nature.

Table 4.2 Estimates of damage that could be caused to the historic centre of Horta

Total number of buildings – 192	Macroseismic Intensity (EMS-98)							
	V	VI	VII	VIII	IX	X	XI	XII
Collapsed	/	/	0	0	17 (7.7%)	101 (52.7%)	/	/
Unusable	/	/	3 (1.8%)	33 (17.3%)	93 (48.2%)	72 (37.3%)	/	/

Number of dead, severely injured and homeless

Total number of inhabitants – 1596	Macroseismic Intensity (EMS-98)							
	V	VI	VII	VIII	IX	X	XI	XII
Dead and severely injured	/	/	0	0	37 (2.3%)	252 (15.8%)	/	/
Homeless	/	/	29 (1.8%)	278 (17.4%)	856 (53.7%)	1184 (74.2%)	/	/

Estimates of repair costs

Total number of buildings – 192	Macroseismic Intensity (EMS-98)							
	V	VI	VII	VIII	IX	X	XI	XII
Replacement costs (millions of Euros)	0.37	2.77	15.77	37.36	60.5	73.03	77.51	77.96
Percentage of GDP (Azores Autonomous Region)	0.01	0.07	0.42	1.0	1.62	1.95	2.07	2.08

Based on the modelling of Ferreira et al. (2017, table 5, p. 2895, table 6, p. 2896 and table 7, p. 2897)

earthquake intensities on São Miguel (i.e. Intensity X EMS-98) have been re-corded (Silveira et al. 2003). The remainder of the *concelho* lies in the Intensity IX zone. Martins et al. (2012a) approach the issue of long-term risk exposure from a contrasting perspective to other writers, by examining how land-use changes over time have shifted populations and economic activities between the two zones. Between 1994 and 2005 urban growth occurred according to a municipal plan, which concentrated expansion on the margins of existing settlements. Although based on well-considered land-planning criteria, the unexpected consequence was that, of the 135.4 ha of new urban development some 91.6% occurred in the Intensity X zone (Martin et al. 2012a, p. 2738), the implication being that in the future land-use and hazard planning require better co-ordination.

The same authors (Martins et al. 2012b), use a GIS-based methodology and geo-referenced information on demography and socio-economic conditions, to show that a strong spatial correlation exists between vulnerable buildings in the centre of the village of Vila Franca do Campo and concentrations of economically and socially disadvantaged people. They also demonstrate that greater vulnerabilities occur in some settlements rather than in others, because of higher illiteracy rates and lower levels of educational attainment, with Água d'Alto and Ponta Garça emerging as being particularly disadvantaged. Both these studies have implications for the development of future policy and are considered in more detail in Chapters 5 and 6.

Volcanic eruptions

As mentioned in Chapter 3, in the 1970s a team from Imperial College London under the leadership of Professor George Walker carried out a detailed study of pre-historic silicic volcanism mostly on São Miguel, and towards the end of this project produced the first long-term assessment of volcanic hazards in the Azores (Booth et al. 1983). Arguing that on São Miguel there have been three principal types of activity: trachytic-explosive; land-based basaltic-Strombolian and marine hydrovolcanic (i.e. Surtseyan), *scenarios* of eruption for differing styles and magnitudes were constructed in order to demonstrate probable spatial impacts across the island. Although a variety of potential haz-ards were considered including lahars and pyroclastic density currents (PDCs), emphasis was placed on the potential damage from pyroclastic fall deposits because, outside the immediate vicinity of an eruption site, roof collapse would be the major threat to people and their activities. Recognising that the area affected by fall deposits would be strongly influenced by wind direction and strength, stratospheric circulation was assumed for larger Plinian events whereas surface winds were modelled for smaller-scale eruptions. Figure 4.3 is an example the hazard maps produced by the Imperial College research team and shows pyroclastic fall dispersal according to a number of scenarios: a pre-historic trachytic eruption from Sete Cidades; the eruption of Fogo in 1563; Furnas in 1630[13] and Fogo A (4.6 ka BP), the latter being the largest eruption to have impacted São Miguel in the past 5000 years.

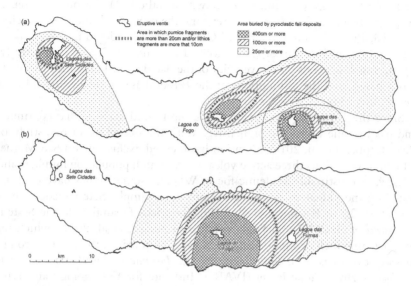

Figure 4.3 Dispersal maps of pyroclastic fall deposits from some trachytic eruptions on São Miguel. a. From west to east: the pre-historic eruption from Sete Cidades; the 1563 eruption of Fogo and the 1630 Furnas eruption. b. The Fogo A eruption (4.6 ka BP). Booth et al. 1983, Fig. 9.1, page 100. Copyright permission Dr Basil Booth.

In the 1990s and as part of the European *laboratory volcanoes* project (see Chapter 3), Furnas on São Miguel was the focus of attention and a methodology similar to that employed by George Walker's team was adopted to evaluate the spatial impacts of two eruption *scenarios*: Furnas 1630 and the much larger Furnas C event (*c.*2000 BP) (Guest et al. 1999; Cole et al. 1999). One innovation introduced into this project was the probabilistic assessment of the recurrence of future explosive eruptions from Furnas. This was based on statistical modelling using the *Poisson* distribution and employed a data set comprising both historic and pre-historic radio-metrically dated events (Jones et al. 1999).[14] Probabilities of eruption were calculated at: 3% in 5 years, 5% in 10 years, 26% in 50 years and 53% in 100 years, figures that emphasise the risk exposure of the population especially those living within and adjacent to Furnas Caldera (see Chapters 3 and 5).

More recently Caniaux (2005) has extended statistical modelling to the whole archipelago, expressing his results as recurrence intervals. According to this analysis the most likely future eruptions are from; Pico Volcano (*Montagne du Pico*) Pico Island; the Picos Fissural Volcanic System (*Région des Picos*); Sete Cidades on São Miguel; the linear complex of Planalto da Achada fissure zone (*São Roque-Piedade*) on Pico Island and the Capelo Fissural System of Faial with, recurrence intervals, respectively, of 62, 152, 161, 206 and 225 years (terminology of Caniaux is shown in italics). At the other extreme, eruptions from the fissure zone of Terceira, the central volcano (i.e. Caldeira) on Faial, Pico Alto on Terceira and Graciosa are considered highly unlikely with recurrence intervals of 1,067, 1,300, 1,337 and 2,900 years, respectively. Again, much depends on the veracity of the dated record on which the analysis is based, and the assumptions of the *Poisson*-based model being fulfilled (see endnote 14).

Since the 1990s *long-term* prediction using hazard mapping has continued and has become more comprehensive. With the exception of one study of Pico (Cappello et al. 2015), attention has focused exclusively on São Miguel which, in view of its three active volcanoes and high population numbers and density, is a justifiable academic focus. Whereas some studies have concentrated on individual volcanoes or regions, for example, Sete Cidades (Queiroz et al. 2008; Kueppers et al. 2019), the Picos Fissural Volcanic System (Pedrazzi et al. 2015) and Fogo Volcano (Medeiros et al. 2021), which for the first time included modelling of impacts on tourism, the most thorough investigations has been undertaken by the *Instituto de Investigção o em Vulcanologia e Avaliação de Riscos* (IVAR – Institute for Volcanology and Risk Assessment Research – Gaspar et al. 2015b). Since the Imperial College and *laboratory volcanoes* projects were completed and published, knowledge of volcanological processes has improved and these insights have been incorporated by IVAR into their assessments. Much has been learnt from studies of *long-term prediction* at other volcanoes and this has also informed research in the Azores. For instance, the IVAR team is now concerned not just with risks posed by pyroclastic fall deposition, as was the case in earlier modelling, but the whole gamut of hazards presented by eruptions on São Miguel, i.e. lava flows, domes, pyroclastic density currents (PDCs) and lahars. Furthermore, risks posed by volcanoes when they are not erupting, such as degassing and landform instability, have been addressed in additional studies published both by IVAR and other researchers (see below).

Taking lava flows as an example, Gaspar et al. (2015b) used historical and radiometrically dated pre-historic records to conclude that during the past 5,000 years there had been as least 36 basaltic eruptions in the western part of São Miguel, but only two in the east representing recurrence intervals, respectively, of 139 years and *c.*2,500 years. To assess the varying spatial susceptibility to inundation by lava flows, a computer-based GIS tool first developed by Felpeto et al. (2007) in the Canary Islands was employed, which makes the assumption that ground contours are the principal factor controlling both the direction of flows and those areas that are topographically protected from inundation by lava. All the basaltic vents on the island were mapped and, using a digital elevation model (DEM), a circle with a radius of 500 m was traced around each eruptive centre. The DEM was divided into cells of 50 m, the assumption being that any point within a circle could be an eruptive centre and that all cells within circles had the same probability of acting as source areas for a lava flow eruption. A Monte Carlo simulation[15]

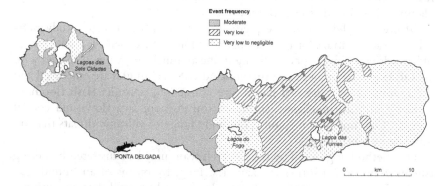

Figure 4.4 São Miguel: Areal susceptibility to basaltic lava flows and the eruptive frequency of Hawaiian/Strombolian eruptions over the last 5 ka. Based on Gaspar et al. 2015, Fig. 12.3, page 159. Copyright permission Geological Society of London.

was performed to model the progression of flows from one cell to another using 4,000 iterations per vent and assuming a maximum flow length of 20 km. The results are reproduced as Figure 4.4, which shows the spatial susceptibility of being affected by basaltic lava flows generated by either Hawaiian or Strombolian activity. Maps were also published for threats posed by: trachytic domes and associated lava flows; scoria fall from Strombolian eruptions; pumice fall from VEI 4 and 5, sub-Plinian events; PDCs generated by VEI 4 and 5 eruptions and lahars from Fogo and Furnas volcanoes.

Multiple and related hazards

Since 2002 IVAR and its predecessor, CVARG (*Centro de Vulcanologia e Avaliação de Riscos Geológicos* or Centre for Volcanology and Geological Risk Assessment), has been building a GIS-based data-bank to encompass all geo-referenced information relevant to risk analysis (Gaspar et al. 2004, 2011; Goulart et al. 2011; Wallenstein et al. 2015). Known as AZORIS, an acronym for AZOres RISk, the building-blocks of the system are nine sets of data relating to: the geography of the islands and, more particularly their socio-economic characteristics; civil protection measures; mapped geological and geomorphological information; landslide occurrence; volcanic phenomena; historic earthquake activity; geodesy; fluid geochemistry and meteorology. Although recognising that not all relevant data are capable of being geo-referenced, AZORIS nevertheless provides a facility for such information to be viewed alongside spatial data. Research on improving and refining AZORIS is ongoing and, although of relevance to *long-term prediction*, it does not involve new research on hazard assessment, being focused on providing information in order effectively to respond to emergencies. For these reasons AZORIS is more fully discussed in subsequent chapters as are more planning-based studies (e.g. Borges et al. 2011, 2014).

Of greater relevance in the prediction of hazards is a major study published under contract for the European Union and focused on a wide range of hazards faced by the people living in the archipelago (Anon 2016). As with AZORIS geo-referenced information and GIS-techniques are employed and the following categories of hazard are considered: seismic; flash flooding; tsunamis and storm surges; landslides; soil erosion; lava flows and coastal erosion. One notable omission is the wide range of volcanic threats that are unrelated to lava flows.

The methodology used by the research team may be illustrated by taking tsunamis and storm surges as examples. First, by means of an 'inundation scenario' (i.e. maximum wave-height) based on historical data and using a detailed DEM, the spatial variation in hazard severity for coastal areas is evaluated and frequencies of occurrence calculated, which allows five hazard categories to be recognised, i.e. very low, low, medium, high and very high (Anon 2016, p. 54). Most inhabitants of the Azores are located near to the coast and the authors examine the spatial patterning of population density, infrastructure and buildings at risk and land-use all of which are incorporated into a GIS to produce, not only maps of assets at risk, but also tables summarising the exposure of: population; built-up areas and land-cover/land-use. Table 4.3 is an example of these summations and shows population at risk across the islands. Finally, risk abatement measures are suggested which are a mixture of engineering solutions (e.g. the reinforcement of port facilities and the construction/improvement of breakwaters and seawalls) and planning policies (e.g. improvements in evacuation procedures, better land-use zoning and the avoidance of further development in hazard prone locations) (see Chapter 5). This research highlights the degree of exposure both of individual islands, by means of the tables, and particular areas within the islands through detailed computer-based maps. With regard to the former it is salutary to reflect that, based on the data in Table 4.3, some 8% of São Miguel's population shows a degree of exposure, with c.1% being at very high risk; and that Graciosa shows an exposure of just under 8%, Faial c.7% and Pico c.5%.

Table 4.3 The Azores: Population at risk from tsunamis and storm surges

Population										
	Corvo	*Flores*	*Faial*	*Pico*	*S. Jorge*	*Graciosa*	*Terceira*	*S. Miguel*	*S. Maria*	*Total*
V. Low	5	5	237	265	121	195	515	3236	12	4591
Low	0	3	211	219	68	103	298	1574	3	2479
Medium	0	1	375	188	92	72	164	3182	2	4076
High	0	5	230	160	31	20	86	1118	7	1657
V. High	0	2	0	20	0	0	0	1382	0	1404

Based on information in Anon (2016, table 15, p 57).

As well as the study carried out under the auspices of the *European Union*, there has been other research on volcanic and seismic-related phenomena that pose a threat to the Azores. Potentially hazardous intra-eruption and intra-earthquake activities are in fact major areas of vulnerability in the Azores (Wallenstein et al. 2005, 2007, 2015). These include tsunamis; floods, crater-lake breaching; landsides/mass wasting and degassing which may either coincide with extreme geophysical events, or may be triggered by a range of processes that operate in the intervals between earthquakes and eruptions.[16]

As far as tsunamis are concerned the historical catalogue published by Andrade et al. (2006, p. 181) concludes with a cautionary note that 'the documentary record is fragmentary and incomplete; in many cases, accounts of flooding and associated damage are restricted to property, goods and fatalities that affected a limited (wealthy) social class.' Andrade et al. (2006) also argue that incomplete recording is long-standing and castigate municipal authorities in the past for not documenting the effects of even large tsunamis such as those generated by the 1755 Lisbon earthquake. Where damage has been recorded, data have often been biased towards losses experienced by larger settlements and property owned by elites[17] and even as late as the twentieth century there were insufficient tide-gauges to allow comprehensive information on tsunamis to be collected. Despite these and other cautionary caveats, Andrade et al. (2006) present an example of how *long-term prediction* might proceed, by the construction of a map of Angra do Heroísmo on Terceira, which shows the estimated extent of inundation under a '1755-type scenario' assuming a 11–15m run-up height. Such a tsunami would see flooding of some 75,500 m^2 of the historic town centre, much of which is a World Heritage Site; destruction of the passenger terminal, light cargo and fishing harbours and severe damage to many historic buildings and centres of administration.

Inundation by storms-waves and tsunamis may become more acute in the future with climate change, not only raising sea levels but also probably increasing the size and recurrence of large storms, and there is a need for more comprehensive zoning of littoral areas because the population of the islands is concentrated at and near to coastlines (Santos et al. 2004; Calado et al. 2011, see Chapters 1 and 6).

River floods and lahars may occur, either as a result of seismic or eruption-induced failure of crater lake rims, or by a wide range of climatic and geomorphological causes. As discussed briefly in Chapter 1, the archipelago is geologically young and several islands attain considerable altitudes over short distances from their coastlines which, combined with high precipitation totals and pronounced seasonality, make the islands subject to flooding, mass-movements in the form of mudflows, landslides and flashy river-regimes. Instability is heightened by the unconsolidated nature of many volcanic products and incautious planting in the past of exotic shallow-rooted trees particularly the Japanese red-cedar (*Cryptomeria japonica*). In the past, large landslides have been triggered by volcanic and seismic events and have produced extensive debris flows, including coastal platforms that are known in the Azores as *fajãs* and which are formed of lava, pyroclastic deposits and landslide debris. Instability is also associated with sub-aerial mass-wasting especially during and after heavy rainfall (Malheiro 2006; Marques et al. 2008, see Chapter 3), and one example already mentioned in Chapter 1 is the major landslide that occurred on 31 October 1997 in the Furnas area of São Miguel (Cunha 2003). Twenty-nine people were killed in the coastal settlement of Ribeira Quente, leaving 114 homeless, the village cut off from the rest of the island for more than 12 hours and causing economic losses estimated at 20 million Euros (1997 prices). As discussed in Chapter 6, this disaster also had a profound influence on policy formation in the years which followed.

Crater lakes found on Fogo and Furnas volcanoes (São Miguel) represent the greatest threat and hazard maps have been published by Gaspar et al. (2015b, pp. 163–164). These are constructed using a program named LA-HARZ and which was developed by the United States Geological Survey. It employs ARCH/INFO[18] GIS technology to calculate the areas that would be inundated by lahars and floods of differing volume (Schilling 1998) and a digital elevation model (DEM) is constructed using geo-referenced information, with a resolution of 10 m, contours at the same 10 m interval and spot heights extracted from 1: 25,000 scale maps.[19] Varying estimates of damage are calculated for lahars with volumes ranging from 10^4 to 10^7 m^3. In the case of Fogo, maps show major impacts on the village of Praia, on a large hotel (the Bahia Palace), which is located on a delta and on the coastal road that links the southern settlements of the island. In the case of Furnas, damage would be severe in the villages of Furnas and Ribeira Quente (Figure 4.1). Additional factors exacerbating flooding are episodes of high rainfall instigating reductions in soil-moisture storage capacity; narrow valleys forcing high rates of discharge to be accommodated by increased stage-height and the fact that both gorges are densely vegetated and have steep unstable slopes, which would cause temporary dams to be constructed only to fail with high volumes of material periodically debouching downstream (Chester et al. 1995).

Large-scale mass-wasting occurred in the Azores and other Atlantic Islands in the Pleistocene and at least eight giant landslides have been recognised on Flores, Pico, Santa Maria and São Miguel, which shaped coastlines and deposited debris as fans across the ocean floor (Costa et al. 2014, 2015; Blahût et al. 2018; Hildenbrand et al. 2018). At a smaller-scale in Povoação *concelho* (i.e. county) on São Miguel, research by Rui Marques and his colleagues in 2008 represented the first attempt at forecasting landslides (Marques et al. 2008, p. 486, see Figure 4.1). This involved plotting historical records of rainfall intensity (mm/day) against rainfall duration (D days) and demonstrating that intensity (I) increases exponentially as duration decreases, according to a regression equation $I = 144.06\,D^{-0.5551}$. The regression line specifies thresholds above which instability becomes more common and the historical catalogue indicates that landslides are related to both: short duration (one to three days) rainfall, with high mean intensities of 78–144 mm/day); and longer (one to five month) precipitation periods, with lower mean intensities of between 9 and 22 mm/day. On São Miguel rainfall regimes with these characteristics are frequently encountered between October and March and landslides occur in São Miguel during most winters. In Povoação *concelho* between 1918 and 2002 some 85% of landslides occurred between October and March and, of 40 cases reviewed, only one-quarter could be classified as 'minor' (Marques et al. 2008, p. 484).

A more comprehensive study, including the construction of a landslide sus-ceptibility map has been published for São Miguel (Marques et al. 2015). First, a detailed historical catalogue was constructed of 193 events that occurred between 1900 and 2008. These landslides, most of which were triggered by rainfall, caused significant demographic and socio-economic impacts and were responsible for 67 deaths, 20 cases of serious injury, the destruction of many homes and much homelessness. Second, 12 pre-disposing features were identified which include altitude; several factors relating to slope characteris-tics; insolation (i.e. incoming solar radiation); wetness; distance from streams; drainage density; geology and land use, and these were used probabilistically to assess spatial variations in landslide susceptibility. Finally, four classes of susceptibility were recognised: Low; Moderate; High and Very High, and these were both mapped and tabulated for the six *concelhos* (i.e. counties) that comprise São Miguel (see Figure 4.1), separate tabulations being produced for areal units, roads and buildings. The conclusions drawn from the analysis by Marques and his colleagues have implications for land-planning and include the following:

1   The principal pre-disposing conditions for landslips are steep slopes (i.e. >30°); pumiceous deposits older than 5 ka BP; low insolation rates; altitudes above 700 m (reflecting higher precipitation); areas denuded and/or covered with low-lying vegetation and wetness. Povoação and Nordeste *concelhos* emerge as the most exposed, whereas Lagoa shows the lowest levels of exposure.
2   Slope-angle and geology emerge as the major causal factors and water-saturation is also of importance. Lower insolation levels mean reduced rates of evaporation and greater moisture within substrates.
3   Degraded slopes and shallow-rooted vegetation are factors exacerbating instability.
4   Some 23% of the land area of São Miguel has high or very high levels of susceptibly to landslides, with Nordeste (*c.*47%) and Povoação (*c.*43%) *concelhos* emerging as the most exposed.

Since the 1990s geochemical studies have been carried out on São Miguel in villages where there are anomalously high rates of degassing (Baxter et al. 1999). More recently this work has been extended to other islands including

Terceira, Graciosa, Faial and Pico, although the greatest vulnerability as measured by population at risk is on São Miguel (Viveiros et al. 2010, 2015a, 2016, 2017; Andrade et al. 2020). Volcanic gases pose threats to populations not just in periods of eruption and unrest, but also when volcanoes are quiet (Wallenstein et al. 2007). Soil-diffuse degassing is an insidious threat because dangerous gases, including CO_2 and radon ^{222}Rn, are often released without the inhabitants being aware of the dangers. CO_2 is an odourless and colourless gas and adversely affects the population when concentrations are above 15% vol., making death by asphyxiation more likely as volumes increase (Viveiros et al. 2015a). CO_2 is also a dense gas and can build-up lethal concentrations when 'ponded' in poorly ventilated basements and there are numerous examples of people being killed by asphyxiation in volcanic areas throughout the world (Hansell et al. 2006). In 1992, two visitors were killed in a cave in the floor of the caldera of *Vulcão Central* on Graciosa when CO_2 concentrations exceeded 15% by volume (Gaspar et al. 1998 and Chapter 2), and other residents have been deleteriously affected by high concentrations of gas across the archipelago. In addition to CO_2, other gases are emitted including hydrogen sulphide (H_2S), hydrogen (H_2), nitrogen (N_2); with oxygen (O_2), methane (CH_4) and argon (Ar) which are detected in lower concentrations. H_2S, in particular, causes irritation of the respiratory tract and at concentrations of over 700 ppm (0.07% vol.), may be lethal (Viveiros et al. 2015a, p. 186).

To illustrate this aspect of forecasting in the Azores the research of Viveiros et al. (2009, 2015a) may be used as an example. On São Miguel and Faial some people have had to be moved because of high indoor levels of CO_2 and, on the former island, several villages located on or near to its three active volcanoes have been identified as being exposed to actual or potential public health risks: Mosteiros (NW flank of Sete Cidades); Ribeira Seca and Caldeiras da Ribeira Grande (northern flank of Fogo) and Furnas and Ribeira Quente (Furnas). In these villages soil CO_2 emissions are recorded on regular basis (Figure 4.5) and additionally ^{222}Rn is monitored. As Figure 4.5 shows, in Furnas CO_2 concentrations may be very high in some areas and, indeed, well above the danger level of 15% vol. Further research by Silva et al. (2015a) shows that ^{222}Rn values demonstrate a strong spatial correlation with CO_2 levels, with some dwellings in Furnas and Ribeira Quente experiencing anomalously high readings in excess of 100 Bq m^{-3}: the maximum recommended by the World Health Organisation (WHO 2010). In addition to measuring CO_2 within villages, individual houses have also been monitored and a disturbing finding is that concentrations may be dangerously high even when outdoor readings are much lower.

Monitoring and mapping of gaseous concentrations has major implications for planning. This not only involves choosing the right areas in which to build new housing, but also ensuring that building codes make adequate ventilation a mandatory provision (see Chapter 6).

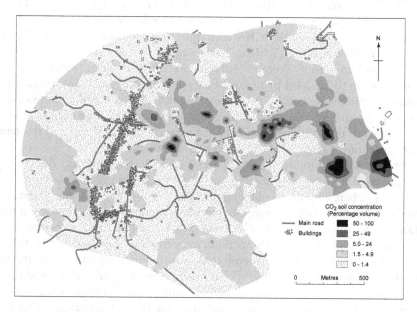

Figure 4.5 Concentration of CO_2 in Furnas Village. Based on Viveiros et al. 2015a, Fig. 14.2, page 187. Copyright permission Geological Society of London.

Short-term prediction: progress in research

The threats posed by gaseous emissions form a convenient segue between *long* and *short-term* prediction because, although hazard maps are produced, monitoring is on-going and contributes to research into understanding activity over more restricted time scales. As discussed at the start of this chapter, monitoring seeks to identify changes in putative precursory phenomena which may presage an imminent eruption, landslide or earthquake. As discussed in Chapter 3, in the Azores the desire better to understand eruption and earthquake precursors is the rationale behind the extensive monitoring networks that have been established from the time of the 1980 Terceira earthquake and with quickening pace over the past 25 years. The stimulus given to the earth sciences in the Azores by the joint European Communities (now European Union) and European Science Foundation's *laboratory volcanoes* project of the 1990s (see Chapter 3) was crucial, as has been the administrative structure that has been put in place to process and evaluate information. This structure grew out of the university's *Departamento de Geociências* and its *Centro de Vulcanologia* supported by the *Instituto Nacional de Investigação Científica* (INIC), with a research centre being established, CVARG (*Centro de Vulcanologia e Avaliação de Riscos Geológicos* – Centre for Volcanology and Geological Risk Assessment), which was renamed IVAR (*Instituto de Investigção o em Vulcanologia e Avaliação de Riscos* – Institute for Volcanology and Risk Assessment) in 2017. The role of CVARG/IVAR has been to advise the Regional Government and the Civil Protection Authority (*Serviço Regional de Proteção Civil Bombeiros dos Açores* – SRPCBA) through an organisation named CVISA (*Centro de Informação de Vigilância Sismovulcânica dos Açores* – Centre for Information and Seismo-Volcanic Surveillance of the Azores), which was established jointly in 2008 by the university and the Regional Government to 'assure the implementation, development and maintenance of... monitoring systems for the Azores... based on geophysical, geodetic, geochemical and meteorological continuous real-time data acquisition networks and discrete data sampling' (Gaspar et al. 2011, no page numbers).

In the Azores monitoring by CIVISA is comprehensive and details are summarised in Table 4.4. Data from the networks are either transmitted to CIVISA, where they are used for research or, in times of crisis, are forwarded to the IVAR/CIVISA Emergency Operations Centre, which is housed in the *Universidade dos Açores* in Ponta Delgada (São Miguel). Critical information is also shared with the Civil Protection Authority and the Regional Government. Geo-referenced records collected from monitoring stations are stored in the AZORIS database (see above) where they may be used should they be required during a seismic or volcanic emergency.[20] Progress that is being made may be illustrated by examining four areas of investigation which involves studies of seismicity, ground deformation, geochemical monitoring and infrasound.

Seismic monitoring

> High seismicity recorded in the Azores and the impacts caused by the main destructive earthquakes demonstrate the importance of monitoring the active volcano-tectonic systems. Therefore, it is crucial to have seismic monitoring networks and a permanent information and surveillance service, which allows technical and scientific advice to be given to the civil protection authorities and decision-makers and allow the production of essential information for land-use and emergency planning.
>
> (Silva et al. 2020, p. 20)

In the absence of a major crisis since the 1998 earthquake on Faial to test the effectiveness of the seismic network, research has been on-going to evaluate the potential of seismic monitoring. The dense network of seismic stations has, for instance, enabled a correlation to be established between heightened seismic activity and volcanic unrest. Between May and September 2005 on São Miguel and concentrated beneath Fogo volcano, intense seismic swarms were linked, not just to ground deformation and surface rupture, but also to the triggering of more than 250 landslides (Marques et al. 2008; Wallenstein et al. 2009; Gaspar et al. 2011; Silva et al. 2015b). Results also show the potential that detailed seismological investigations hold for understanding the

Table 4.4 Characteristics of meteorological and geophysical monitoring in the Azores

Network	Characteristics
Seismic	The network is composed of 33 short-period and five additional broadband stations,[1] which are deployed across seven of the nine islands. These are complemented by portable seismic stations and arrays, which are used during times of seismic and volcanic crises. Data collected are used to monitor: tectonic and volcanic earthquakes and tremor; seismic source regions; site-specific seismic characteristics of areas especially those with buildings with differing vulnerabilities and modelling possible impacts of future events.
Geodetic	Regular field surveys support permanent *Global Navigation Satellite Systems* (GNSS) stations and are linked to meteorological sensors to correct readings for atmospheric effects. Geodetic monitoring allows crustal deformation studies to be undertaken, not only to identify the locations of point source changes in elevation, but also to track temporal and spatial variations of magmatic intrusions. At a more local scale, the network may be used to monitor slope instability.
Fluid geochemistry	This includes permanent stations to monitor CO_2 and H_2S fluxes and ^{222}Rn activity. These stations are coupled to meteorological sensors to assess the influence of external factors on processes of degassing. Additionally, temperatures and water levels are monitored in wells to appraise local stresses caused by crustal deformation and/or sallow intrusions.
Meteorology	In addition to being linked to the fluid geochemistry and geodetic networks, in the last ten years a new hydro-meteorological network has been installed across the islands to assist in the prediction of flooding and landslides. This allows meteorological information to be linked to geomorphological, geological and hydrological data.
Infrasound	Infrasound monitoring is based on Graciosa and is one of the stations set up to monitor compliance with the *Comprehensive Nuclear-Test-Ban Treaty*. Data are sent to the *International Monitoring System* (IMS) in Vienna. Infrasound measurements may also be used to monitor eruption at locations far removed from the Azores. Current research has grown out of a European Union funded project, *ARISE 2*, (2015–2018) which involved 24 institutes and universities, ten European member states and some external collaborators.

Based on Gaspar et al. (2011); Matos et al. (2017); Silva et al. (2020) and additional information.
1 Short-period seismographs are sensitive to seismic waves that vibrate several times per second and are used to record local earthquakes. Long period seismographs respond to lower frequency waves and are used to record distant earthquakes. Broadband seismographs perform both functions.

dynamics of magma chambers. Using results from seismic monitoring together with studies of geochemistry and petrology, Zanon et al. (2020) have modelled the evolution of a shallow reservoir (5.6–6.8 km) beneath Pico volcano, mapping possible ascent paths magma would take to reach the surface.

In terms of future prediction, a report by Silva et al. (2020) is of interest. Analysing 90,000 earthquakes recorded since 1997, they point out that most occur in three areas and coincide with zones from which major historic earthquakes have been generated: (1) Fogo and the Congro Fissural Volcanic System on São Miguel; (2) in the ocean SE and south of São Miguel and (3) from beneath Faial and locations north and north east of the island. These observations allowed the authors to identify significant 'seismic gaps' north of São Miguel and north of São Jorge, areas where there is less recent seismic activity than might be expected and from where the 1852 and 1757 earthquakes were, respectively, generated.[21] Another promising line of enquiry is recently published research by Rodrigues and Oliveira (2020), who argue that the traditional consensus that the location of earthquake epicentres across the islands is independent of previous events may not be secure. The transfer of stress across seismic zones may be manifested as 'epicentre migration' which may be capable of being modelled statistically.

Ground deformation

Using a Global Positioning System (GPS) and data collected from São Miguel for the period 2008–2013 and more particularly an episode of seismic unrest between 2011 and 2012 in the area between Congro and Furnas, Okada et al. (2015) proposed a hypothesis to explain this instability. Drawing parallels between this episode in the Azores and similar periods of unrest elsewhere in the world, such as the Matsushiro earthquakes swarm (Japan) and the Campi Flegrei (Italy) bradyseismic crisis in the 1980s, the authors propose the following model to account for observed patterns of seismicity and inflation/deflation. First, inflation is related to the lateral diffusion of fluids which is guided by existing cracks and fissures and is associated with the occurrence of intense high-frequency earthquakes. Second, an increase in fluids increases pressure in cracks and fissures and generates lower-frequency earthquakes and, finally, the dispersal of fluids allows pressure to drop, deflation to occur and eventually a return to seismic quiescence. In the future the link established between seismic activity, ground inflation and magma movement may prove an important tool in the *short-term prediction* of eruptions.

Geochemical monitoring

In the Azores geochemical observations involve periodic sampling of fumarolic gases and the continuous measurement of soil diffuse gas emissions. The principal gases analysed are CO_2 and ^{222}Rn, and significant progress has been made in soil diffuse CO_2 monitoring with permanent stations being installed across the islands as part of the IVAR/CIVISA seismo-volcanic network. The main aim of this research is to recognise signals within the data that suggest early-warnings of magmatic or hydrothermal activity (Viveiros et al. 2015b, p. 271). Such a task is not without its difficulties because a range of environmental factors are known to be responsible for anomalies in gas flux over the short-term and these include variations in soil water content, barometric pressure, wind speed, air temperature and relative humidity. Only when the effects of these are removed, may deeper-seated and longer-term anomalies be recognised within the time-series of observations (Oskarsson et al. 1999; Hernández et al. 2004; Granieri et al. 2010).

One example of such research in the Azores is a study of degassing at Furnas and Fogo volcanoes on São Miguel. Multiple regression techniques have been used to filter or 'clean' environmental signatures from the time-series data of CO_2 flux, so leaving residuals that may be compared with similar time-series produced by monitoring other anomalies such as seismic activity. Although this research is ongoing its aim is for the 'filtered real-time series … (to be) … used as part of the warning system for seismic-volcanic activity at Furnas and Fogo volcanoes' by establishing base-line data for what is a quiescent phase in the volcanic histories of these two volcanoes (Viveiros et al. 2015b, p. 286).

Infrasound

Infrasound research in the Azores differs from that on seismic, ground deformation and geochemical monitoring, because it allows volcano surveillance to take place over great distances. It also differs by not allowing the identification of signs of volcanic unrest because it is only effective in the open-vent stage of volcanic activity once an eruption has begun. Infrasound allows the detection of subtle variations in pressure which enables observers to anticipate changes in the eruptive behaviour. As Table 4.4 shows, scientists in the Azores are part of a large international network of collaborating bodies which work on the principal of reciprocity, with each institution monitoring distant volcanoes in partner countries and exchanging information (Wallenstein 2019). Station IS42 sited on Graciosa has already made a contribution and Matos et al. (2017, 2019, 2020) report the results of joint research carried out by IVAR and the *Dipartimento di Scienze della Terra* in the *Università di Firenze* in Italy. Two eruptions were monitored Grimsvötn (Iceland) involving a hydro-magmatic eruption occurring in May 2011 and summit activity on Mount Etna (Italy), which took place in May 2106. In both cases, it was possible to correlate local observations with remote infrasonic indications which shows the future potential of what is at present a research frontier and a work in progress.

Concluding remark

Portugal is an economically developed state which is ranked 42nd out of 185th countries and territories in terms of GDP per capita by the International Monetary Fund (IMF 2020) and this, combined with a strong Portuguese tradition of pure and applied investigations in the earth sciences and political will and commitment at both national and regional levels to fund mitigation-based research, has meant that considerable progress has been made in hazard evaluation over the past four decades. Regrettably only a future large earthquake and/or eruption will be able fully to test the islands' preparedness in terms of measures of both *long and short-term prediction* which has been discussed in this chapter, and with respect to policies of disaster risk reduction which are discussed in Chapter 6.

Notes

1 A geographic information system (GIS) is an IT- based set of procedures designed to archive, manipulate and analyse spatial (i.e. geo-referenced) data. It allows different 'layers' of data, for example, on hazard, population, land-use and transport, to be compared and for spatial inter-relationships to be established. In hazard research it has become progressively more prominent over the past three decades. In disasters GIS is particularly useful because risk maps can be rapidly modified as new data become available (see Carrara & Guzzeti 1995).

2 Some scholars call this approach *short-term prediction* and others make a distinction between prediction and forecasting (Anon 2011, p. 319). This distinction is not followed in the present chapter and prediction and forecasting are used inter-changeably.

3 What was thought at the time to be a causal mechanism was the hypothesis known as *dilatancy-diffusion,* in which it was observed in the laboratory that highly stressed rock showed an increase in volume (i.e. dilatancy), which could be correlated to a range of pre-cursory phenomena such as: seismic P wave velocity; ground uplift and tilt; radon emission; electrical properties and the number of local earthquakes (Main et al. 2012). Other procedures concentrated on: the ratio of P wave to S wave velocities (i.e. V_p/V_s), again related to dilatancy (Scholz et al. 1973) and radon emissions on their own (Cicerone et al. 2009).

4 Even some of the claimed successes, such as the Parkfield earthquake in California of 2004 have been questioned by subsequent research. In the case of Parkfield the prediction was firm that the earthquake would actually occur between 1988 and 1993 (Hough 2005).

5 Infrasound is defined as waves below 20 Hz where most of a volcano's sound energy is concentrated (McNutt et al. 2015).

6 As discussed in Chapter 3, the 1st January 1980 Terceira earthquake was recorded by the three stations in the Azores which were maintained by the Portuguese *Instituto de Meteorologia*. In the immediate aftermath of the disaster the *Universidade dos Açores* established a temporary network and a few months later had six stations operating in Terceira and eight on São Miguel (Fontiela et al. 2018).

7 The technique is known as *Kriging*. Originating in mining geology, it is named after a South African mining engineer, Danie Krige, and uses a limited set of points to estimate the value of a variable - in this case Imax - over a continuous spatial field. Detailed discussion is beyond the scope of this volume and for further information reference should be made to Matheron (1963).

8 As discussed in Chapter 3, intensity values are calculated on the basis of the damage caused to different types of building. EMS-98 (Grünthal 1998) classifies buildings into six categories: A - being the least resistant and F - the most resistant. Most traditional buildings on São Miguel fall into Class A, making them extremely vulnerable to seismic shaking and in this study it is assumed that all buildings fall into this category. Although it is known that Class B and C buildings are to be found in the central areas of some of the principal settlements,

the data available did not allow these to be identified either by number or by the degree to which they were affected (Silveira et al. (2003, p. 617).

9   Figure based on the 2011 census (SREA 2012).

10 Seismic retrofitting is the modification of buildings to make them more resistant to earthquakes. Traditionally many buildings were designed without adequate seismic protection, and many countries have policies and codes to encourage retrofitting. Such codes need regular revision because new earthquakes often reveal further design inadequacies (Chuang & Zhuge 2015).

11 This project was entitled *URBSIS: Assessing Vulnerability and Managing Earthquake Risk at the Urban Scale*, and was commissioned by the *Câmara Municipal* (County Council) of Horta on Faial Island (Ferreira et al. 2017).

12 The *Vulnerability Index* is based on 14 parameters divided into four groups: Structure of the Building (i.e. type of resisting system, quality of resisting system, strength, maximum distance between walls, number of floors, location and soil conditions); Irregularities and Interactions (i.e. position and interaction, plan configuration, height regularity and wall facade, openings and alignments); Floor Slabs and Roofs (i.e. horizontal diaphragms and roofing type) and Conservation Status and Other Elements (i.e. fragilities and conservation status and non-structural elements). Weightings assigned to the different elements was based on expert opinion introducing a level of subjectivity. It is argued, however, that this has proved to be robust across a number of applications.

13 Wrongly dated and named as 'Furnas C 1630' by Booth et al. (1983, p. 100)

14 The *Poisson* distribution is widely used in the statistical prediction of future extreme natural events, but its application is not without problems because of two assumptions which it makes about time-series data. The first is that the events in the time-series are stationary: that is they show no linear trend over time and, second, that successive occurrences are independent of each other (Cox and Lewis 1966; Chester 1986; Davis 2002). Although this may be assumed for many climatically induced extreme events (e.g. storms and droughts), this might not be the case for earthquakes and volcanic eruptions where non-independence (i.e. temporal auto-correlation) may be present, since if magma supply is constant than one major eruption may preclude similar large eruptions for a considerable period of time. Despite these misgivings, *Poisson*-based modelling is widely employed in volcanology and seismology because it is considered, not only statistically robust, but also pragmatically the best technique that is currently available (Lockwood & Hazlett 2010, p. 428). It is also assumed that all the historic and, more particularly the radiometric, dates are accurate.

15 Monte Carlo simulations use probability to assess the most likely of several possible outcomes. In the case of the lava flow simulation, the calculation of possible paths for a flow begins with the cell chosen as the eruption vent and its propagation into an adjacent cell is calculated probabilistically. This is then repeated for the next cell until a flow has attained its maximum pre-set length, in this case 20 km (Felpeto et al. 2001, 2007).

16 As discussed in Chapter 3, although it is possible to calculate an average recurrence interval for damaging tsunamis of *c.*70 years, this figure may be misleading. Tsunamis have been generated either, from within or adjacent to the islands (55%), or from distant sources (45%) such as the eastern North Atlantic Ocean in the case 1755 Lisbon earthquake. The mechanisms involved are numerous and include earthquakes; slope-failure (sub-aerial and marine); pyroclastic flows into the sea, submarine eruptions; volcanic explosions and collapse; caldera formation and lahars entering the ocean, and in many instances records of events are not well constrained by historical data, this being particularly the case for the 45% of tsunamis that originate from distant source areas (Santos et al, 2004; Calado et al. 2011).

17   Andrade et al. (2006), also note that these points were first made more than 150 years ago by the Azorean historian and polymath, Francisco Ferreira Drummond (1850–64).
18   ARCH/INFO (now archinfo) is a Geographical Information System (GIS) produced by Esri (Environmental Systems Research Institute) of Redlands, California.
19   These are the 1: 25,000 scale maps (*Carta Militar de Portugal*) produced by the *Instituto Geográfico do Exército*.
20   In the future IVAR may be able to call upon the assistance of several national and international research and technology networks. In addition to the Comprehensive Nuclear-Test-Ban Treaty (CTBTO - see Table 4.4), IVAR is associated with: the Network of European Regions Using Space Technologies (NEREUS) and participates with other volcano observatories in the European Observing System (EPOS).
21   A seismic gap is a segment of an active fault known to produce significant earthquakes which has not been active for a long time, compared with other known earthquake generating areas. A gap is considered to be a possible location for future earthquakes (McCann et al. 1979)

References

Andrade, C., Borges, P and, M.C. Freitas, M.C. 2006. Historical tsunami in the Azores archipelago (Portugal). *Journal of Volcanology and Geothermal Research* **156**, 172–185.
Andrade, C., Cruz, J.V., Viveiros, F. and Coutinho, R. 2020. CO_2 emissions from Fogo intracaldera volcanic lakes (São Miguel Island, Açores). *Journal of Volcanology and Geothermal Research* **400**, 1–10, paper 106915.
Anon. 2011. *Operational Earthquake Forecasting: State of Knowledge and Guidelines for Utilization. Report by the Commission on Earthquake Forecasting for Civil Protection.* Department of Civil Protection, Instituto Nazionale di Geofisica e Vulcanologia, Rome.
Anon. 2016. Multiple natural hazard risk assessment- Planning and Recovery. Technical Report EMSN-018 (Contract 259811). European Commission - Joint Research Centre, Institute for the Protection and Security of the Citizen, Global Security and Crisis Management Unit, Ispra (VA), Italy.
Baxter, P.J., Baubron, J-C. and Coutinho, R. 1999. Health hazards and disaster potential of ground gas emissions at Furnas Volcano, São Miguel, Azores. *Journal of Volcanology and Geothermal Research* **92** (1–2), 95–106.
Blahůt, J., Klimes, J., Rowberry, M. and Kusák, M. 2018. Database of giant landslides on volcanic islands – first results from the Atlantic Ocean. *Landslides* **15**, 823–827.
Bolt, B.A. 2004. *Earthquakes*. W.H. Freeman, New York.
Booth, B., Croasdale, R. and Walker, G.P.L. 1983. Volcanic hazard on São Miguel, Azores, *In*: Tazieff, H. and Sabroux, J-C. (eds) *Forecasting Volcanic Events*, Elsevier, Amsterdam, 99–109.
Booth, B., Croasdale, R. and Walker, G.P.L. 1978. A quantitative study of five thousand years of volcanism on São Miguel, Azores. *Philosophical Transactions of the Royal Society of London* **228A**, 271–319.
Borges, P., Mg, K. and Calado, H. 2011. Coastal hazards in the Azores Archipelago – Coastal storms and flooding. *Journal of Coastal Research* **SI 61**, 474.
Borges, P. Phillips, M.R., Ng, K., Medeiros, A. and Calado, H. 2014. Preliminary coastal vulnerability assessment for Pico Island (Azores). *Journal of Coastal Research* **SI 70**, 385–388.

Calado, H., Borges P., Ng, K. and Alves, F. 2011. The Azores archipelago, Portugal: Improved understanding of small island coastal hazards and mitigation measures. *Natural Hazards* **58**, 427–444.

Caniaux, G. 2005. Analyse statistique de la fréquence des éruptions volcaniques aux Açores: Contribution à l' évaluation des risques. *Bulletin de la Societe Geologique de France* **176** (1), 107–120.

Cappello, A., Zanon, V., Del Negro, C., Ferreira, T.J.L. and Queiroz, M.G.P.S. 2015. Exploring lava-flow hazards at Pico Island, Azores Archipelago (Portugal). *Terra Nova* **27**, 156–161.

Carrara, A. and Guzzeti, F. 1995. *Geographical Information Systems in Assessing Natural Hazards.* Springer, New York.

Carvalho, A., Sousa, M.L., Oliveira, C.S., Campos-Costa, A. and Forjaz, V.H. 2001. Seismic hazard for the Central Group of the Azores Islands. *Bollettino di Geofisica ed Applicata* **42** (1–2), 89–105.

Cashman, K.V. and Sparks, R.S.J. 2013. How volcanoes work: A 25 year perspective. *Geological Society of America Bulletin* **125** (5/6), 664–690.

Chester, D.K. 1986. Comments on "a statistical analysis of flank eruptions on Etna volcano" by Mulargia, F., Tinti, S. and Boschi, E. *Journal of Volcanology and Geothermal Research* **28**, 385–395.

Chester, D.K. 1993. *Volcanoes and Society.* Edward Arnold, London.

Chester, D.K., Dibben, C. and Coutinho, R. 1995. Report on the evacuation of the *Furnas District*, São Miguel, Azores, in the event of a future eruption. CEC Environment – ESF Laboratory Volcano Open File Report 4, University College, London.

Chuang, S.W. and Zhuge, Y. 2015. Seismic retrofitting of unreinforced masonry buildings – a literature review. *Australian Journal of Structural Engineering* **6** (1), 25–36.

Cicerone, R.D., Ebel, J.E. and Britton, J. 2009. A systematic compilation of earthquake precursors. *Tectonophysics* **476** (3–4), 371–396.

Cimellaro, G.P. and Marasco, S. 2018. Seismic hazard analysis. *In*: Cimellaro, G.P. and Marasco, S. (eds) *Introduction to Dynamics of Structures and Earthquake Engineering.* Springer International, Switzerland, Chapter 11, 263–280.

Cole, P.D., Guest, J.E., Queiroz, G., Wallenstein, N., Pacheco, J.M., Gaspar, J.L., Ferreira, T. and Duncan, A.M. 1999. Styles of volcanism and volcanic hazards on Furnas volcano, São Miguel, Azores. *Journal of Volcanology and Geothermal Research* **92**, 39–53.

Costa, A.C.G., Marques, F.O., Hildenbrand, A., Sibrant, A.L.R. and Catita, C.M.S. 2014. Large-scale catastrophic flank collapses in a steep volcanic ridge: The Pico-Faial Ridge, Azores triple junction. *Journal of Volcanology and Geothermal Research* **272**, 111–125.

Costa, A.C.G., Hildenbrand, A., Marques, F.O., Sibrant, A.L.R. and Santos de Campos, A. 2015. Catastrophic flank collapses and slumping in Pico Island during the last 130 kyr (Pico-Faial ridge, Azores Triple Junction). *Journal of Volcanology and Geothermal Research* **302**, 22–46.

Cox, D.R. and Lewis, P.A.W. 1966. *The Statistical Analysis of Series of Events.* Methuen, London.

Cunha, A. 2003. The October 1997 landslides on São Miguel Island, Azores. *In*: Javier, H. (ed) *Lessons Learnt from Landslide Disaster in Europe.* European Commision Joint Research Cetre, ISPRA, Document 20558EN, Brussels.

Davis, J.C. 2002. *Statistics and Data Analysis in Geology.* Wiley, New York, 3rd Edition.

Drummond, F.F., 1850–64. *Anais da ilha Terceira, vols. I a IV.* Edição fac-similada de 1981, Governo Autónomo dos Açores, Angra do Heroísmo.

Fagundes, C., Bento, R. and Cattari, S. 2017. On the seismic response of buildings from the Azores. *Structures* **10**, 184–196.

Felpeto, A., Arana, V., Ortiz, R., Astiz, M. and Garcia, A. 2001. Assessment and modelling of lava flow hazard on Lanzarote (Canary Islands). *Natural Hazards* **23**, 247–257.

Felpeto, A., Marti, J. and Ortiz, R. 2007. Automatic GIS-based system for volcanic hazard assessment. *Journal of Volcanology and Geothermal Research* **166**, 106–116.

Ferreira, T.M., Maio, R., Vicente, R. and Costa, A. 2016. Earthquake risk mitigation: The impact of seismic retrofitting strategies on urban resilience. *International Journal of Strategic Property Management* **20** (3), 291–304.

Ferreira, T.M., Maio, R. and Vicente, R. 2017. Seismic vulnerability assessment of the old city centre of Horta, Azores: Calibration and application of a seismic vulnerability method. *Bulletin of Earthquake Engineering* **15**, 2879–2899.

Fontiela, J., Bezzeghoud, M., Rosset, P. and Rodrigues, C. 2017. Maximum observed intensity maps for the Azores Archipelago (Portugal) from 1522 to 2012 seismic catalog. *Seismological Research Letters* **88** (4), 1178–1184.

Fontiela, J., Oliveira, C.S. and Rosset, P. 2018. Characterisation of seismicity of the Azores Archipelago: An overview of historical events and a detailed analysis for the period 2000–2012. *In*: Kueppers, U. and Beier, C. (eds) *Volcanoes of the Azores, Active Volcanoes of World*, Springer-Verlag, Germany, 127–153.

Gaspar, J.L., Ferreira, T., Queiroz, G., Baubron, J.C. and Baxter, P. 1998. High levels of CO_2 in the atmosphere of Furna do Enxofre lava cave (Graciosa Island, Azores): a case of public health risk. *EC Advanced Study Course 1998. Volcanic Hazard Assessment, Monitoring and Risk Mitigation.* European Commission, Environment and Climate Programme (abstract).

Gaspar, J.L., Goulart, C., Queiroz, G., Silveira, D. and Gomes, A. 2004. Dynamic structure and data sets of a GIS database for geological risk analysis in the Azores volcanic islands. *Natural Hazards and Earth System Sciences* **4**, 233–242.

Gaspar, J.L., Queiroz, G., Ferreira, T., Amaral, P., Viveiros, F., Marques, R., Silva, C. and Wallenstein, N. 2011. Geological Hazards and Monitoring at the Azores (Portugal). *Earthvine.* https://earthzine.org/geological-hazards-and-monitoring-at-the-azores-portugal/

Gaspar, J.L., Queiroz, G., Ferreira, T., Medeiros, A.R., Goulart, C. and Medeiros, J. 2015a. Earthquakes and volcanic eruptions in the Azores region: Geodynamic implications from major historical events and instrumental seismicity. *In*: Gaspar, J.L., Guest, J.E., Duncan, A.M., Barriga, F.J.A.S. and Chester, D.K. (eds) *Volcanic Geology of São Miguel Island (Azores Archipelago)*, Geological Society, London Memoir **44**, 33–49.

Gaspar, J.L., Guest, J.E., Queiroz, G., Pacheco, J.M., Pimentel, A., Gomes, A., Marques, R., Felpeto, A., Ferreira, T. and Wallenstein, N. 2015b. Eruptive frequency and volcanic hazard zonation in São Miguel Island, Azores. *In*: Gaspar, J.L., Guest, J.E., Duncan, A.M., Barriga, F.J.A.S. and Chester, D.K. (eds) *Volcanic Geology of São Miguel Island (Azores Archipelago)*, Geological Society, London Memoir **44**, 155–166.

Goulart, C., Gaspar, J.L. and Queiroz, G. 2011. Azoris: An infrastructure to improve scientific responsiveness to natural disasters in the Azores. *Proceedings of* the Geo-information for Disaster Management (Gi4DM) Annual Conference, OP 28, Antalya, Turkey.

Granieri, D., Avino, R. and Chiodini, G. 2010. Carbon dioxide diffuse emission from the soil: Ten years of observations at Vesuvio and Campi Flegrei (Pozzuoli), and linkages with volcanic activity. *Bulletin of Volcanology* 72, 103–118.

Grünthal, G. 1998 *European macroseismic scale 1998*. Conseil de L'Europe Cahiers du Centre Européen de Géodynamique et de Séismologie vol. 15, Luxenbourg.

Guest, J.E., Gaspar, J.L., Cole, P.D., Queiroz, G., Duncan, A.M., Wallenstein, N., Ferreira, T. and Pacheco, J-M. 1999. Volcanic geology of Furnas Volcano, São Miguel, Azores. *Journal of Volcanology and Geothermal Research* 92, 1–29.

Hansell, A., Horswell, C. and Oppenheimer, C. 2006. The health hazards of volcanoes and geothermal areas. *Occupational and Environmental Health* 59, 628–639.

Hernández, P., Pérez, N., Salazar, J., Reimer, M., Notsu, K. and Wakita, H. 2004. Radon and helium in soil gases at Cañadas Caldera, Tenerife, Canary Islands, Spain. *Journal of Volcanology and Geothermal Research* 131, 59–76.

Hildenbrand, A., Marques, F.O. and Catalão, J. 2018. Large-scale mass wasting on small volcanic islands revealed by the study of Flores Island (Azores). *Scientific Reports* 8, 13898. doi: 10.1038/s4198-018-32253-0.

Hipólito, A., Madeira, J., Carmo, R. and Gaspar, J.L. 2013. Neotectonics of Graciosa Island (Azores): A contribution to seismic hazard assessment of a volcanic area in complex geodynamic setting. *Annals of Geophysics* 56 (6), S0677. doi:10.4401/ag-6222.

Hough, S.E. 2005. Earthquakes: Predicting the Unpredictable. *Geotimes* 50 (3). http://www.geotimes.org/mar05/feature_eqprediction.html

IMF. 2020. *World Economic Outlook Database*. International Monetary Fund, Washington, DC. https://en.wikipedia.org/wiki/List_of_countries_by_GDP_(PPP)_per_capita

Joffe, H., Rossetto, T., Bradley, C. and O' Connor, C. 2018. Stigma in science: The case of earthquake prediction. *Disasters* 42 (1), 81–100.

Jones, G., Chester, D.K. and Shooshtarian, F. 1999. Statistical analysis of the frequency of eruptions at Furnas Volcano, São Miguel, Azores. *Journal of Volcanology and Geothermal Research* 92, 31–38.

Kilburn, C.R.J. 2018. Forecasting volcanic eruptions: Beyond the failure forecast method. *Frontiers in Earth Science* 6, article 133, 15 pages.

Kueppers, U., Pimentel, A., Ellis, B., Forni, F., Neukampt, J., Pacheco, J., Perugini, D. and Queiroz, G. 2019. Biased volcanic hazard assessment due to incomplete eruption records on Ocean Island: An example of Sete Cidades Volcano, Azores. *Frontiers in Earth Science* 7 (article 22), 1–17.

Lockwood, J.P. and Hazlett, R.W. 2010. *Volcanoes: Global Perspectives*. Wiley-Blackwell, Oxford.

Madeira, J.E. 1998. *Estudos de Neotectónica nas ilhas do Faial, Pico e S. Jorge: uma contribuição para o conhecimento Geodinâmico da junção tripla dos Açores*. Dissertação apresentada à Universidade de Lisboa para a obtenção do grau de doutor em geologia, na especialidade de Geodinâmica Interna. Faculdade de Ciências da Universidade de Lisboa, Departamento de Geologia. pp 428.

Main, I.G., Bell, A.F., Meredith, P.G., Geiger, S. and Touati, S. 2012. The dilatancy-diffusion hypothesis and earthquake predictability. *Geological Society of London, Special Publication* 367 (1), 215–230.

Maio, R., Estêvão, J.M.C., Ferreira, T.M. and Vicente, R. 2017. The seismic performance of stone masonary buildings in Faial Island and the relevance of implementing effective seismic strengthening policies. *Engineering Structures* **141**, 41–58.

Maio, R., Ferreira, T.M., Vicente, R. and Costa, A. 2019. Is the use of traditional seismic strengthening strategies economically attractive in the Renovation of urban cultural heritage assets in Portugal? *Bulletin of Earthquake Engineering* **17**, 2307–2330.

Malheiro, A. 2006. Geological hazards in the Azores archipelago: Volcanic terrain instability and human vulnerability. *Journal of Volcanology and Geothermal Research* **156**, 158–171.

Marques, R., Zêzere, J.L., Trigo, R., Gaspar, J and Trigo, I. 2008. Rainfall patterns and critical values associated with landslides in Povoação County (São Miguel, Azores): Relationships with the North Atlantic Oscillation. *Hydrological Processes* **22**, 478–494.

Marques, R., Amaral, P. Araújo, I., Gaspar, J.L. and Zêzere, J.L. 2015. Landslides on São Miguel Island (Azores); susceptibility analysis and validation of rupture zones using a bivariate GIS-based statistical approach. *In*: Gaspar, J.L., Guest, J.E., Duncan, A.M., Barriga, F.J.A.S. and Chester, D.K. (eds) *Volcanic Geology of São Miguel Island (Azores Archipelago)*, Geological Society, London Memoir **44**, 167–184.

Martins, V.N., Cabral, P. and Sousa e Silva, D. 2012a. Urban modelling for seismic areas: The case study of Vila Franca do Campo (Azores Archipelago, Portugal). *Natural Hazards and Earth System Sciences* **12** (9), 2731–2741.

Martins, V.N., Sousa e Silva, D. and Cabral, P. 2012b. Social vulnerability to seismic risk using multi-criteria analysis: The case of Vila Franca do Campo (São Miguel Island, Azores Portugal). *Natural Hazards* **62**, 385–404.

Matheron, G. 1963. Principles of geostatistics. *Economic Geology* **58** (8), 1246–1266.

Matos, S., Wallenstein, N., Ripepe, M. and Marchetti, E. 2017. *Long-Range Infrasound Detections of Volcanic Activity by IS42 Station*. Comprehensive Test Ban Treaty (CTBT) Science and Technology Conference, Vienna, Paper T1.1-P16. https://ctnw.ctbto.org/ctnw/abstract/22288

Matos, S., Wallenstein, N., Campus, P., Marchetti, E. and Ripepe, M. 2019. *Analysis of multiple detections of Mount Etna eruptive activity at different IMS infrasound stations compared with near source observations*. CTBT: Science and Technology 2019 Conference, Viena, Paper T1.1-P3. https://events.ctbto.org/sites/default/files/2020-05/20190614-2019%20Book%20Of%20Abstracts%20Web%20Version%20with%20front%20cover%20-%20edited.pdf

Matos, S., Wallenstein, N., Marchetti, E. and Ripepe, M. 2020. Location of Stromboli volcano July 2019 paroxysm event based on long-range infrasound detections in several IMS stations, EGU General Assembly 2020, Online, May 4–8. doi: 10.5194/egusphere-egu2020–10156

McCann, W.R., Nishenko, S.P., Sykes, I.R. and Krause, J. 1979. Seismic gaps and plate tectonics: Seismic potential for major boundaries. *Pure and Applied Geophysics* **117**, 1082–1147.

McNutt, S.R., Thompson, G., Johnson, J., De Angelis, S. and Fee, D. 2015. Seismic and Infrasonic Monitoring. *In*: Sigurdsson, H., Houghton, B., Mc Nutt, S., Rymer, H. and Stix (eds) *The Encyclopedia of Volcanoes*. Elsever, Amsterdam, 1071–1099, 2nd Edition.

Medeiros, J., Carmo, R., Pimentel, A., Vieira, J.C. and Queiroz, G. 2021. Assessing the impact of explosive eruptions of Fogo volcano (São Miguel, Azores) on the tourism economy. *Natural Hazards and Earth System Sciences* **21**, 417–437.

Nunes, J.C. 1986. *Sismicidade Histórica e Instrumental do Arquipélago dos Açores. Catálogo Preliminar, 1444–1980.* Relatório INMG/LNEC, Lisboa.

Nunes, J.C. 1999. *A actividade vulcânica na Ilha do Pico do plistocénica ao holocénico: mecanismo eruptive e hazard vulcânica.* PhD thesis, Dept. Geociências, Universidade dos Açores.

Nunes, J.C., Forjaz, V.H. and Oliveira, C.S. 2000. Catálogo sísmico da região dos Acores, Vol. I (1980–88), Vol. II (1989–98) PPERCAS Project, Azores Univesity, Ponta Delgada, Portugal.

Nunes, J.C., Forjaz, V.H. and Oliveira, C.S. 2004. *Catálogo Sísmico Região dos Açores Versão 1.0 (1850–1998).* SÍSMICA 2004–6th Congresso Nacional de Sismologia e Engenharia Sísmica, Guimarães, 14, 15 e de Abril 2004.

Okada, J., Sigmundsson, F., Ófeigsson, B.G., Ferreira, T.J.L. and Rodrigues, R.M.M.T.C. 2015. Tectonic and volcanic deformation at São Miguel, Azores, observed by continuous GPS analysis 2008–13. In: Gaspar, J.L., Guest, J.E., Duncan, A.M., Barriga, F.J.A.S. and Chester, D.K. (eds) *Volcanic Geology of São Miguel Island (Azores Archipelago),* Geological Society, London Memoir **44**, 239–256.

Oskarsson, N., Pálsson, K., Ólafsson, H. and Ferreira, T. 1999. Experimental monitoring of carbon dioxide by low power IR-sensors: Soil degassing in the Furnas Volcanic Centre, Azores. *Journal of Volcanology and Geothermal Research* **92**, 181–193.

Pallister, J. and Mc Nutt, S.R. 2015. Synthesis of Volcano Monitoring. In: Sigurdsson, H., Houghton, B., Mc Nutt, S., Rymer, H. and Stix (eds) *The Encyclopedia of Volcanoes.* Elsevier, Amsterdam, 1151–1171, 2nd Edition.

Paolucci, R. 2002. Amplifcation of earthquake ground motion by topographic irregularities. *Earthquake Engineering and Structural Dynamics* **31** (10), 1831–1853.

Pedrazzi, D., Cappello, A., Zanon, V. and Del Negro, C. 2015. Impact of effusive eruptions from the Equas-Carvão fissure system, São Miguel, Azores Archipelago (Portugal). *Journal of Volcanology and Geothermal Research* **291**, 1–13.

Queiroz, G., Gaspar, J.L., Cole, P.D., Guest, J.E., Wallenstein, N., Duncan, A.M. and Pacheco, J. 1995. Erupções vulcânicas no vale das Furnas (Ilha de S. Miguel, Açores) na primeira metade do seculo XV. *Açoreana* **8**, 159–165.

Queiroz, G., Pacheco, J.M., Gaspar, J.L., Aspinall, W.P., Guest, J.E. and Ferreira, T. 2008. The last 5000 years of activity at Sete Cidades volcano (São Miguel Island, Azores: implications for hazard assessment. *Journal of Volcanology and Geothermal Research* **178**, 562–573.

Rodrigues, M.C.M. and Oliveira, C.S. 2020. Considering spatial memory to estimate seismic risk: The case of the Azores Archipelago. *GEM International Journal on Geomathematics* **11**, 16. https:/doi.org/10.1007/s13137-020-00152-0

Santos, F.D., Valente, M.A., Miranda, P.M.A., Aguiar, A., Azevado, E.B., Tomé, A.R. and Coalho, F. 2004. Climate change scenarios in the Azores and Madeira Islands. *World Resources Review* **16** (4), 493–491.

Schilling, S.P. 1998. *LAHARZ: GIS Programs for Automated Mapping of Lahar-Inundation Hazard Zones.* Open File Report, US Geological Survey, Washington, DC.

Scholz, C.H., Sykes, L.R. and Aggarwal, Y.P. 1973. Earthquake. prediction: A physical basis. *Science* **181** (4102), 803–810.

Silva, M. 2005. *Caracterização da sismicidade histórica dos Açores com base na reinterpretação de dados de macrossísmica: contribuição para a avaliação do risco sísmico nas ilhas do Grupo Central.* Dissertação de Mestrado em Vulcanologia e Riscos Geológicos, Universidade dos Açores.

Silva, C., Viveiros, F., Ferreira, T., Gaspar, J.L. and Allard, P. 2015a. Diffuse soil emanations of radon and hazard implications at Furnas Volcano, São Miguel (Azores) *In*: Gaspar, J.L., Guest, J.E., Duncan, A.M., Barriga, F.J.A.S. and Chester, D.K. (eds) *Volcanic Geology of São Miguel Island (Azores Archipelago)*, Geological Society, London Memoir **44**, 197–211.

Silva, R., Medeiros, A., Carmo, R., Luís, R., Wallenstein, N., Bean, C. and Sousa, R. 2015b. Seismic activity on São Miguel Island volcanic-tectonic structures (Azores archipelago). *In*: Gaspar, J.L., Guest, J.E., Duncan, A.M., Barriga, F.J.A.S. and Chester, D.K. (eds) *Volcanic Geology of São Miguel Island (Azores Archipelago)*, Geological Society, London Memoir **44**, 227–238.

Silva, R., Carmo, R. and Marques, R. 2020. Characterization of the tectonic origins of historical and modern seismic events and their societal impact on the Azores Archipelago, Portugal. *In*: Dilek, Y., Ogawa, Y. and Okubo, Y. (eds) *Characterization of Modern and Historical Seismic-Tsunamic Events and Their Global-Societal Impacts*. Geological Society, London, Special Publication, 501. http://doi.org/10.1144/SP501-2019-106.

Silveira, D., Gaspar, J.L., Ferreira, T. and Queiroz, G. 2003. Reassessment of the historical seismic activity with major impact on S. Miguel Island (Azores). *Natural Hazards and Earth System Sciences* **3**, 615–523.

SREA. 2012. *Censos 2011.* Serviço Regional de Estatística dos Açores, Açores.

Vicente, R., Parodi, S., Lagomarsino, S., Varum, H. and Mendes da Silva, J.A.R. 2011. Seismic vulnerability and risk assessment: Case study of the historic city centre of Coimbra, Portugal. *Bulletin of Earthquake Engineering* **9** (4), 1067–1096.

Viveiros, F., Ferreira, T., Silva, C. and Gaspar, J.L. 2009. Meteorological factors controlling soil gases and indoor CO_2 concentration: A permanent risk in degassing areas. *Science of the Total Environment* **407**, 1362–1372.

Viveiros, F., Cardellini, C., Ferreira, T., Caliro, S., Chiodini, G. and Silva, C. 2010. Soil CO_2 emissions at Furnas volcano, São Miguel Island, Azores archipelago: Volcano monitoring perspectives, geomorphologic studies, and land use planning application. *Journal of Geophysical Research* **115**, B12208. doi:10.1029/2010JB007555.

Viveiros, F., Gaspar, J.L., Ferreira, T., Silva, C., Marcos, M. and Hipólito, A. 2015a. Mapping of soil CO_2 diffuse degassing in São Miguel Island and its public health implications. *In*: Gaspar, J.L., Guest, J.E., Duncan, A.M., Barriga, F.J.A.S. and Chester, D.K. (eds) *Volcanic Geology of São Miguel Island (Azores Archipelago)*, Geological Society, London Memoir **44**, 185–195.

Viveiros, F., Ferreira, T., Silva, C., Vieira, J.C., Gaspar, J.L., Virgili, G. and Amaral, P. 2015b. Permanent monitoring of CO_2 degassing at Furnas and Fogo volcanoes (São Miguel Island, Azores. *In*: Gaspar, J.L., Guest, J.E., Duncan, A.M., Barriga, F.J.A.S. and Chester, D.K. (eds) *Volcanic Geology of São Miguel Island (Azores Archipelago)*, Geological Society, London Memoir **44**, 271–288.

Viveiros, F., Gaspar, J.L., Ferreira, T. and Silva, C. 2016. Hazardous indoor CO_2 concentration in volcanic environments. *Environmental Pollution* **214**, 776–784.

Viveiros, F., Marcos, M., Faria, C., Gaspar, J.L., Ferreira, T and Silva, C. 2017. Soil CO_2 concentration in volcanic environments. *Environmental Pollution* **214**, 776–784.

Wallenstein, N. 2019. A utilização de infrassons na monitorização de eventos extremos – aplicação aos Açores. *NO Revista*, junho de **2019**, 28–29.

Wallenstein, N., Chester, D. and Duncan, A.M. 2005. Methodological implications of volcanic hazard evaluation and risk assessment: Fogo Volcano, Sao Miguel, Azores. Zeitschrift für *Geomorphologie* **140**, 129–149.

Wallenstein, B., Duncan, A.M., Chester, D.K. and Marques, R. 2007. Fogo Volcano (São Miguel, Azores): a hazardous landform. *Géomophologie: Relief, Processus, Environment* **3**, 259–270.

Wallenstein, N., Silva, R., Riedel, C., Lopes, C., Ibáñwz, J., Silveria, D and Montalvo, A. 2009. Recent developments in seismic studies in the Fogo-Congro area, São Miguel Island (Azores). *In:* Bean, C.J., Braiden, A.K., Lokmer, I., Martini, F. and O'Brian, G.S.I. (eds) *The Volume Project: Volcanoes: Understanding Subsurface Mass Movement.* School of Geosciences, University of Dublin, 207–216.

Wallenstein, N., Chester, D., Coutinho, R., Duncan, A. and Dibben, C. 2015. Volcanic hazard vulnerability on São Miguel Island, Azores. *In:* Gaspar, J.L., Guest, J.E., Duncan, A.M., Barriga, F.J.A.S. and Chester, D.K. (eds) *Volcanic Geology of São Miguel Island (Azores Archipelago),* Geological Society, London Memoir **44**, 213–225.

Walker, G.P.L. 1974. Volcanic hazards and the prediction of volcanic eruptions. *In:* Funnell, B.M. (ed) *Prediction of Geological Hazards.* Geological Society of London, Miscellaneous Paper **3**, 23–41.

WHO. 2010. *WHO Guidelines for Indoor Air Quality: Selected Pollutants.* Word Health Organisation, Denmark.

Zanon, V., Pimentel, A., Auxerre, M., Marchini, G. and Stuart, F.M. 2020. Unravelling the magma feeding system of a young basaltic oceanic volcano. Lithos 352–253. http://doi.org/10.1016/j.lithos.2019.105325.

5 Coping with disasters in the Azores

From the middle of the 1930s the American scholar G.F. White[1] began to study the impact and mitigation of flooding in the USA (White 1936, 1945). Subsequently and with the assistance of colleagues, research was extended to embrace the totality of environmental extremes that affected people globally (White 1974; Burton et al. 1978). So strong was the influence of this body of research on policies adopted by the United Nations and national governments, that it became known as the *dominant paradigm*.[2] The *dominant paradigm* accepts that factors of wealth, systems of belief, experience of previous hazardous events and psychological factors profoundly influence responses to disasters, but their effects may be mitigated by so-called *adjustments* (i.e. measures that societies may adopt, in order to cope with environmental extremes) (Macdonald et al. 2012, p. 129; see Alexander 2000, pp. 23–53 and Chester 2005, pp. 414–421, for further details). It is argued that the physical magnitude of an event is the principal factor causing losses and that the higher the level of economic development the wider the range of *adjustments* that are available and from which a society may choose. *Adjustments* comprise either policy options (e.g. land use zoning), or engineering and technologically based mitigation measures (e.g. flood forecasting).

In the 1980s, Kenneth Hewitt a one-time associate of White edited a volume, *Interpretations of Calamity* (Hewitt 1983), which marked the start of a period of intense criticism of the *dominant paradigm*. It was argued, especially with respect to long-onset disasters such as droughts in economically less developed countries, that pre-existing factors, including the marginalisation of sections of a society and widespread poverty, were more important causes of death and economic dislocation than episodes of low or non-existent rainfall. Although disasters may be triggered by environmental extremes, it is human beings who make them 'unnatural hazards' (Kelman 2018, p. 254).

Evolution in approach from a physically deterministic brief to one centred on human vulnerability and resilience first emerged during the United Nations' *International Decade for Natural Disaster Reduction* (IDNDR) during the 1990s and became more prominent during the subsequent *International Strategy for Disaster Reduction* (ISDR), which has guided disaster reduction research and policy since 2001.[3] Today research is focused on the uniqueness

DOI: 10.4324/9780429028007-5

of place, not only in terms of the geophysical threats they may face, but also upon the ways in which environmental extremes may interact with those demographic, economic and cultural features that make societies what they are (Chester et al. 2019a). Disasters represent the effects of extreme events on vulnerable people and their activities, vulnerability being those features of societies that make them susceptible to harm.[4] The extent and duration of a disaster depends on the balance between vulnerability and resilience, the latter being those characteristics which enhance the ability of a society to cope. Resilience also reflects the capacity 'of the social system to re-organize, change and learn in response to a threat' (Wisner et al. 2004; Cutter et al. 2008, p. 599), with disasters acting as 'social laboratories' revealing the manner in which people evolve mechanisms of survival (García-Acosta 2002, p. 65). Several scholars have devised typologies to examine human *vulnerability* (e.g. Alexander 1997; Zaman 1999; Degg & Homan 2005; Cutter et al. 2008) and one tailor-made to the situation in the Azores has been proposed by the authors (Wallenstein et al. 2015). This is reproduced as Table 5.1 and will be used in this chapter. Over the past three decades greater emphasis has been placed on the fact that disasters bring into sharp focus deep-seated economic, social and cultural issues that have long remained dormant within societies, and these will also be examined in the context of the Azores.

One feature of the *dominant paradigm* of continuing relevance is Gilbert White's distinction between *pre-industrial* (or *folk*) and *industrial* (or *modern technological*) societies, and the characteristic ways in which they respond to disasters through boosting resilience (White 1974, p. 5). The list of typical responses found within *pre-industrial* and *industrial* societies have been expanded by later researchers (Chester et al. 2012a) and are summarised in Table 5.2. Instances where responses to earthquakes and volcanoes have been successful are usually ones where policies of hazard reduction are in place before disaster strikes. Such policies are termed *proactive*. In contrast in *pre-industrial* societies and sometimes in *industrial* ones too, responses are hurriedly put together

Table 5.1 Typology of human vulnerability to volcanic, volcano-related and seismic events on São Miguel

Type of vulnerability	Characteristics in the Azores
Physical and demographic	Building quality and population: numbers; distribution and location of settlements
Economic	Economic characteristics, dependent socio/economic cohorts within the population and occupational structure
Social and cultural	The social structure, spiritual and cultural *milieu* of the people at risk
Perceptual and informational	Accurate and inaccurate perceptions of risk. The lack of accurate information

Based and modified on the basis of information in: Alexander (1997); Zaman (1999); Degg and Homan (2005); Cutter et al. (2008) and Wallenstein et al. (2015).

Table 5.2 Responses to disaster in societies at different levels of development

Type of Society	Typical hazard responses according to Gilbert White and later writers
Pre-industrial (or folk)	a A wide range of initiatives by individuals or small groups
	b The State is only peripherally involved
	c Harmonisation with nature, rather than people seeking to use technology to control or modify natural processes
	d Low cost
	e Variation in responses over short distances
	f Flexibility, so that attempts at mitigation may abandoned if they prove to be unsuccessful
	g Losses are explained as being caused by a vengeful God, who is assumed to control human fate
	h The slow accumulation of knowledge from previous events of a similar magnitude and type
	i The effectiveness of local leadership may be critical in determining the effectiveness of responses
	j There is an absence of pre-disaster hazard reduction plans and policies are reactive
Industrial (or modern technological)	a A more restricted range of adjustment
	b Actions require co-ordination by government
	c A bias towards technological control over nature
	d High costs that have to be met by the State
	e More uniformity and inflexibility in responses
	f A belief that losses can be reduced by government action, science technology and planning
	g The notion of the Act of God is replaced by attitudes that place responsibility on human action or inaction
	h Responses are is still predominantly reactive, although there may be elements of proactive planning.

Based on information in: White (1974, p. 5) and later writers see Chester et al. (2012b).

during an emergency and in its immediate aftermath and are described as being *reactive* (Chester 2002). As will be clear in the discussion that follows and in Chapter 6, until recently the 'reactive model' has held sway in the Azores.

In the 1970s, White (1974, p. 5) proposed an 'ideal' future situation, in which there would be a fusing of the *resilience* mechanisms found within *pre-industrial* and *industrial* societies, so as to maximise the benefits of localised and state-based approaches.

Pre-industrial responses

Since the first settlement of the Azores in the fifteenth century, most responses to earthquakes, eruptions and related phenomena have been *pre-industrial*. Indeed, it is only from the time of the 1957/58 eruption and earthquake on Faial (Coutinho et al. 2010), that the State became a major player in responding to emergencies and its role has become progressively more important with each successive disaster. Although the *pre-industrial* era ended in the

1950s characteristic elements of it are still to be found during and following later events. In the aftermath of the 1998 earthquake on Faial, for instance, voluntary un-planned emigration was still taking place to adjacent islands and even to the Portuguese mainland (Oliveira et al. 2012).

As discussed in Chapter 1, socio-economic conditions in the Azores improved greatly from the fifteenth century to the mid-1950s, but throughout the *pre-industrial* era the archipelago lagged behind Western Europe and mainland Portugal by a wide margin. Late *pre-industrial* times in the Azores may be assessed by examining the situation leading up to the 1957/58 emergency, when all four elements of vulnerability listed in Table 5.1 were not just evident but long-standing. Some features of vulnerability were mitigated by deep-seated facets within Azorean society that boosted resilience in an unplanned or accidental fashion (e.g. the coastal orientation of settlement, extended family networks and spirituality), whilst others only emerged in times of crisis (e.g. the quality of local leadership and the limited role of the state).

Physical and demographic vulnerability

In terms of physical vulnerability Faial is isolated in the middle of the Atlantic Ocean and in the 1950s had seaborne connections to mainland Portugal and to the other islands in the archipelago. In addition and from the time of the Second World War, the United States Air Force base at Lajes (originally Lagens) on Terceira and the international airport on Santa Maria had improved communications. Faial also had first-class telecommunications, being an important node on the international telegraph network.[5] Faial in common with some of the other islands in the Azores is physically vulnerable because of its small size (173 km^2) which would provide limited scope for intra-island evacuation in the event of an emergency. This is an aspect of vulnerability that Faial shares with Graciosa (61 km^2) and São Jorge (244 km^2), although there is more scope for internal migration and re-settlement of evacuees on Pico (445 km^2), Terceira (400 km^2) and São Miguel (745 km^2). A further aspect of physical vulnerability concerns buildings which were and are highly susceptibility to failure and damage in a major tectonic earthquake or, indeed, more localised seismic activity associated with eruptions. As discussed in Chapter 3 and in spite of a history stretching back many centuries, traditional rubble-stone single-story rural domestic dwellings and more substantial buildings in towns and villages of a similar construction, perform badly in earthquakes (see Chapter 3). Even as late as the mid-1950s few improvements had been made, with the housing stock showing similar levels of vulnerability as it had in previous centuries. In a 1955 symposium to mark the 200th anniversary of the Lisbon earthquake (Ordem dos Engenheiros 1955; Azevedo et al. 2009 and see Chapter 6), it was evident that Portugal's buildings were highly exposed to seismic-induced losses and a new building code was enacted three years later. This was too late, however, to mitigate losses during the 1958 earthquake.

Through most of the *pre-industrial* era communications within the islands were poor and road improvement only dates from the middle of the nineteenth century (Costa 2008). Even as late as the 1940s, there was only one major metalled road on Faial (Admiralty 1945) which ran around the island and on the eve of the 1957/58 emergency many settlements still relied on un-metalled roads and tracks. In the absence of good roads and as a consequence of the coastal orientation of most settlements, communication by boat was important and served to prevent the total isolation of communities on more than one occasion during emergencies (see below) and may be viewed as a further feature of unplanned resilience.

As well as being expressed physically, vulnerability was also demographic. High rates of natural increase and emigration have been themes throughout Azorean history and the 1950s were no exception. The 1951 census recorded a population for the Azores of over 318,000 and a density of $c.134$ persons per km^2, the latter being significantly higher in some islands, with a density of $c.213$ persons per km^2 being recorded on São Miguel (SREA 2011). Although emigration to Brazil may be traced back to the seventeenth century, mass migration dates from the middle of the nineteenth century and acted as a 'safety-valve' whereby population pressure was relieved by emigration first to Brazil and later to the USA (Ávila & Mendonça 2008b). For instance, in 1864 Faial's population was at an all-time high of 25,839 falling to 18,917 in 1920 (SREA 2007), whereupon it increased because of immigration restrictions imposed in the aftermath of the First World War by the Brazilian and American governments. By the 1950s it was estimated that there were some 250,000 people of Azorean descent living in the USA,[6] whereas at the same time the population of the Azores was only just over 300,000 (Callender & Henshall 1968, p.19). Whether this 'safety value' was open or closed had an influence on the degree of distress suffered by people following a disaster, whilst aid from diasporic communities boosted resilience in several emergencies in *pre-industrial* and *industrial* times.[7] Although some limited emigration to Canada occurred after 1953 and small numbers still joined their compatriots in the USA, the ending of significant out-migration resulted in an excess of labour and widespread rural poverty across the Azores and this blighted the economy of Faial in the years leading up to the 1957/8 emergency (Ávila & Mendonça 2008c). At the time

> agriculture employed two thirds of the employed population and was characterised by many semi-subsistence farmers who often owned as few as 4–5 head of cattle and frequently cultivated fields of less than 1 hectare in area. In most years there were only $c.90$ days of farm work and, hence, unemployment and mass underemployment were endemic on the island.

as they were, albeit with less severity across the archipelago (Callender & Henshall 1968; Coutinho et al. 2010, pp. 270–271).

It is not just the number of people that make a place vulnerable, but their distribution. In the Azores there is both a strong spatial patterning of risk and of population and one factor influencing vulnerability is the correlation between these distributions. It is because of a combination of the history of settlement, the availability of flat land for cultivation and economic activities (e.g. shipping, overseas commerce, fishing and whaling) and the often-steep character and high-altitudes of island interiors, that the vast majority of people in the Azores both in the past and today live at or near to the coast. For instance, on São Miguel, Sete Cidades, Santa Bárbara, Furnas and Nossa Senhora dos Remédios are the only inland settlements of importance and there is a high concentration of population around Ponta Delgada, the island's capital and principal settlement. In 1950 this city contained 14% of the island's inhabitants, located in the four *freguesias* (i.e. parishes) which comprise the capital in official statistics, with this figures rising to *c.*44% for Ponta Delgada *concelho* as a whole (Wallenstein et al. 2015; SREA 2020a). These features of coastal orientation and urban primacy[8] are repeated across the archipelago, though a coastal focus is less well marked – though still present – on the less vertiginous islands of Santa Maria and Terceira.

With some exceptions such as Ribeira Grande and Vila Franca do Campo, on São Miguel the overall distribution of settlement is fortuitously well adjusted to volcanic risk because most the land that would be affected by the three active volcanic *areas* is rural and many *freguesias* show low population densities. In many volcanic regions of the world low population densities serve to hamper successful evacuation because it may be difficult to locate people, but this is not the case in the Azores because people have been historically concentrated within the principal settlement (*povoação sede de freguesia*) of each parish, with dispersed rural settlement being uncommon. In contrast on Faial, the island's capital is highly exposed to seismic risk (Chapter 3) and in 1950 Horta contained *c.*36% of the island's population and nearly 60% if adjacent *freguesias* are added (SREA 2020a).[9]

Economic vulnerability

On the eve of the emergency of 1957/8, the economic situation of the people of Faial and the Azores was insecure, with precariousness being the lot of Azoreans for most of their history (see Chapter 1). At the time the economy of Portugal was comparable to that found in other semi-industrialised countries of southern Europe and Latin America with: per capita incomes below $300 (US); a predominance of an unskilled, often illiterate workforce; low productivity; a high proportion of the labour force employed in agriculture and other primary activities and technological backwardness (Baklanoff 1992). The economic policies of Salazar and the *Estado Novo* (i.e. New State – see Chapter 1) were focused around a balanced budget, which meant limited investment and rates of growth in *Gross Domestic Product* (GDP) that were low in comparison with those achieved in other Western European countries, averaging only *c.*1% per annum between 1934 and 1947 and just over 2% between 1948 and 1958 (Lains 2003). By the start of the 1950s the government was alarmed, as living standards in Portugal were not rising on a per-capita basis because of a swelling population, a situation that was even more acute in the Azores (Anon 1956 and see below). The *Plano de Fomento 1953–1958* (i.e. Develop-ment Plan) (Feraz 2020) was a programme that involved the investment *c.*17% of GDP each year (at 1950 values) with a focus on infrastructure (Neves 1994) and, although this raised growth rates in GDP to over 4%, by the time of the 1957/8 emergency little of this wealth had trickled down to the Azores which remained a poor and peripheral region (Lains 2003 and see Chapter 1).

In 1957, Faial in common with the rest of the Azores was dominated by the production of primary products both for subsistence and export. In earlier cen-turies victualing ships was important, but with the advent of steam propulsion in the nineteenth century (Costa 2008b) the agricultural economy of the islands had to change to become one strongly focused on fodder crops and exports of live cattle, tinned butter and cheese which became the staples of agricultural production. Cereals – especially maize and to a lesser extent wheat – were grown for local commercial consumption and for subsistence. Whaling and fishing were also important industries (Callender & Henshall 1968).

Social and cultural vulnerability

In Chapter 1 the cultural *milieu* of the Azores, including the spirituality of the Azoreans, was discussed and from this several points relevant to disaster vulnerability may be highlighted. First, throughout much of the history of the islands administration has been directed by members of three social groups: the *nobreza* (i.e. nobles); and more locally by people who were either *proprietários* (i.e. owners of property and land, shop keepers and small-scale entrepreneurs) or members of the clergy. In the nineteenth century, leaders began to be recruited from what has been termed 'established educated' backgrounds, individuals who frequently filled positions as: teachers; medi-cal practitioners and politicians within *concelhos* and *freguesias* (Chapin 1989). As will be discussed below, in the *pre-industrial* era the quality, or otherwise, of local leadership was often an important factor in shaping responses to eruptions and earthquakes and the degree to which they were successful.

Second, although the islands of the archipelago vary economically, de-mographically and in their degree of historic isolation which is reflected in the development of distinctive dialects, what has been termed *Açorianidade* (Azoreanness),[10] is a key feature of Azorean identity. It is based on a shared adherence to a singular expression of Catholic spirituality and social teaching (see below), and on community solidarity (Almeida 1980; Silva 2008). Espe-cially in times of emergency *Açorianidade* may boost resilience through:

a   Extended families networks, which include members of the Azorean di-aspora, and who have often provided support – moral and material – in times of need.

b A worldview that is notable for its stoic pessimism. Comfort is taken from the realisation that in the past families and communities survived situations of extreme hardship brought about, not just by poverty, isolation and separation over vast distances because of emigration, but by emergencies occasioned by environmental extremes. As discussed in Chapter 1 this worldview embraces *saudade*,[11] an attitude of mind deep-seated within Azorean psyche, which implies a nostalgic longing for supposed better times in the past when families were not separated by great distances (Teletin & Manole 2015; Anon 2016a).

c Localism is a well-established trait. This is expressed across the islands, not just in a distinctive Azorean culture, but by differences within the archipelago where there is strong atavistic identification with one's island of origin and/or residence.

A third element of Azorean culture concerns religion, and several aspects of the spiritual have continued to shape hazard responses from *pre-industrial* times through to the present day. Table 5.3 which summarises the discussion of religion in Chapter 1 is self-explanatory, but it needs to be stressed that the distinctive character of Azorean Catholicism with its emphasis on pastoral care, a highly nuanced theodicy and the principles espoused by the *Culto do Espírito Santo* (i.e. Cult of the Holy Spirit) have been significant factors in boosting resilience, especially in the *pre-industrial* era. In addition, for many people in the Azores there is a strong sense of belonging to a popular Catholic culture in turns of care and praxis, without attending mass on a regular basis, accepting official dogma or, indeed, the fundaments of Christian orthodoxy (Davie 1990; Chester et al. 2019b).

Perceptual and informational vulnerability

In his important treatise on disaster theory David Etkin (2016, p. 66), begins his discussion of risk perception with the observation that 'people do not respond directly to the risk they are exposed to; rather they respond to their perceptions of those risks.' Indeed, perceptual vulnerability and informational vulnerability have been identified as being important from the earliest days of disaster research and this recognition continues to the present day. For instance, White (1974, p. 5) argues that the choice of adjustment is a function of an individual's perception: of risk; choices open to them; their command of technology; relative economic efficiency of the alternatives and links to other people.

The established methodology used to evaluate the ways in which people perceive environmental risk is based on social surveys usually using questionnaires and/or interviews. In the Azores surveys have only been undertaken during the last few decades and these are discussed in Chapter 6, but for historic earthquakes and eruptions recourse has to be made to archival sources. These documents and published accounts have to be used with caution because outside observers, in particular, may misunderstand local culture

Table 5.3 Azores: Aspects of religion and spirituality

Features of religion Expression	Details
Similarities with Southern European Catholicism	Over 90% of Azoreans have a confessional allegiance to Roman Catholicism. In societies with historically high levels of illiteracy and in common with other areas of southern European Catholicism, faith is expressed in lavishly decorated church interiors, highly choreographed patronal festivals and processions to mark important points in the liturgical year, e.g. *Festas do Senhor Santo Cristo dos Milagres* (Festival of the Christ of Miracles) held in Ponta Delgada (São Miguel) on the fifth Sunday after Easter. In common with southern Italy, processions and *romarias* (religious pilgrimages) are often linked to sites that experienced historical earthquakes and eruptions.
Unique features of Azorean Catholicism	a **Christianisation** The Azores was evangelised by missionaries from mainland Portugal most of whom belonged to religious orders, particularly the Franciscans and Jesuits. From the beginning it was not only a highly 'learned' priesthood, but also one that emphasised pastoral care. At the turn of the nineteenth century, there was an estimated one priest for every 156 people in the Azores so there was no shortage of potential clerical leaders. From the time of settlement, clergy concentrated not just on the cure of souls, but on praxis through emphasising empathy with the people regardless of background. The accent was on self-sacrificial ministry to a parish, monastery, school or hospital. This ministry of service continued after the Jesuits were banned in 1760. In 1834 all convents, monasteries, colleges and hospices were dissolved and their property was confiscated (Robinson 1977).
	b **Theodicy** (i.e. theological explanations of suffering). The most prominent model in societies with a southern European Catholic ethos was that disasters represented an expression of divine wrath for sins committed by individuals and communities. In southern Italy and some other areas, this is still encountered today (Chester et al. 2008). This model of suffering was (and is) not the only one encountered in the Azores. From the sixteenth century, educated priests such as Gaspar Frutuoso, many Jesuits (Udías 2009) and some lay inhabitants began to share the Aristotelian view that disasters were purely natural phenomena (Pinto 2003). A theodicy that became increasingly attractive following the 1755 Lisbon earthquake, involved a reworking by Leibniz of a model first suggested by St. Irenaeus (ca.115–190). Known as a 'the best of all possible worlds' model, this holds that the earth represents the best that could be created by means of the physical processes that operate within, and laws which control, the universe.

Culto do Divino *Espírito Santo* (Cult of the Holy Spirit)	This theologically heterodox cult was introduced into the archipelago by Portuguese Franciscans and Flemish settlers. The cult comprises: egalitarian 'brotherhoods' of the Divine Holy Spirit (*Irmandades do Divino Espírito Santo*), which are notable for the equality of status of their members, who include men and women, young and old; and an emphasis on hope (*esperança*), with the Holy Spirit being the font of knowledge and order. The Holy Spirit is omniscient and omnipresent and promotes values of solidarity, charity and fortitude. Although growing out of Catholicism and involving some church attendance, because of its heterodox character the cult has always been led by lay people and each brotherhood has its own small building that looks like a chapel and is known as an *Império* (i.e. empire or theatre). The leader (*mordomo*) of each brotherhood is chosen by lot and a crown is placed on the head of an adult or small child to represent equality of status. Members are expected to live their lives according to principles of concern, solidarity with neighbours and charitable giving (Anon 2020c).

Based on information in Chapter 1 and the references cited in the table.

---

and make prejudiced judgements about how local people responded during emergencies. The British Royal Naval Captain Edward Boid's strongly anti-Catholic rant when writing of Faial, for example, sums up local culture as comprising an ignorant populace led by an equally ignorant, repressive clergy (Boid 1835). Slightly earlier the American Consul General, John Bass Dabney,[12] in his account of the 1808 eruption on São Jorge, reports that residents were 'perfectly panick (*sic*) struck wholly given up to religious ceremonies and devotion' and that 'consternation of and anxiety were for some days so great among the people, that even their domestic concerns were abandoned; and, amidst plenty, they were in danger of starving' (Dabney 1808, pp. 46 & 48). In contrast and by reviewing the totality of historic documents and published accounts, the picture of clerical leadership may be more subtly drawn and, although there were some failures, there were also many successes. There is also little evidence of panicked responses and Dabney's report proves to be the exception rather than the rule. Experience from many disasters around the world indicates that panic, which may be defined as 'irrational, groundless, or hysterical flight that is carried out with complete disregard for others' (der Heide 2004, p. 342), is largely missing from records of Azorean disaster responses. Records shows that people were often frightened – sometimes terrified – but panic is absent with most people acting rationally and with regard to the safety of family members and others within their communities.

Poverty, low educational achievement and high rates of illiteracy during the *pre-industrial* era and in fact well into the 1960s, should not be taken to imply that people had no understanding of geophysical processes and what extreme events would mean to them and their families. Although prior to 1957/8 the last major eruption in the Azores was in 1808, several submarine eruptions occurred, some islands had active fumaroles and throughout most of the history there was a living memory of damaging earthquakes and widespread experience of lower magnitude felt, though not destructive, ground shaking. In addition and from earliest times, many educated clergy had severe doubts about models of theodicy which invoked divine punishment (Table 5.3), while from the nineteenth century the diffusion of newspapers and their public reading did much to spread scientific knowledge about the natural world and to dispel ignorance of extreme events (Simões et al. 2012 and see Chapter 1).

Examples of pre-industrial responses

Records of *pre-industrial* responses to disasters are both incomplete and vary greatly in the information they convey. It might be expected that records become progressively more complete from the fifteenth century onwards, but this is only true to a degree and details of responses to some events that occurred in the late nineteenth and even the early decades of the twentieth centuries are sparse. The principal reason why records have survived from the earliest centuries of settlement is because of interest in the natural world and diligence shown by the priest and chronicler: Gaspar Frutuoso (1522–1591, Figure 5.1a, see Chapters 1 and 3). Additionally present-day authors are greatly indebted to the nineteenth century literary figure Ernesto do Canto (1831–1900; Figure 5.1b) who in his periodical, *Archivo dos Açores*, brought together and edited many published and unpublished accounts of eruptions and earthquakes, and the ways in which people responded to them. In the account that follows the features listed in Table 5.2 and which characterise *pre-industrial* responses will be discussed in turn.

Individual, family and small group initiatives

Unplanned evacuation by families and larger groups during and immediately following earthquakes and eruptions has been documented on many occasions. Following the 1522 earthquake on São Miguel, for instance, many inhabitants who survived the destruction left the ruined town of Vila Franca do Campo and relocated to Ponta Delgada (Forjaz 2008). Similar spontaneous evacuations are recorded on São Miguel in 1591 (Halikowski-Smith 2010), from Vila da Praia da Vitória (now named Praia da Vitória) on Terceira in 1841 (Anon 1841a, 1841b) and on Faial and Pico in 1926 (Anon 1926b). In some cases, people are recorded as camping in town squares as well as living with members of extended families. Evacuation during and following

Figure 5.1 Pioneers of Azorean volcanology.

a. Statue of Gaspar Frutuosa in Ribeira Grande (São Miguel) his place of birth c.1522. He was the son of a local landowner and trained for the priesthood at the University of Salamanca (Rodrigues 1991).
b. Photograph of Ernesto do Canto in 1884. Canto was a notable Azorean man of letters and was a son of one of the most influential Azorean family (Diãs 1933).
Authors' photographs. Originally published in: Chester, D., Duncan, A., Coutinho, R., Wallenstein, N. and Branca, S. 2017. Communicating information on eruptions and their impacts from the earliest times until the 1980s. *In:* Bird, D., Fearnley, C., Haynes, K., Jolly, G. and McGuire, B. (eds) *Observing the Volcano World; Volcano Crisis Communication.* Springer, Berlin, Fig. 3, pp. 431 & 432. They are published under a *Creative Commons Attribution 4.0 International License (http://creativecommons. org/licenses/by/4.0/).*

eruptions was often for longer periods of time because, in addition to the destruction and damage to buildings that occurs in earthquakes, large areas of productive land were effectively sterilised for many years. Nunes (1999) records that, following the long-duration eruption on Pico in 1562–1564, people not only relocated to other parts of the island but also to other areas in the Azores, particularly to São Jorge, Faial and Terceira. This evacuation involved a high proportion of the total population and may have impeded the recovery of Pico for several years. During the 1672–1673 eruption of Faial, seismic activity caused inhabitants from the villages of Praia do Norte and

Capelo spontaneously to abandon their homes and live in huts made of straw (Madeira 1998). Recovery was protracted because the villages had to be re-built in new locations (Figure 3.11): Capelo to the west of its former site and Praia do Norte beyond the NE margin of the flow (Nunes 1999). Ribeira Brava was never re-established. The use of temporary accommodation is also noted following the 1808 eruption on São Jorge when, at least initially, peo-ple were forced to live in tents and even in the open (Dabney 1808). Later in this emergency and following the eruption of 11 May, many people fled from Urzelina and found accommodation in the villages of Manadas, Calheta and Rosais (Canto 1884), for locations – see Figure 2.3 in Chapter 2).

Individual actions were sometimes marked by altruism and community solidarity. Many acts of selflessness were performed by members of the clergy (see below), but others were undertaken by lay people. According to Gaspar Frutuoso (Canto 1880), during the 1580 eruption on São Jorge 15 men banded together and returned by boat to collect their possessions from a farm build-ing located in vineyards that had been threatened. Madeira (1998) argues that the location was on the western margin of the Ponta da Queimada lava delta below the Ribeira do Almeida cones (Figure 3.16). Some of the salvagers stayed in the boat, whilst others went to the farm to collect their possessions. A large cloud, which the present authors have argued (Wallenstein et al. 2018) was a pyroclastic density current (PDC), engulfed the farmhouse (see account in Chapter 3). Some of the men startled by the shadow cast by the glowing cloud (i.e. *nuée ardente*)[13] tried to run back to the boat, but the 'the air of the cloud burned all of them, with skin falling from their bodies' (Canto 1880, p. 189 our translation). Those who remained in the house were killed and others were severely injured. Following this eruption limited self-evacuation was allowed by the authorities (Wallenstein et al. 2018). A further example of community solidarity is to be found following the 1672–1673 eruption on Faial when 70 families who were cut off had to be supplied by boat from neighbouring parishes (Canto 1881).

Harmonisation with the environment

In a society with limited capital and with some exceptions little help from the State and its officials, recovery from disasters was slow. For much of the *pre-industrial* era harmonisation involved using locally available materials and in-digenous building designs to produce highly vulnerable 'rubble-stone' houses similar to those that had proved to be deficient in previous earthquakes (see Chapters 3 and 4). It is a well-known maxim introduced into hazard studies by the eminent seismologist Professor Nicholas Ambraseys (1929–2012), that 'earthquakes don't kill people; buildings do' (Ambraseys 1972; Anon 2013a, p. 302), and it is notable that it was not just domestic buildings that were reconstructed on traditional lines but public buildings and churches as well, so passing down this aspect of physical vulnerability through the generations

(Arêde et al. 2012).[14] Indeed, harmonisation was not just environmental, but also aesthetic as new and repaired churches were fashioned so as to maintain the traditional appearance of Azorean religious buildings. In this respect the Azores is atypical with respect to the general situation in most *pre-industrial* societies (Table 5.2), because rather than learning from experience the same un-safe building practices continued, showing little change until the second half of the twentieth century.

In terms of land covered by volcanic products, re-colonisation was left to natural processes and, as in other volcanic regions, re-use of land occurred more rapidly on land blanketed by tephra-fall materials, than on terrain covered by PDC deposits. Lava takes the longest time to recover and observations in the field and through remote sensing media, show that land covered by historic lava may still be identified centuries after the eruption occurred on account of less dense vegetation, more limited ground cover and more restricted land uses when compared to adjacent areas which are not affected. Re-colonisation and re-use may be illustrated by examining the Furnas area (São Miguel) in the century and a half following the 1630 eruption. With the exception of the trachytic lava dome, this eruption produced tephra-fall and areally limited PDCs (Chapter 3). Furnas remained desolate until about 1640 when people began to return in an unplanned fashion, initially as foresters and herders from neighbouring parishes. They found the topography dramatically changed, the two small lakes had disappeared and settlers could not find the property or land they had formerly owned or rented (Diãs 1933, 1936). Most land was in fact owned by the Jesuit Order who after 1665 rented cultivation plots, and from this time onwards there was some permanent settlement, although land use intensity remained low. By 1706 there were only 74 inhabitants occupying just 22 dwellings. After the suppression of the Jesuit Order in 1760 land was auctioned and agriculture gradually became well established, but it was only towards the close of the eighteenth century that population numbers reached a level that allowed the area to be officially recognised as a separate *freguesia* (i.e. parish) (Diãs 1936; Martins 1990; Dibben & Chester 1999). The Scottish horticulturalist and botanist Francis Masson (1741–1805) in a letter of 1778 to the Botanical Gardener of his Majesty (i.e. King George III of the UK), describes the valley of Furnas as 'well cultivated, producing wheat, Indian corn, &c. The fields are planted round with a beautiful fort of poplars' (Masson & Banks 1778, p. 604).

Leadership

Following most eruptions and earthquakes and in keeping with the situation across *pre-industrial* societies, leadership was exercised at family, extended family and village level, with government officials playing a minimal role. Occasionally officials are remembered because of their deficiencies, with the most notorious instance being during and immediately following the 1808 eruption on São Jorge. This occurred in the *concelho* (county) of Velas, when the *Presidente da Câmara* (mayor) deserted his post leaving his subordinates to handle the crisis. This they did with commensurate skill with measures that included the distributing of grain to the destitute and which was provided by the Captain General of the Azores (Avellar 1902, pp. 440–441; Costa 2008a).[15]

There are other examples where officials had positive impacts on disaster outcomes. In 1522, the capital of São Miguel and the seat of successive *capitães dos donatários* (captains of the donataries) was destroyed in what was the greatest earthquake loss of life to have affected the Azores. Caused by a combination of direct seismic activity and the triggering of a destructive landslide (Chapter 3), this event killed the captain's eldest son and much of his family along with an estimated 5,000 inhabitants in Vila Franca do Campo and other settlements mainly along the south coast of São Miguel. The hereditary captain at the time, Rui Gonçalves da Câmara II (c 1490–1535), together with his wife and second son, were absent at the time residing on their estate in Lagoa. The captain's leadership was clearly effective in the aftermath of the disaster because he organised rescue work,[16] later established a new capital at Ponta Delgada which became a city in 1546 and arranged for Vila Franca to be re-built on the western edge of the landslide debris (Bento 1993; Gerrard et al. 2021). On the basis of archaeological investigations, a joint Portuguese and British research team have argued that rapid reconstruction is suggestive of a resilient community (Anon 2014; Forlin 2016), but against this has to be set the fact that the town did not regain anything like its former prosperity until the eighteenth century when it became a centre of orange production and later of pineapples (Anon 2015). Indeed the 'footprint' of the landslide debris remained as open land until urban development occurred after a gap of around 200 years (Gerrard et al. 2021).

Perhaps the best recorded example of impressive leadership is that of José Silvestre Ribeiro (1807–1891)[17] and his role in managing recovery following the 1841 earthquake on Terceira. This earthquake is known as the *Caída da Praia*[18] (the fall of Praia) and caused the collapse of 2,100 houses in the east and NE of the island, with the greatest damage being centred on Praia da Vitória[19] and to a lesser extent Agualva, Fonte do Bastardo, Lajes and São Sebastião (Ribeiro 1844). Details of this earthquake and recovery from it are reported in detail in *Archivo dos Açores* and have been both summarised and supplemented by Sousa Martins (2015). At the time, Ribeiro was a career civil servant and *Administrador-geral do Distrito de Angra do Heroísmo* (General Administrator of the District of Angra do Heroísmo), serving from 1839 to 1844. Amongst the initiatives with which he was involved included, *inter alia*: setting up local relief committees (*Comissões de Soccorros*), chaired by parish priests; ensuring that contractors re-built to a higher standard through exercising control over what was constructed; responding to the needs of the poor; seeking funds, not just from the Horta District, but from the central government, Ponta Delgada District, expatriate communities in Brazil and Europe and by means of charitable donations; and reducing the number thatched properties, which at the time carried the sigma of destitution (Sousa Martins 2015). Ribeiro was interested in both architectural style and urban planning, developing a distinctive style in which reconstructed buildings were symmetrical, uniform and proportional. Finally Ribeiro hoped, albeit in vain, that the lessons learnt from this disaster could be of used in subsequent emergencies and in 1844 he published a volume; *Collecção dos escriptos administrativos e litterarios do ill.mo e ex.mo snr. José Silvestre Ribeiro* (Collection of administrative and literary texts by illustrious and Most Excellent Mr José Silvestre Ribeiro) (Ribeiro 1844). It is fitting that Ribeiro is commemorated today by a prominent statue in Praia da Vitória.

Religion and spirituality

As discussed above and in Chapter 1, popular southern European Catholicism was the glue that held Azorean society together in the *pre-industrial* era, by explaining how life should be lived and how death should be bourn. There is much evidence that between settlement in the fifteenth century and the 1950s many responses to earthquakes and eruptions accepted a theodicy of divine retribution for sins supposedly committed and for which the only responses were confession, absolution, amendment of life and propitiation through liturgical actions especially procession during and after disasters. Several examples of many may be cited. First during the 1562–1564 eruption on Pico, Frutuoso describes people waving crosses in front of the advancing lava flow seeking to arrest its progress (Pinto 2003, p. 223) and, in the immediate aftermath of this event, the term *mistério* (i.e. mystery of God) was employed for the first time to describe new land; in this case a lava delta formed at the coast. Ever since this eruption *mistério* has been used to describe terrain created by lava (Forjaz 2008, pp. 28–29; Chester et al. 2019b, p. 44). A second example occurred during the 1563 eruption on São Miguel when the church at Ribeira Seca was surprisingly not destroyed by lava. Thus began an annual procession involving a group of horsemen (the *Cavalhadas de São Pedro*) riding round the church in commemoration of both safe deliverance and to ward off any recurrence (Figure 5.2). Gaspar Frutuoso documented responses to many early earthquakes and eruptions including *inter alia*: the 1522 and 1547 earthquakes (São Miguel); and eruptions in 1562–1564 (Pico), 1563 (São Miguel) and 1580 (São Jorge). He reports that the people generally explained these phenomena in "religious terms as punishment from God … (and)… make frequent mention of processions and other (ways) of showing regret for sins committed, all intended to placate… Divine wrath" (Pinto 2003, p. 223). In the 1761 eruption on Terceira processions were again a noteworthy feature of the response' (Chester et al. 2019b, p. 45).[20] In 1580 on São Jorge widespread civil disturbance was quelled when a Jesuit priest visited the area, giving both spiritual reassurance and promise of eternal life through the agency of the sacrament

Figure 5.2 The annual procession of horsemen (the *Cavalhadas de São Pedro*) leading a procession around the church at Ribeira Seca to commemorate its safe deliverance after it had been threatened by lava during the 1563 eruption. Authors' photograph.

(Canto 1880). Before the nineteenth century, there was also a belief across the Azores that churches were protected from the effects of eruptions but this proved to be illusory when, with the exception of its tower and belfry, the parish church of São Mateus – Urzelina (see Figure 3.17 in Chapter 3) was destroyed during the 1808 eruption on São Jorge (Madeira 1998; Scarth & Tanguy 2001, p. 152).

As discussed in Chapter 1 and summarised in Table 5.3, from the sixteenth century alternative models of theodicy were adopted by some clergy and educated lay people, the life of the locally born cleric, Gaspar Frutuoso (*c.*1522–1591), being particularly informative. Frutuoso was educated at the University of Salamanca and studied both theology and the natural philosophy and was later appointed to be priest in his home town of Ribeira Grande on São Miguel. He made frequent visits to other parts of the archipelago and produced the first major academic work by an Azorean writer, which is entitled *Saudades da Terra*.[21] This multi-volume work provides details of eruptions, earthquakes and responses to them from the time of Portuguese settlement of the islands up to Frutuoso's death in 1591, and it is today the principal source of information about early post-settlement earthquakes and eruptions (Pinto 2003). The present-day Jesuit and eminent geophysicist,

Agustín Udías (Udías 2009, see Table 5.3), has posited that naturalistic explanations of environmental extremes were embraced by many Catholic writers and theologians, particularly members of the Jesuit Order, centuries before the Eighteenth-Century Enlightenment. Intellectual questioning also followed the 1755 Lisbon earthquake (Kendrick 1956; Paice 2008), when both naturalistic explanations of disaster associated with the writings of Jean-Jacques Rousseau (1712–1778) and a 'best of all possible worlds' theodicy became more prominent (Dynes 2003, 2005; Bassnett 2006). Views broadly similar to Rousseau's were held by the Marquês de Pombal, under whose leadership a national recovery programme was organised and which included the reconstruction of Lisbon and the development of Portugal's first national earthquake building code (França 1977 and Chapter 6). Views of disasters as social constructs slowly diffused through Portuguese intellectual circles and secular elites greatly assisted by the anti-clerical character of liberal governments in the nineteenth century and during the time of the first Republic (1910–1926). Portuguese universities had also been outside religious jurisdiction since 1772 and Salazar's *Estado Novo* (see Chapter 1), although close to Catholicism and Catholic social teaching more particularly, maintained a clear distinction between church and State (Carneiro et al. 2013; Chester et al. 2019b). As shall be discussed below, by the start of the *industrial* era this view of disasters as purely natural phenomena was widespread.

Gaspar Frutuoso whilst proclaiming an 'official' *retributive* theodicy, 'clearly had severe objections to it. In his writing he embraces an Aristotelian understanding of natural systems and goes even further by carrying out albeit rudimentary experiments on lava and even showing an inchoate uniformitarian understanding that the present could be the key to the past. In the view of his biographer, Frutuoso "admitted he could be wrong and might need to be corrected, should the Holy Church... think so. This was a subtle, elegant way of disagreeing with the Biblical doctrine, into which his ideas about the volcanic origin of the islands did not fit, (in) the sense that they defied the account of the harmony of creation taught by some powerful churchman of the time'" (Pinto 2003, p. 225; Chester et al. 2019b, p. 46). Just how widespread were Frutuoso's opinions cannot be determined with any accuracy, but until it was suppressed by the Marquês de Pombal in 1759 the Jesuit Order practised a learned 'disaster' ministry in the Azores and the clergy from this order and more generally focused their efforts during emergencies on pastoral care and praxis rather than on dogmatic statements about judgement and Divine punishment. Examples are legion and include an incident during the 1580 eruption on São Jorge when the people greatly appreciated the arrival of a Jesuit to minister and comfort the people; and in 1808 when, as well as leading processions hopefully to appease God's supposed wrath most – but regrettably not all – clergy are remembered for remaining in their parishes and sharing the suffering of their parishioners. In addition, more practical help was forthcoming as they helped to salvage personal affects from threatened buildings and assisted in recovery in many ways not least by raising morale (Canto 1880, 1884).

The *Culto do Espírito Santo* (Table 5.3 and Chapter 1) was already well established when Gaspar Frutuoso was writing *Saudades da Terra* and he makes general reference to its charitable purposes in early eruptions and earthquakes (Anon 2020c). During the 1761 eruption on Terceira, an *Império* was specially constructed to emphasise the charitable values associated with the cult, acts of this kind being 'very common in the Azores during seismic and volcanic swarms' (Forjaz 2008, p. 37). On Faial during the eruption of 1672 a chapel, known as the *Império dos Nobres* (Império of the Nobles), was constructed in Horta to commemorate acts of charity by members of the cult (Macedo 1871; Canto 1881). As will be discussed later in this chapter, the charitable role of the *Culto do Espírito Santo* has continued to the present day.

Finally, before leaving this discussion of religion and spirituality it is important to stress that, even when disasters were viewed as being in God's hands, there is no evidence that such a belief provoked a debilitating fatalism amongst the people of the Azores and their religious leaders, which either inhibited responses or increased vulnerability by offering an 'alternative reality.' What is more it is increasingly apparent that this is a finding which transcends faith, tradition, time and culture and has emerged in many empirical studies of religious responses to disasters (Chester & Duncan 2007; Chester et al. 2012a). Holding together two apparently contradictory positions at the same time is known in the older literature on hazards as *cognitive dissonance* (Burton & Kates 1964, p. 433), although today the more usual term is 'parallel practice.'[22]

Limited State involvement

As noted in Table 5.3, one characteristic of *pre-industrial* responses to disasters is minimal involvement from central government. For much of the period under review European governments generally, and those in Portugal more particularly, did not view the provision of welfare as part of their remit. In Portugal 'social security in rural society was based on charity (and the) giving of alms to the poor was ingrained in Catholic society' (Birmingham 1993, p. 130), and in the Azores the church made itself responsible for hospitals and schools as well pastoral care within parishes (Chester et al. 2019b, Chapter 1). From the mid-nineteenth century public works programmes and very limited welfare provision under the *Estado Novo* (Birmingham 1993), did little to change this situation until after the Portuguese Revolution of 1974. Before the advent of steam ships and the electric telegraph, even if the central government wanted to help, there was little that could be done in the short-term because it took weeks for messages to be sent to Lisbon, responded to and replies sent back to the islands suggesting a particular course of action.

Records show that central government actions with respect to disasters were few in number and had limited impact. Sometimes initiatives were the result of successful petitioning by local officials. For instance, following the 1841 earthquake on Terceira, José Silvestre Ribeiro managed to obtain limited re-building funds from Lisbon (see above) and after the April 1852 earthquake on São Miguel the civil governor obtained funds from government to carry out a survey of damage. The resulting *Comissão da Câmara* (Council Commission) concluded that age and poor maintenance were both factors exacerbating building damage but, as far as the authors are aware, nothing was done to respond to the commission's findings (Silveira 2002).

Another example occurred following the 1672/3 eruption on Faial when it became apparent, because of the loss of productive land due to lava inundation, that the affected part of the island (Figure 3.11) could no longer support its population. The municipality of Horta concluded that the excess population could not be successfully absorbed within the island and successfully petitioned the government in Lisbon to facilitate the emigration of around over 400 people to Brazil (Canto 1879, 1881, 1883, 1888). The plan involved two phases of emigration of 100 'couples' to Maranhâo (actually 453 people), the first in 1676 (234 people) and the second in 1677 (219) (Canto 1879, pp. 371–372, 1883, pp. 265–266), this being the first documented example of an organised permanent relocation of population following a volcanic eruption or earthquake (Gaspar et al. 2015).[23]

Later after telegraphic communications were established, governors could inform Lisbon immediately and, following the 1926 earthquake that affected Faial and Pico, the government dispatched a naval vessel, the *Carvalho Araújo*, to assist in local relief efforts and no doubt to raise morale (Anon 1926a).

Industrial (or modern technological) responses

The era of *industrial or modern technological* responses to disasters has coincided with five significant earthquakes (one major) and one eruption. There have been earthquakes in: 1958 on Faial, also impacting Pico; 1964, effecting São Miguel; 1973, causing damage to northern settlements on Pico; 1980, producing deaths and extensive losses on Terceira, with effects also felt on São Jorge, Graciosa, Pico and Faial and the 1998 disaster on Faial, when more limited impacts were inflicted on Pico and São Jorge. In addition, the 1958 earthquake was associated with a major eruption which began in 1957 and in 1964 seismic activity coincided with submarine volcanic activity off the coast of São Jorge. With the exception of the 1973 earthquake, which is not considered further because it is not well documented,[24] information on how Azorean society coped with and recovered from these emergencies is far more extensive than for earlier events and, in addition to written records, there are many contemporary archival photographs and newsreel films. All events in the *industrial* era have also occurred within the living memories of Azoreans who can provide eyewitness testimonies.

As Table 5.2 shows, the principal distinguishing feature of the *industrial* era is the scale of government involvement, but actions were still predominantly *reactive* being prompted by a combination of: local officials contacting, at times lobbying central government; the globalisation of concern as knowledge of emergencies diffused rapidly, not only within the Azorean diaspora especially in that cohort residing in North America, but also to non-lusophone countries by means of the news media. Indeed, the growth and increasing speed of communications from the 1950s enabled the government in Lisbon to monitor the nature of a disaster impact and respond rapidly and much more effectively if it chose to do so. Internationalisation allowed the spotlight of public opinion to be shone more brightly on the action of Portuguese government. *Proactive* planning remained inchoate and did not emerge until after the start of the current millennium, but some progress did occur and by the time of the 1998 earthquake important initiatives had been introduced administratively by the Regional Government, and academically by the University of the Azores and other researchers both in Portugal and abroad.

The 1957/8 Capelinhos eruption and earthquake on Faial

Until well into the twenty-first century, there was a marked contrast between detailed scientific accounts that had been published on this eruption and earthquake and the comparative dearth of outputs that were concerned with how Faial was impacted by and recovered from the effects of these disasters. Under the *Estado Novo* regime (Chapter 1), scientific study was encouraged and was indeed financially supported (e.g. Castello Branco et al. 1959; Machado 1959; Zbyszewski & Veiga Ferreira 1959)[25] but, in contrast, there were many newspaper reports but few in-depth publications on impacts and responses. Critical studies of policy were not welcomed by Salazar's authoritarian government and its civil service who placed a premium on secrecy (Coutinho et al. 2010).

In common with other authoritarian regimes in Europe, Portugal had a highly effective bureaucracy and detailed records of government responses to the emergency are preserved in the archives and were interrogated by the present authors more than a decade ago (Coutinho et al. 2010). These include, *inter alia*: official reports (e.g. Junta-Geral Horta, 1958a, 1958b); a valuable published compilation of correspondence and telegrams between the civil governor in Horta and the central government in Lisbon (Araújo 2007) and a plethora of data on demographic, economic and social indicators that allow an assessment to be made of the quantitative impacts of the volcanic and seismic crises. These have been published by the *Serviço Regional de Estatística dos Açores* (Regional Statistics Service of the Azores). A comprehensive land use survey of Faial was undertaken by the Oxford Women's Expedition in 1960 (Callender & Henshall 1968). This comprises not just a 1: 25,000 scale map, but detailed information on the land affected by volcanic products and the extent of any recovery in cropping that had occurred in the two years which elapsed between the eruption and the land use mapping. The accompanying memoir provides a near contemporary account of the economy and society on Faial, together with nine interviews with some of those who were involved in the events of 1957/8. To mark the fiftieth anniversary of the Capelinhos emergency two important works were published: *Vucão dos Capelinhos Memórias 1957–2007* (Memories of Capelinhos Volcano 1957–2007) (Forjaz 2007b) and *Capelinhos, A Volcano of Synergies* (Goulart 2008). The former is a collection of works, some of which are obscure, that were published at the time with additional material about how the emergency was handled, whereas the latter concentrates on families within North American diaspora communities and includes first-person testimonies about the emergency. Finally, eyewitness recollections have been collected by the present authors and are referred to in the account which follows and in Table 5.4.

Table 5.4 The impacts of the 1957/8 Capelinhos eruption and 1958 earthquake on the people of Faial, and the responses of the authorities to the disaster

Events	Human impacts and responses
Phase 1 – *Surtseyan* activity. September 16 to November 7 1957	September 16–27. Many pre-eruption earthquakes caused concern, but produced no damage (Anon 1957a). Much evidence of prayer and more frequent church attendance was noted (Anon 1957b).
	September 27. Whalers and their families abandoned Porto do Comprido whaling station (Lobão 2008; see Fig. 3.19 Chapter 3). Before the eruption there over 100 whalers were employed and operated around 20 boats (Garcia 2005).
	October 1. Fisherman were forced to leave the area because of contamination of the water. The lighthouse was abandoned (Anon 1957c; Cunha 1959).
	On October 7. The Governor, Dr António de Freitas Pimentel, called a meeting of the leaders of the principal civil, political and military organisations on the island.
	October 6–7. There was a change of wind from SE to NE and most of the island suffered ash fall. The Governor mobilised medical aid and transport and was advised by Professor Orlando Ribeiro (see below) to evacuate the affected area. Governor Pimentel consulted the *Presidente da Câmara Municipal da Horta* (Dr. Sebastião Goulart, i.e. chairman of the county council).[1] Around 1,500 people were evacuated from the worst affected *freguesias,* by the *bomberios* (i.e. fire fighters) and the *Legião Portuguesa.*[2]
	Some people were evacuated to relatives' homes in other settlements on Faial (Garcia 2005). Rainfall, ash and high winds impeded the evacuation and people were transported by trucks, cars and ambulances in an *ad hoc* fashion. Ash fall caused roofs to collapse. Most people were accommodated in Praia do Norte (Fig. 5.4), food was supplied to the victims (Pimentel 1957a) and the authorities restricted access to the areas affected (Anon 1957d, 1957e).

(Continued)

Events	Human impacts and responses
	October 5–December 30. A technical mission was sent from the Ministry of Public Works in Lisbon to Faial, headed by Professor Orlando Ribeiro, who arrived in Horta aboard the warship *S.Tomé* (Campos et al. 1960; Lobão 2008). Its mission was focused on: geology and meteorology; financing the emergency and reconstruction; assisting in evacuation; authorisation and planning of public works: the removal of ash from roofs and roads and the re-planting of land affected by ash fall. It was proposed that that:
	a For a period of two to three years, the unemployed should be occupied in public works.
	b The agricultural service (*Direcção Geral dos Serviços Agrícolas*) should carry out research to understand how land covered by ash could be brought back into cultivation.
	c The *freguesia* of Capelo should have its land taxes remitted.
	d Research should be commissioned to investigate how agricultural on Faial may in the longer term provide greater security.
	e Aid from State should be provided to enable the repair of housing, roads and the lighthouse.
	f Loans had been advanced to allow livestock to be purchased and their terms and conditions should be eased.
	October 20. A specialised geological mission arrived on Faial led by Dr Castello Branco. During the eruption members of the Ministry of Education were employed to assist in the observation of the eruption (Forjaz 2007a). Committees were established to organise: transport; food sourcing and distribution; lodging and veterinary care/animal assistance (Lobão 2008).
	October 28. Between the start of the eruption and 28 October, 150 families had been evacuated and Governor Pimentel wrote to the Minister of the Interior stating that he was seeking to provide accommodation in unoccupied houses. Some 150 men and a few women had been found work and total evacuees now numbered *c*.1,700. Many farm animals (mostly cattle) had been removed from the areas affected. Men were paid *c*.70 cents (US) per day to clear roads (Scofield 1958) and woman *c*.30 cents to clear roofs (Fraião 2005).[4] Governor Pimentel interceded with the government for taxes to be suspended in *freguesia* do Capelo (Pimentel 1957b). Council workers, fire fighters and a small number of soldiers assist the evacuees and photographs show that a variety of vehicles were used (Rosa & Pereira 2008).
Phase 2 – *Surtseyan* activity. November until the middle of December 1957	December 16. Torrential rain damages buildings and wet ash causes roof failure and further evacuation is required. In a telegram to the Minister of the Interior, the Governor reports that the people of Capelo (Figure 5.4) are now threatened (Pimentel 1957c). In reply the Minister of the Interior formed a committee of officials in Lisbon, headed by Engineer Viriato Campos, to assist with evacuation and the committee left for the Azores by air to Santa Maria (Ministry of the Interior 1957).

December 17. Azorean and Portuguese expatriates in California donate *c.* $800 (US) to provide Christmas presents to the victims of the eruption. In his reply the Governor notes that many people have been found work, and that food, clothes and toys have been distributed (Anon 1957f). The clothing is reported to have been in a poor condition (Garcia 2005).

Phase 3 –
Surtseyan
activity.
December
29 1957
to May 12
1958

December 30. The technical mission leaves Faial for Lisbon. By this time the *freguesia* of Capelo has been reduced 'to ash and sand' and those who are too old and/or cannot work receive an allowance of food (Pimentel 1957d): the ration comprising two loaves of bread, *c.*1 litre of milk, some cheese and vegetables (Scofield 1958, p. 753).

Phase 4 –
Seismic
Crisis
Monday May
12 to Friday
October
24,1958

May 12–13. During the night an earthquake destroyed the villages of Praia do Norte, Ribeira do Cabo (near to Capelo) and partly destroyed the smaller settlements of Areeiro and, Cruzeiro (both near Capelo) and Espalhafatos (Figs. 3.19 & 5.4). In line with a recommendation of Eng. Frederico Machado, the Governor ordered an immediate evacuation. Later the Governor, together with Dr. Linhares de Andrade (*Presidente da Junta Geral,* i.e. chairman of the council) and Eng. Frederico Machado (*Director das Obras Públicas,* i.e. Director of Public Works), visited the caldera and found that gases were being emitted and feared a major eruption. The local press and wall posters were used to reassure the people (Anon 1958a, 1958b), and Governor Pimentel and Eng. Machado ordered the evacuation of the population of Capelo and Praia do Norte. Some 500 homes had been destroyed (Anon 1958c; Anon 2008).

May 15. The Minister of Public Works, Eng. Arantes e Oliveira, arrived in Horta and announced a comprehensive *Plano de Recuperação Económica e Reconstrução da Ilha do* Faial (Plan for Economic Recovery and Reconstruction of the Island of Faial); ordered that ships should be made available should rapid evacuation of the island be required and announced that the Under-Secretary of National Education (*Subsecretário Educação Nacional*), Sr. Rebelo de Sousa, would be provide grants.

May 26. Governor Pimentel and the *Junta Geral* discussed a plan to assist farming and the Governor encouraged families to emigrate to Angola (Anon 1958e).[3]

End of May. A US government representative discussed immigration into the USA. This proved to be much more popular than the Angolan option (Anon 1958f).

June 1. USA forwarded $30,000 US worth of food aid.[4] It was provided by a charity CARE (Cooperative (for) American Remittances to Everywhere) (Anon 1958d; 1958g).

August 18 Crops sown after the initial eruption were covered once more by ashes (Anon 1958h).

Phase 5 After
October
1958

A delegation from the Portuguese community in California visited Faial and donated $140, 364 US (Lobão 2008, pp. 52).

(*Continued*)

Events	Human impacts and responses
	40% of economically active population emigrated from Faial following the eruption (Anon 2001). Around 5,000 people left the District of Horta under the Azorean Refugee Acts of 1958 and 1960, most of whom lived on Faial. In 1965 a new immigration law abolished the quota system, and between 1966 and 1980 *c*.80,000 Azoreans emigrated to America, the majority going to the USA (Chapin 1989; Ávila & Mendonça 2008a).
	By 1960 people had started to return to the villages of Capelo and Norte Pequeno and began to replant their fields with maize. The plants rarely reached a height of 0.3 m, but provided some fodder (Callender & Henshall 1968, p. 18). Reconstruction of Praia do Norte began in 1960 (Lobão 2008), which was closed by the authorities between 1958 and 1961 (Neves 2008).

1 Faial was (and is) administered as a single *concelho* with 13 *freguesias*. This meant that during the eruption the two critical figures where the Civil Governor (Dr Pimentel) and the *Presidente da Câmara Municipal* (Dr. Goulart).
2 The *Legião Portuguesa* (Portuguese Legion) was a militia created by the Salazarist government in 1933 following prompting from the political right. It aimed to support the political control of the *Estado Novo* regime (Rodrigues 1996).
3 Official policy at the time encouraged emigration to the Portuguese colonies, especially Mozambique and Angola.
4 One dollar US in 1957 has a purchasing power of just over $9 in 2020. Hence, today $30,000 is equivalent to *c*. $270,000 and $140, 364 is equivalent to over $1.2 million. 70 cents was a small daily wage at the time and far below average incomes: 30 cents was miserly. Average annual incomes at the time were around $300 per annum (Baklanoff 1992, p. 4).

Based on Coutinho et al. (2010, Figure 3, p. 272).

The impact of the 1957/8 events on people and housing have been discussed in Chapter 3, but additionally much land was also devastated. The maximum area covered by volcanic products was estimated at 4,652 hectares (i.e. *c*.25% of the Faial), with about 14% being covered by a tephra blanket of more than 5 cm in thickness (Campos et al. 1960; Cunha 1962; Rosa & Pereira 2008). As Figure 5.3 shows, the effects of this loss of land were exacerbated because much of the island could not be cultivated or even grazed intensively because of altitudinal constraints.

Table 5.4 summarises the impacts of and responses to the Capelinhos eruption and earthquake and shows just how far the State was involved in comparison to its much more limited role in earlier emergencies. There are several features in the table that require emphasis and additional discussion. Some adjustments typical of *pre-industrial* times were still in evidence. First, prayer and liturgical actions remained essential to the manner by which many people conceptualised disasters and coped with the depredations they caused, but even for those who accepted a degree of divine responsibility, *cognitive dissonance* (see above) enabled people to work enthusiastically with the authorities

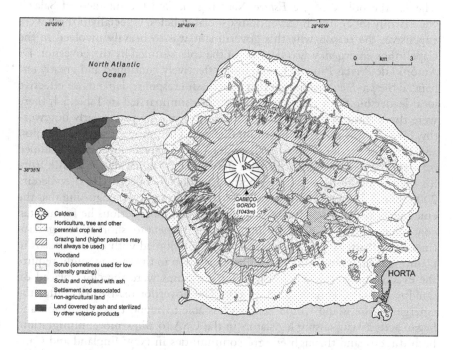

Figure 5.3 The land use of Faial in 1960. The area covered by volcanic products was formerly used by cattle rearing, and maize and vineyard production. From Coutinho et al. 2010, Fig. 4, p. 271 and based on Callender and Henshall 1968, endpaper and field work by the authors. Copyright Elsevier.

in recovery programmes (Garcia 2007; Coutinho et al. 2010). Theologically much of the population had moved away from the acceptance of a theodicy of divine punishment,[26] and the principal roles of both the 'official' church and the Cult of the Holy Spirit were pastoral care and community support. A second feature typical of the *pre-industrial* was unplanned evacuation of affected settlements early in the eruption as families abandoned the whaling station at Porto do Comprido and other villages, with some being recorded as lodging with members of extended families in other villages on the island (Table 5.4).

Notwithstanding these *pre-industrial* elements, responses to the events in 1957 and 1958 were dominated by government action. In reviewing the role of government there have been differing opinions expressed in the literature Beier and Kramer (2018, p.10). Ramos Villar (2006, p. 14) has argued that recession and the outbreak of insurrections in the African colonies reduced the help that could be provided by the State, with Oliveira (2008, p 101) going further and stating that 'the Portuguese government did not feel obliged to provide any help.' In contrast the present authors have posited that 'response, recovery and rehabilitation were generally highly successful' (Coutinho et al. 2010, p. 265). These apparent contradictions are more apparent than real for whereas the policy of the *Estado Novo* regime and the instincts of Salazar were firmly set against interventionist social policies, especially if these were expensive, the reason why this government was so heavily involved in the Capelinhos emergency was because of the role assumed by the governor, Dr António de Freitas Pimentel, who most effectively continued and greatly enhanced the *pre-industrial* notion that a successful response depends on effective local leadership.[27] In addition to the actions summarised in Table 5.4, there were three features of Pimentel's leadership that were particularly noteworthy. First, he was an excellent communicator. The governor stayed at his post and through the local press, wall posters and personal appearances calmed fears and showed his solidarity with the people by donating fresh milk from his own resources and by other gestures of social concern (Rosa & Pereira 2008, p. 62). Pimentel was, second, a man of foresight. Recognising that the emergency could develop into something much more serious he formulated contingency plans, successfully lobbying government for short and long-term finance and, after the earthquake in May 1958, joined with other officials in planning for the possibility that a major eruption in the caldera might require the evacuation of the whole island by sea. Ships were therefore placed on standby (Rosa & Pereira 2008, p. 60). The governor realised that after the emergency the island would be even less able to support its population and began activity canvassing legislators in the USA to allow more immigration, both directly and through émigré communities in New England and California.[28] Finally and trumping all others actions, he used subtle political skills to apply pressure on the Salazarist regime to make Capelinhos the first example in the Azores of a government-led *industrial*-type response to a disaster. Governor Pimentel had the advantage of a strong personal friendship with the Minister of Public Works, Eng. Eduardo Arantes e Oliveira (Table 5.4) and good working relationships with other ministers (Forjaz 2007c). Arantes e Oliveira in turn was able to influence additional ministers not least Salazar himself. The archives make it clear (Araújo 2007) that Governor Pimentel kept up a constant stream of telegraph traffic passing on details of the event and making a carefully argued case for more support. The most difficult task was to obtain the support of Salazar who had personally to approve all major items of new government expenditure. On one notable occasion and using considerable guile, Pimentel reminded the dictator that he was the only governor in the whole of Portugal who has expanded his territory as a result of new volcanic products being added to the western tip of Faial. Salazar, a normally quiet and reserved man, laughed at this joke and signed the papers authorising the additional expenditure (Forjaz 2007c).

A further contrast with *pre-industrial* times is that government involvement continued after the close of the eruption on 24 October 1958. As Table 5.4 shows many measures had already been put in place, including a comprehensive *Plan for Economic Recovery and Reconstruction of the Island of Faial* and policies for emigration to the USA. There were also additional education grants, but one feature not fully captured in Table 5.4 are the efforts that went into agricultural recovery. These included plans to conserve land covered by tephra until it could be returned to its owner; assistance in land preparation; grants for machinery, technical assistance, cropping and husbandry advice and the provision of seed (Campos et al. 1960). Right up to the end of his term of office in 1973 Governor Pimentel was able to supervise recovery and reconstruction and attracted further inward investment in the form of paved roads, plans for reforestation and agricultural co-operatives, improved hospitals services and an airport. So severe was the impact of the emergency that even well into the post-revolutionary era from 1974, the island had not fully recovered with the 1991 census showing a 62% population reduction on the figure for 1950 and much land in the west still in a state of dereliction. Counterfactual history can easily degenerate into baseless speculation, but it could be argued that economic decline on Faial would have been even more severe if it were not for the disaster concentrating the attention of the national government, communities within the diaspora and foreign governments particularly that of the USA.

1964 São Jorge (Rosais) earthquake

One of the reasons why the 1957/58 emergency was well handled by the authorities was because events unfolded slowly allowing the Governor Pimentel and the government in Lisbon to put mitigation measures in place before distress became acute. Before the May 1958, earthquake attention was already focused on the island because of preceding volcanic activity and, when the earthquake occurred, contingency plans could be devised for a whole island evacuation should this have proved to be necessary. The situation in 1964 and later in 1980 and 1998, could not have been more different, because earthquakes are short onset events. Their impacts and responses to them reveal the weakness of a *reactive* planning regime especially when the disaster is located far removed from the seat of central government, a situation that is exacerbated in Azores because of the isolation of the islands in the middle of the Atlantic Ocean.

The February 1964 earthquake (M 4.8 – maximum Intensity EMS VIII – Gaspar et al. 2015), was associated with offshore volcanic activity (see Chapter 3) and was very destructive. Reflecting the vulnerability of Azorean housing to even a moderate magnitude event, a subsequent survey by the Ministry of Public Works showed that in excess of 900 homes had been damaged, around 400 destroyed and over 250 required extensive repairs. More than 5,000 people were made homeless (Machado and Forjaz 1966, pp. 25–26). On 15 February, there were 179 felt earthquakes and 125 on 16 February with epicentres being located near to Urzelina and Manadas on the SW coast (see Figure 2.3). Ground shaking initiated the spontaneous evacuation of people to Velas, who were so concerned to leave the affected area that they abandoned many personal possessions and livestock. Thereafter the epicentral location migrated to the NW causing extensive destruction in the town of

Velas and village of Rosais and an estimated 5,000 homeless people then made their way on foot to the port of Calheta, a distance of *c.*24 km from Velas (Figure 2.3). The earthquakes occurred in winter and newsreel films show a distressed, apprehensive and poorly clothed population waiting in Calheta for seaborne evacuation. As in 1957/58 the Civil Governor, in this case Teotónio Machado Pires (1902–1993), was the critical figure in organising the response of the national government. Based in Terceira, but with responsibility for São Jorge and Graciosa, the governor arranged for shipping in the vicinity of the Azores to approach the affected island and remove evacuees to Angra do Heroísmo (Terceira).[29] At the same time a battalion of troops was moved in the reverse direction from Terceira to São Jorge and brought tents, medicines and food. Numbers differ (Anon 1964a; Machado and Forjaz 1966), but before the evacuation was called off on 20 February in excess of 1,000 people reached Angra do Heroísmo and Praia da Vitória, and were accommodated both in barracks and with local families. One evacuee, António Loureiro a road maintenance official, spoke for many when he told his rescuers that he had had nothing to eat for three days (Anon 1964b). Indeed, both the evacuation and caring for those who remained on São Jorge (*c.*5,000), were severely hampered by poor weather which culminated in a severe storm on 15 February at the height of the rescue operation. Major assistance was also offered and accepted from the United States Air Force (USAF) base at Lajes on Terceira, aircraft being grounded for much of the time because of bad weather. The acting commander, Colonel Hugh E. Wilde, obtained information on available shipping in the North Atlantic from the New York Coast Guard and used small craft to ferry evacuees from larger ships into the harbour at Angra do Heroismo (Anon 1964a, 1964c, 1964d).

Following the earthquake immediate relief was provided by both the Portuguese authorities and by the USAF. The Civil Governor visited the island to co-ordinate relief supervising the materials brought to the island by Portuguese troops and received tents, food and the first temporary buildings[30] donated by the USAF (Anderson 1964). Later legal decrees enacted in Lisbon put the long-term relief effort on a more secure footing (Portuguese Government 1964a, 1964b). A total of 21.25 million Escudos[31] was released to be spent on: urban reconstruction; public works (e.g. repair of roads, restoration of water supply and other services and the erection of tents); economic regeneration, including agriculture and various forms of assistance (e.g. food supply, medicines, domestic equipment, clothes and transport to places of work). Despite this programme, which in the context of the fiscally austere standards of the *Estado Novo* regime, was generous but emigration remained a long-term consequence of the emergency. The government encouraged emigration to the African colonies, especially to Angola, but the majority of people eventually ended up in North America (Williams & Fonseca 1999), and by 1970 the population of the island had fallen rapidly to just under 13,000 from a figure of *c.*6,000 in 1960 (SREA 2020b).

1980 Terceira earthquake

The 1 January 1980 earthquake had a devastating impact which was concentrated in the north western part of Terceira and around Angra do Heroísmo, the principal settlement of Terceira. Other towns and villages on Terceira were also affected together with areas of São Jorge, Graciosa, Pico and Faial (see Chapter 3 & Table 3.4). The lethal combination of a Magnitude 7.2 earthquake, the 'soil-effect' (see below and Chapter 3) and vulnerable buildings which are to be found across the Azores produced: a death toll of 59; caused injuries to more than 100 people; either destroyed or severely damaged 15,000 buildings and displaced thousands of inhabitants (Gaspar et al. 2015, p. 36). With an initial estimated cost of c.100 million US dollars, equivalent to around 102% of the 1979 budget of the Regional Government (Pell 1980),[32] the earthquake exacted a severe financial burden on what was at the time a very poor region of Portugal in which per-capita incomes ranged across the islands from 21% to 48% of the national average (Shapleight et al. 1983, p. 213; also see Chapter 1).

This was the first major disaster to have affected the Azores since the 1974 Portuguese Revolution. The 1980 emergency was also the first time in which both immediate responses and longer-term recovery were largely the responsibility of the autonomous Regional Government, although the central government in Lisbon reserved to itself some important functions and provided most of the financial support (see below). At the time, the centre-right Social Democrat Party was in power both regionally and nationally and, following political instability in the immediate aftermath of the revolution, the system had stabilised by 1980 (Shapleight et al. 1983). One feature that had not changed since the days of the *Estado Novo* regime was that planning remained *reactive*, there being no civil protection authority, no emergency plans and no command structure to respond urgently when a disaster struck: all had to be improvised.

The combination of a more open and accountable government, global interest and developing scholastic concern with disaster mitigation both internationally and in Portugal, caused the 1980 earthquake to spawn a vast academic literature, in addition to extensive coverage by newspapers of record and newsreels at home and abroad. For the first time, some film stock was in colour. Particularly useful sources are a report prepared for the Senate of the United States of America and authored by Senator Claiborne Pell (Pell 1980);[33] an edited two volume work, *Problemática da Recostrução* (Problems of Reconstruction), published in 1983 by the *Instituto Açoriano de Cultura* (Azorean Institute of Culture) in Angra do Heroísmo (Anon 1983) and a ten-year retrospective study, *Monografia 10 anos após o sismo dos Açores de 1 de Janeiro de 1980* (Monograph ten years after the earthquake of 1 January 1980), produced by members of *the Laboratório Nacional de Engenharia Civil* (National Civil Engineering Laboratory) in Lisbon (Oliveira et al. 1992).

Table 5.5 Immediate responses to the 1980 earthquake

Responses	Details
First Few Days	When the disaster struck on 1 January, the area was almost completely cut off from the outside world, with just one Portuguese military radio link remaining to Lisbon and a single USAF link to Washington DC (Anon 1980a; Jolliffe 1980). In Angra do Heroísmo in order to fill the power and information vacuums, public officials, military and ecclesiastical leaders met on the day of the disaster and established a command post within the police headquarters. The following day the President of the Republic, António Ramalho Eanes, flew to Terceira to demonstrate support, supervise operations and to declare three day of national mourning. He also brought medical personnel and supplies (Jolliffe 1980). In spite of a total lack of preparedness, action was quickly taken to restore minimal living conditions. The hope was not just humanitarian, but to discourage unplanned out-migration. These measures included the following:
	a Rescuers (Portuguese and American) searched through collapsed and damaged buildings looking for survivors.
	b Homeless people were settled in open spaces near to the city.
	c Equipment and household affects was moved to less-damaged properties.
	d The debris was removed to allow roads to be opened and to enable supplies to be moved.
	e Technical staff checked all damaged building to ensure that they could be safely inhabited.
	Much assistance was provided from the USAF base at Lajes. This included help in clearing roads; digging through rubble and helping families to shelter; the gymnasium at the base was used as a reception point; food was provided on a 24 hour basis; people were accommodated in dormitories; USAF medical staff were deployed and generators were donated, because 90% of Terceira's power had been cut off. (Anon 1980c.)
	By 3 January, Portuguese military helicopters and naval ships were evacuating people from inaccessible areas on São Jorge (Anon 1980b, 1980d).
First three months	Later the homeless were settled in tents, then in pubic buildings, schools, garages and in the homes of friends and neighbours. By 15 January, 150 families were housed on the USAF base (Anon 2013b). Tents could only be a temporary solution and temporary metallic dwellings were pre-fabricated and erected in urban areas. In rural areas construction timber was supplied to build huts and to enable the homeless to return to work. The Regional Government established the *Gabinete de Apoio e Reconstrução* (Office of Assistance and Reconstruction)

or GAR, which co-operated with local authority leaders in *concelhos* (i.e. counties) and *freguesias* (i.e. parishes) and encouraged an 'open door' management style. From the start the emphasis of the GAR was on financially subsidised self-reconstruction, but this was initially hampered by: a lack of an urban planning structure; a dearth of professional staff and only a small workforce of skilled construction workers. There was a strong desire to 'build-back-better' and eventually this was translated into a planned and highly regulated policy of reconstruction (see text).[1] By the end of March economic life had resumed and fears of mass migration proved unfounded.

National and International Support (January to March)	a The Portuguese Government provided $4.4 million and two lines of credit over 15 and 30 years, plus non-financial aid. Pre-fabricated housing was funded by the Regional government (estimated at $2–4 million) and from the USA ($2.3 million)

b International Red Cross.

c US disaster funds $650,000.[2] Some $25,000 was donated in emergency relief, plus material and non-material aid from the Lajes base (see above).

d Rhode Island residents in the USA collected $35,000 and over 50,000 kg of supplies. There was also support from other diaspora communities.

e Foreign Governments: Canada; France; Greece; Japan; S. Korea; UK and West Germany – total of $218,000.

f Private Sector *c.* $900,000.

g International organisations and charities

i Canadian International Development Society and Canadian Red Cross $30,000.

ii US Disaster Relief Organisation $150,000.

iii US CARITAS $320,000.

iv European Economic Communities $150,000

v European Investment Bank $36,000.

vi Austrian CARITAS $40,000.

---

1 It is notable that a 'build-back-better' policy presaged by more than three decades one of the principal objectives (Priority 4) of the United Nations' *Sendai Framework for Disaster Risk Reduction* (United Nations 2015).

2 In all cases $ refers to US dollars. In March 1980 1$ US was equal to 44 Escudos. 1$ US in 1980 is roughly equivalent to 3$ US in 2020.

Based on information in: Pell (1980); Shapleight et al. (1983); Oliveira et al. (1992); Feio et al. (2015) and other references cited in the table.

As Table 5.5 shows and notwithstanding the total lack of preparedness, the State and, in particular, the Regional Government was able quickly and effectively to mobilise in marked contrast to the highly centralised Lisbon–centric approach of the former government. Elements of the *pre-industrial* practices were still in evidence, the church and its clergy provided succour to distressed people and were involved at the local level in organising the response (Table 5.5), but for the first time the church was viewed not just as a provider of spiritual comfort, but as an important stakeholder in maintaining community solidarity. The earthquake exacted a severe toll on church buildings. Two churches and seven chapels were destroyed, with many other places of worship – including the cathedral in Angra – being extensively damaged, yet the clergy laboured successfully to keep worshipping communities together during the emergency (Mendonça 1983). Also families, extended families and friendship bonds continued to be vital assets in making available temporary accommodation and in providing economic support.

Three features of the response, nascent in the weeks immediately following the disaster, featured prominently during the course of recovery. First, there was a role for the USA; its military personnel and infrastructure. Reading Senator Pell's report (Pell 1980), it is clear that the principal reason for the generosity of USA was a mixture of humanitarian concern and a desire to support the large Azorean diaspora communities resident in North America, some of whom were Claiborne Pell's Rhode Island constituents. Following the revolution there was also a desire to respond strategically to prevent Portugal tilting towards communism. By the end of March 1980 and in addition to material aid, the USA provided cash sums of c.675,000 US dollars, equivalent to over two million US dollars at 2020 values (Pell 1980). Indeed by the end of May 1982 total US public and private aid was valued at over six million US dollars: worth over 19 m US dollars in 2020. The equivalent Portuguese spend over the same period was c.31 million dollars US, making the US contribution around 19% of the Portuguese total (Annex B 1983, p. 281).[34] A second feature that continued was international interest and this was also characterised by generosity. Aid not only reached the Azores in cash and non-monetary donations from the communities of the diaspora, but also from a large number of foreign governments and charities, this generosity reflecting international media attention and the actions taken by the Regional Government in publicising their predicament, so different from the secretive norms of the *Estado Novo*. For instance, Dr Álvaro Monjardino (President of the Legislative Assembly of the Regional Government), visited Rhode Island and lobbied successfully for funds to support house building and rehabilitation (Pell 1980). Finally, out-migration contrasted with what had been observed in earlier disasters. In 1980 emigration from Terceira involved 1,752 persons, twice the total of previous years, but fell back to 574 in 1981 and in this respect the policy to stem emigration was deemed to have been successful (Shapleight et al. 1983, p. 270).

Longer-term recovery has been discussed in detail by Feio et al. (2015) and the following account is based on their research. Reconstruction was led by the *Gabinete de Apoio e Reconstrução* (Office of Assistance and Reconstruction) or GAR (see Table 5.5), with funds from international, national and local sources being consolidated into a *Fundo de Apoio à Reconstrução* (Fund for Assistance and Reconstruction). To implement a policy of self-construction the GAR provided financial assistance and technical support and this policy quickly captured the public mood. In fact, some re-building began within days of the disaster. Reconstruction projects had to be licensed, supervised and the GAR provided support to local architects, planners and introduced training for construction workers. Planning considerations were also to the fore and the opportunity was grasped to plan the future expansion of Angra and other settlements. One major consideration was preserving the architectural heritage of the city of Angra do Heroísmo which had been so severely damaged by the earthquake. Reconstruction and preservation of historical buildings was mandatory and just three years after the disaster and whilst reconstruction was still ongoing, the *United Nations Educational, Scientific and Cultural Organisation* (UNESCO) paid the Azores the signal honour of designating the city a World Heritage Site.

A feature new to the Azores was detailed review and critique of policy. The overall conclusion of most commentators has been that, even though the response was *reactive* and planned from scratch, it was highly successful. The conclusion of Feio et al. (2015, p. 90) with respect to Angra do Heroísmo, but applying with equal weight across the devastated area, is typical.

> Decision makers were capable of implementing locally adequate reconstruction actions, avoiding the expected emigration problems, dealing with the lack of resources, and taking advantage of the population's will to rebuild. Moreover, they where also able to use reconstruction as an opportunity to improve buildings and quality of life as well as to plan for the future development of the city. Therefore, 'Angra do Heroísmo is a successful example, notwithstanding a complete absence of earthquakes preparedness.'

This last comment is too critical because reconstructed and rehabilitated buildings were clearly more seismically resilient than they were before the earthquake and there was also some retrofitting of earthquake resistant elements (see below), but the opportunity for root and branch redesign was not taken up. Programmes of reconstruction have to strike a balance between preserving the heritage, on the one hand, and improving safety, on the other, and the post-1980 rehabilitation of Angra do Heroísmo emphasised the former.

In Chapter 3, it was argued that in 1980 and especially in Angra do Heroísmo building losses were due to a combination of earthquake waves, vulnerable buildings and the so-called 'soil-factor'; differing seismic responses being observed across varying surface geologies. Buildings constructed since the 1940s and incorporating reinforced concrete columns, beams and other strengthening elements performed better than older structures built using rubble stone (Teves-Costa et al. 2007). In the early 1980s, little was known about soil-amplification, but subsequent studies have shown that Angra do Heroísmo and other settlements on Terceira are still extremely vulnerable. Two papers, for example (Teves-Costa & Veludo 2013; Veludo et al. 2013), show that under a number of earthquake scenarios the city is still very hazard exposed, although since 1980 the building stock in the historic centre of Angra presents greater seismic resilience due to 'widespread retrofit works carried out in that zone after the earthquake' (Veludo et al. 2013, p. 447).

1998 Faial earthquake

Responses to the 1998 earthquake diverged from what had gone before. One contrast was that, whereas the 1980 disaster became an international media event, the 1998 earthquake was largely ignored outside Portugal and, although some newspapers of record noted its occurrence, there was virtually no in-depth reporting.[35] Direct United States involvement and the desire to keep post-revolutionary Portugal favourably disposed towards western strategic interests were both important considerations in 1980, but given the impact of and damaged caused in 1998 it is still difficult to explain this international lack of interest, especially from within diaspora communities. A second area of contrast was that the emergency occurred on the cusp of major refocusing of civil protection policies that were in the next two decades to become strongly *proactive* (see Chapter 6). During and following the 1998 earthquake elements of both *reactive* and *proactive* approaches may be discerned in the policies adopted by the regional government.

The 9 July 1998 Magnitude 5.8 earthquake (maximum Intensity VIII MM56) had a devastating impact on Faial and western Pico, with São Jorge also being affected to a more limited degree. The earthquakes caused nine deaths, injuring over 100 people, destroying more than 1,500 buildings and displaced around 2,500 residents (see Chapter 3). Figure 5.4 shows the settlements most severely impacted and, as in earlier earthquakes, intensities are high given the relatively modest magnitude of the event (M 5.8), because of the location of the proximity of the epicentre to Faial and Pico (c.10 to 20 km), the large number of vulnerable buildings and the amplification effects that were witnessed due to the soil-factor (Senos et al. 1998 and Chapter 3). Although the Portuguese Building Code for new buildings was strengthened in 1983 (see Chapter 6), by 1998 few Azorean domestic dwellings complied with these standards and the affected islands remained redoubts of what David Alexander has termed 'residual un-ameliorated vulnerability' (Alexander 1997, p. 292), with the housing stock in many settlements being dominated by old and vulnerable homes. In addition to housing, the earthquake also damaged roads, water and sewage systems, communications and also destroyed and caused harm to many valued heritage buildings including numerous churches (Anon 2020d). Of the 13 churches on Faial, four collapsed (wholly or partially), four suffered serious damage and some 58% of the population was denied the opportunity to worship. On Pico there were no collapses, but three churches suffered serious damage and 40% of the population was adversely affected (Oliveira et al. 2012). As discussed above and in Chapter 1, Azorean Catholicism is characterised by many people feeling a strong sense of belonging, even though they may be theologically sceptical and/or heterodox

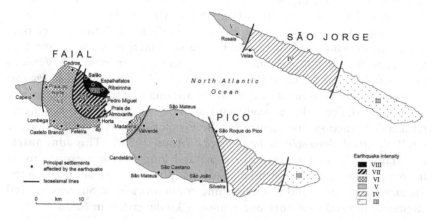

Figure 5.4 The 1998 earthquake: seismic intensities and the locations of settlements most badly impacted. On Faial seismic wave amplification was noted on the 'soft' substrates and deep soils upon which the villages of Castelo Branco, Flamengos are built. The settlements showing the greatest damage were Espalhafatos, Flamengos and Ribeirinha. On Pico damage generally decreased from west to east in accordance with seismic attenuation, but with some anomalous high values being observed because of local amplification due to the effects of thick layers of soft sediment. On São Jorge, Rosais was the only village showing major damage to its housing stock. Based on information in Matias et al. 2007; Marques et al. 2014 and additional sources.

believers, with the church acting as a strong social as well as a religious focus within communities (Chester et al. 2019b). The total direct costs of the emergency have been estimated at 300 million Euros at 1998 values (*c.*432 million Euros at 2020 values) (Oliveira et al. 2012).[36]

Scientists based at the *Universidade dos Açores* (University of the Azores) the *Instituto de Meteorologia* (Institute of Meteorology) were prepared in advance and, using the greatly expanded seismic network that had been developed following the 1980 catastrophe, were able closely to monitor seismic activity. In the immediate aftermath of the earthquake – between 9 July and 6 August – the *Sistema de Vigilância Sismológica dos Açores* (Azores Seismological Surveillance System or SIVISA),[37] was able to establish a network of seven stations across the three islands and began monitoring aftershocks (Senos et al. 1998; Matias et al. 2007), several of which caused additional damage to already weakened structures (Malheiro 2006). Initially *reactive* planning reigned and there were still some aspects of typical *pre-industrial* responding with the Portuguese Air Force ferrying in tents for around 1,000 homeless people and other refugees finding accommodation with friends and/

or relatives in less badly affected villages. As in 1980 a senior member of the National Government in Lisbon, Prime Minster António Guterres,[38] arrived both to show support to the people affected and to the Regional Government whose task it was to handle the emergency and to meet its costs. The Regional Government not only had the expertise of SIVISA to call upon, but also the Civil Protection Service (*Serviço Regional da Proteção Civil dos Açores*) established in 1983 in the aftermath of the 1980 emergency (Cabral 2015), fire fighters (*bombeiros*) and a body set-up to promote longer-term reconstruction, the Centre for the Promotion of Reconstruction, or CRP (*Centro de Promoção da Reconstrução*) (Anon 2020d). This administrative framework allowed effective short and longer-term measures to be introduced. In the short-term emphasis was placed on rescue, housing the displaced population and re-establishing communications. Surveys by civil engineers showed that only one school, a kindergarten in Salão had suffered major collapse of an external wall and two schools in Espalhafatos and Ribeira Funda (Figure 5.4) were considered unsafe and demolished, so allowing the remaining school buildings to be used safely as temporary accommodation (Proença 2004; Ferreira et al. 2008). In addition, other public buildings were inspected before they were brought back into use, many of which were constructed to more recent seismic-resistant designs, and were added to the stock of temporary public housing.[39] Some 670 pre-fabricated houses were also supplied by the Regional Government (Oliveira et al. 2012; Anon 2016c).

The process of rehabilitation began with a building survey covering Faial, western Pico and a few houses on São Jorge (Neves 2008; Neves et al. 2012). At the time of the earthquake some 57% of buildings were traditionally constructed and considered dangerous and the survey allowed the CRP to 'triage' housing as:

1   *New construction* is required because of the severity of the damage.
2   *Rehabilitation* is considered a viable and cost-effective option, but requires a budget in excess of 15,000 Euro per dwelling and involves the seismic reinforcement of structural elements within the design.
3   *Repair* is possible with an expenditure of less than 15,000 Euros (Oliveira et al. 2012).

Housing was prioritised over other reconstruction (e.g. churches)[40] and the peak years for building were 2001–2005, creating much needed employment. Even as early as 2001 there were 2.9% more houses on Pico and Faial than there had been in 1991 and by 2006 monies expended on housing reconstruction, rehabilitation and repair were immense, being estimated at 160 million Euros on Faial and 60 million Euros on Pico. In addition, 47 million Euros was expended on the repair and reconstruction of churches; 30 million Euros on infrastructure, 10 million Euros on miscellaneous expenditure by municipalities and four million Euros on schools (Oliveira et al. 2012). Overall 72%

of public spending came from the budget of the Regional Government, with the rest being provided by the National Government in Lisbon.[41]

Notwithstanding some hyperbole on the part of political leaders, most commentators have pointed out that the plan to 'build-back better' has been successful and that the housing stock of the Faial and western Pico is now, not only more seismically resistant, but homes have also greatly improved amenities (Anon 2009, 2016c). In addition, the earthquake brought about innovation and modernisation in sanitation and water supply so benefitted the islands as a whole. There was no major wave of emigration as a result of the effects of the earthquake, the population of Faial remaining constant at around 15,000 between 1991 and 2011 (SREA 2020b). More negatively, Oliveira et al. (2012) have noted that some aspects of the response were still too *ad hoc*. In advance of disaster the Regional Government had not: drawn up plans to collect detailed data on damage; mistakes were made in the allocation of new housing and insufficient attention had been paid to vernacular (i.e. non-religious) historic building that had been damaged by the earthquake, with many deteriorating in the years following the disaster because of a lack of investment.

Concluding remarks: moving forward

The 1998 earthquake was the last major geophysical event to have impacted the Azores. Since this emergency struck the central islands the thrust of applied seismology and volcanology, on the one hand, and disaster planning, on the other, has been to develop a fully *pro-active* system of civil protection for the Azores, the seeds of which were sown in the generally successful response to the 1998 emergency. The ways in which this system has been developed over the past two decades forms the basis of Chapter 6.

Notes

1 Gilbert Fowler White (1911–2006) was a prominent American scholar and humanitarian activist, who was motivated both by the interventionist policies of the 1930s represented by President Roosevelt's *New Deal* and by his Quaker beliefs. These influences gelled into a life that was devoted to human betterment. This not only involved an academic commitment to hazard reduction, but was also expressed in service aiding refugees in the Second World War and in shaping environmental policy in both the USA and Africa (Hinshaw 2006). For many years White was Professor of Geography at the University of Chicago, but in 1970 moved to the University of Colorado. He was still publishing when he was aged over 90.

2 Sometimes the dominant paradigm it is termed the *Chicago school*, because many of its practitioners had at some time during their careers worked at, or had been were associated with, the University of Chicago.

3 Contemporary research aims are more fully articulated in United Nations' Hyogo and Sendai Frameworks for Action. The *Hyogo Framework for Action* (2005–2015) was inaugurated at the World Conference on Disaster Reduction which was held in Japan in January 2005. The *Sendai Framework for Disaster Risk Reduction*

(2015–2030) is a revised version and was published during a second World Conference on Disaster Risk Reduction. This was also held in Japan ten years later (United Nations 2006, 2014, 2015).

4 Fuller definitions of vulnerability and resilience are provided by Cutter et al. (2008, p. 599). 'Vulnerability is the pre-event, inherent characteristics or qualities of social systems that create the potential for hazard. Vulnerability is a function of the exposure (who or what is at risk) and the sensitivity of the system (the degree to which people and places can be harmed). ... Resilience is the ability of a social system to respond and recover from disasters and includes those inherent conditions that allow the system to absorb impacts and cope with an event, as well as post-event, adaptive processes that facilitate the ability of the social system to re-organise, change and learn in response to a threat.'

5 The Azores were important in the early years of transatlantic aviation. Pan-American Clippers (using Boeing 314 seaplanes) called at Horta in the 1930s. From the late 19th century, Faial was an important centre for transatlantic telegraph and telephone cables and acted as a home for expatriate technical personnel and their families (Callender & Henshall 1968, p. 19).

6 Major concentrations of Azorean immigrants in North America are: California (San Jose and Oakland), New England (Massachusetts and Rhode Island), New Jersey, New York and Florida in the USA; and Ontario (Toronto) in Canada. After the 1957/8 most immigrants went to New Bedford or Fall River in Massachusetts, or to California (Chapin 1989; Sá 2008). In 2015 there were over 1.3 million Americans of Portuguese descent, many of whom traced descent from the Azores (Anon 2016b).

7 Examples include aid from Brazil and North America following in the 1841 Terceira earthquake (Sousa Martins 2015) and from a wide range of expatriate communities following crises in 1957/8 and 1980 (Pell 1980; Coutinho et al. 2010). Communities of the North American diaspora also supported their compatriots on a regular basis in order to alleviate endemic poverty. In 1950 an Azorean parliamentarian, Armando Medeiros, pointed out to Salazar that some 37,000 bags of second hand clothes has been received in the Azores and that he had overheard the comment made 'if it were not for America, we would have nothing to wear' (Fernandes 2019, p. 107).

8 Urban primacy is the concentration of urban population in only one settlement and is especially characteristic of many developing countries and regions. In the opinion of some scholars it prevents the balanced development of a country or region as a whole. Political and economic power is frequently concentrated in these large urban centres (Meyer 2019).

9 The three Horta *freguesias* are Angústias, Conceição and Matriz and those adjacent to the capital are Feteira, Flamengos and Praia do Almoxarife.

10 The notion of *Açorianidade* was first coined by the Azorean-born poet and intellectual Vitorino Nemésio (1901–1978) to express how a distinctive identity was forged as people coped with and tried to rationalise life on the islands in the face of extreme physical hardships (e.g. earthquakes, volcanic eruptions and storms), isolation and dispersed families and friends. The term started to be used widely from 1976, a time which coincided with self-government in the Azores. Other literary figures who developed the notion of *Açorianidade* included: Raúl Brandão (1867–1930) and Luis Ribeiro (1882–1955) (Pires 1995).

11 As discussed in Chapter 1, *saudade* is impossible to translate using a single English work, but implies a sad nostalgic longing.

12 John Bass Dabney (1766–1826) was United States Consul General in the Azores and lived in Horta (Faial). He was followed in office by his son Charles William Dabney (1794–1871) and his grandson Samuel Wyllys Dabney (1826–1893) (Anon 2020a)

13 An account at the time in Latin refers to *fumi nubibus flammarum globi* - a globe of flame and smoke (Canto 1880, p. 192, our translation). Parish registers refer to 10 fatalities having been caused by a 'terrible cloud burning like hell' (Canto 1880, p. 192, our translation).

14 The hazard analyst David Alexander (1997) has used the term *residual unameliorated vulnerability* to describe elements of the heritage of an area which exacerbates present-day hazard exposure.

15 The historical records of this incident are rich and have been collected by Azorean historian José Candido da Silveira Avellar (1843–1905) (Avellar 1902, pp. 440–441). The *Presidencte da Câmara* was Dr Antonio Augusto Pereira and the officials who remained were: Captain Amaro Teixeira de Sousa; Sergeant Major José Soares de Sousa; Captain João Ignacio da Silveira and the muncipal solicitor Jorge José Corvilhão. The Captain General at the time was Miguel Antonio de Mello (1766–1836). Dr Pereira was what is termed a *juiz de fora* (judge from outside), a post created by King Afonso IV (reigned 1325–1357). For centuries these officials were appointed directly by the crown and represented central government. They were widely resented by the locally recruited leaders of municipalities (Faria 2014). It is likely that at the time of the emergency, relations between Dr Pereira and his officials had been strained for some time and that his leadership could not survive the crisis.

16 It took more than a year to recover the bodies and Rui Gonçalves da Câmara II recruited relatives and friends of the deceased from across the island (Gerrard 2021). In this strongly Catholic society much effort was spent in recovering the sacred host (or consecrated sacrament), which was believed to be the body of Christ trapped in the rubble (Forlin 2020).

17 José Silvestre Ribeiro was a liberal, took up arms and served time in exile both in England and in France. Later after the Civil (or Liberal) Wars of 1828–1834, he resumed his studies at Coimbra and entered the Civil Service. After his time on Terceira, Ribeiro served in Beja, Faro and Madeira, and later entered parliament and was a member of the government (Anon 2020b).

18 *Caída da Praia* is a somewhat confusing term because it is also used for the earthquake of 1614.

19 Praia da Vitória was known as Vila da Praia da Vitória until 1983.

20 In Catholic practice the Sacrament (or host) may be reserved by the altar after mass and is used for the communion of the sick, and as an aid to prayer and veneration when periodically exposed on the altar. In 1761 the *Santissimo Sacramento* (Blessed Sacrament) was removed to the church in the neighbouring parish of Quatro Ribeiras. A religious procession made its way to the flow front which was advancing slowly and people were able to light their torches from the incandescent lava and prayers were offered for deliverance. Mathias Silveira, a prosperous local resident of Biscoitos, built a small chapel by the road (Canto 1883 - Fig. 3.11).

21 *Saudades da Terra* (Nostalgic Longing for the Earth) - comprises six volumes: Book 1 - Canaries and the Cape Verde Islands; Book 2 - Madeira and Porto Santo; Book 3 - Santa Maria; Book 4- São Miguel; Book 5 - A tale of chivalry and nothing to do with the islands and Book 6 - mostly on Terceira. From 1591 to 1760 the manuscript was held by the College of the Jesuits in Ponta Delgada and was in private hands until donated to the Public Library and Archive also in Ponta Delgada in 1950 (Rodrigues 1991; Pinto 2003). *Saudades da Terra* remained in manuscript form until the late nineteenth century, being published from 1873 onwards (Luz 1996) and did not become widely available outside Portugal until well into the Twentieth Century.

22 In more recent research we have preferred the term 'parallel practice,' because the term *cognitive dissonance* has a far more restricted meaning within psychology (Carroll 1990).

23 An earlier example of unplanned evacuation was from São Jorge following the 1580 eruption, when the authorities allowed evacuation from the island to take place to relieve population pressure (Canto 1880; Wallenstein et al. 2018).

24 It is known that severe damage was caused to housing in northern Pico (see Chapter 3) and that the government in Lisbon sent tents and supplies (Anon 1975). At the time of the earthquake the *estado novo* regime was under pressure, with the Portuguese Revolution occurring in April 1974 (see Chapter 1).

25 Foreign scientists were also welcomed and included the: eminent French volcanologist, Haroun Tazieff; members of the Lamont (now Lomont-Doherty) Geological Observatory at Columbia University, New York and the Cranbook Institute of Science in Michigan. The physical character of the disaster was reported in by *Paris Match, National Geographic, The Times (London)* and the *New York Times*. More recent studies focusing on scientific aspects has included Cole et al. 1996, 2001)

26 Some clergy still viewed the disaster as a divine punishment. In May 1958 the village of Praia do Norte was destroyed and the parish priest, Padre Henrique Pinheiro Escobar, took the theologically questionable decision to absolve all the people on the spot. A later sermon by the Bishop of Angra, Manuel Afonso de Carvalho (1912–1978), in which he preached that the disaster was a punishment of individual human sinfulness, caused such disquiet that many in the congregations spontaneously walked out of church (Fraião 2005).

27 Flores born Dr António de Freitas Pimentel (1901–1981) was a local doctor who served as governor from 1953 until one year before the 1974 revolution. At time the governor's jurisdiction included, not only Faial, but also Pico, Flores and Corvo. Local accounts of the governor's role in the emergency are of a hagiographic character, reflected a strong government influence on the reporting (Lobão 2008, 46, 51), but there is no doubt that his leadership was viewed as heroic by the people of Faial and spontaneously received a procession of thanks and gifts of flowers. (Anon 1958i). He later received honours from the State.

28 The history of post-Capelinhos emigration to the USA is summarised in Table 5.4, but is very complex in terms off legislative instruments and the timing of departure. For amore detailed treatment reference should be made to Coutinho et al. (2010).

29 Twelve vessels were involved: three Portuguese and nine foreign (Machado & Forjaz 1966). Reacting to the farmers' distress over abandoned livestock, a veterinary surgeon was transported to São Miguel by the British cable ship, *Mirror* (Anon 1964a).

30 These were *Quonset* huts, metallic semi-circular constructions invented in the Second World War and meant to be temporary dwellings. One hundred were donated and some were still in use in the late twentieth century (Anderson 1964).

31 At the time there were approximately 29 Escudos to the US dollar, making 21.25 million Escudos equivalent to 732,788 US dollars. This would have an approximate purchasing power of over six million US dollars in 2021.

32 In 1980 the exchange value of one dollar US was 44 Escudos (Pell 1980). 100 million US dollars is approximately equivalent to 320 million in 2021. Bebo (1992) gives a similar estimate (see Chapter 3)

33 Senator Claiborne Pell (1918–2009) served as Senator for Rhode Island from 1961 to 1997. Pell was a former diplomat who was fluent in several languages including Portuguese. He had an interest in the Azores, because members of the diapora were well represented in his State (Miller 2011).

34 By mid 1982 one dollar US was equivalent to 60 Escudos (Shapleight et al. 1983).

35 One notable exception is a fine report in the *Irish Times* (Brough 1998).

36 This is a colossal sum for two small islands. In the 2001 census the combined population of Pico and Faial was 29,239 and 300 mill Euros equates to losses of over 10,000 Euros per person.

37 This was a joint initiative between the *Instituto de Meteorologia* in Lisbon and the Azores-based *Centro de Vulcanologia e Avaliação de Riscos Geológicos*.

38 António Gutteres (born 1949) subsequently had a distinguished career as an international civil servant culminating in him becoming Secretary General of the United Nations in 2017.

39 The socialist government at the time named these public buildings: *Casa do Povo* (houses of the people).

40 In 2009 church at Prianha do Norte was rededicated some 11 years after the earthquake, some 75% of the costs being bourn by the State (Anon 2009).

41 This sums to a total of expenditure of 311 million Euros. An alternative of 330 million Euros is estimated by Ferreira et al. (2017, p. 331).

References

Admiralty. 1945. *Spain and Portugal Volume IV: The Atlantic Islands*. Naval Intelligence Division, Geographical Handbook Series BR 502c, London.

Alexander, D. 1997. The study of natural disasters, 1977–1997: Some reflections on a changing field of knowledge. *Disasters* **21**(4), 284–304.

Alexander, D. 2000. *Confronting Catastrophe*. Terra Publications, Harpenden.

Almeida, O.T. 1980. A profile of the Azorean. *In:* Macedo, D.P. (ed) *Issues in Portuguese Bilingual Education*. National Assessment Discrimination Center, Cambridge, MA, 115–164.

Ambraseys, N.N. 1972. Earthquake hazard and emergency planning. *Build International* (January–February), 38–42.

Anderson, K.K. 1964. Portugal grateful for Navy Seabee - USAF relief sped to the Azores when quake struck. *Navy Civil Engineer* **5** (December), 27–28.

Annex B. 1983. Tables 1 to 5. *In:* Anon (ed) *Problemática da Recostrução - sismo de 1 de Janeiro de 1980*. Instituto Açoriano de Cultura, Angra do Heroísmo, **2**, 281–285.

Anon. 1841a. Earthquake in Terceira. *The Times* July 13, p. 5.

Anon. 1841b. Express from Falmouth. *The Times* July 13, p. 6.

Anon. 1926a. Earthquake in the Azores. *The Times* September 1, p. 12.

Anon. 1926b. The earthquake in the Azores. Casualties over two hundred. A town wrecked. *The Times (London)*, Thursday September 02, p. 10.

Anon. 1956. The cloud on Portugal's horizon. *The Economist*, May 12, 584–586.

Anon. 1957a. Birth of volcanoes: Island that appeared and disappeared in the Azores. *The Times (London)*, November 14, p. 8.

Anon. 1957b. A erupção dos Capelinhos. Correio da Horta, Sept. 28. *In:* Araújo, V. (ed) *O Resgatar de uma Memória: Sub-Prefeitos, Administradores Gerais, Governadores do Distrito Autónomo da Horta (1831–1975)*, Author's edition. Nova Gráfica, Lda. Horta, 697.

Anon. 1957c. The lighthouse was abandoned. Telégrafo, October 1. *In:* Araújo, V. (ed) *O Resgatar de uma Memória: Sub-Prefeitos, Administradores Gerais, Governadores do Distrito Autónomo da Horta (1831–1975)*. Nova Gráfica, Lda, Horta, 698, Author's Edition.

Anon. 1957d. Volcanic ash covers Azores Island. *The Times (London)*, October 9, p. 9.

Anon. 1957e. Report. *The New York Times*, October 9, p.1.

Anon. 1957f. Correspondence between Governor Pimentel and Sr. Joaquim Morrison from California. *In*: Araújo, V. (ed) *O Resgatar de uma Memória: Sub-Prefeitos, Administradores Gerais, Governadores do Distrito Autónomo da Horta (1831–1975)*. Nova Gráfica Lda, Horta, 712–715, Author's Edition.

Anon. 1958a. Report. *New York Times*, May 21, p. 10.

Anon. 1958b. Report. Correio da Horta, May 14. *In*: Araújo, V. (ed) *O Resgatar de uma Memória: Sub-Prefeitos, Administradores Gerais, Governadores do Distrito Autónomo da Horta (1831–1975)*. Nova Gráfica, Lda, Horta, 719, Author's Edition.

Anon. 1958c. Report. Correio da Horta, May 15. *In*: Araújo, V. (ed) *O Resgatar de uma Memória: Sub-Prefeitos, Administradores Gerais, Governadores do Distrito Autónomo da Horta (1831–1975)*. Nova Gráfica, Lda, Horta, 721, Author's Edition.

Anon. 1958d. Report. Diário de Coimbra and the Século, May 15. *In*: Araújo, V. (ed) *O Resgatar de uma Memória: Sub-Prefeitos, Administradores Gerais, Governadores do Distrito Autónomo da Horta (1831–1975)*. Nova Gráfica, Lda, Horta, 722–723, Author's Edition.

Anon. 1958e. Report. Diário de Notícias, May 26. *In*: Araújo, V. (ed) *O Resgatar de uma Memória: Sub-Prefeitos, Administradores Gerais, Governadores do Distrito Autónomo da Horta (1831–1975)*. Nova Gráfica, Lda, Horta, 723–724, Author's Edition.

Anon. 1958f. Statement of the Governo Civil do Distrito Autónomo da Horta. *In*: Araújo, V. (ed) *O Resgatar de uma Memória: Sub-Prefeitos, Administradores Gerais, Governadores do Distrito Autónomo da Horta (1831–1975*. Nova Gráfica, Lda., Horta, 727, Author's Edition.

Anon. 1958g. Report. *The New York Times*, June 1, p.2.

Anon. 1958h. Island volcano in eruption. Tremors in the Azores. *The Times (London)*, August 19, p. 8.

Anon. 1958i. Report. Telégrafo, July 17. *In*: Araújo, V. (ed) *O Resgatar de uma Memória: Sub-Prefeitos, Administradores Gerais, Governadores do Distrito Autónomo da Horta (1831–1975)*. Nova Gráfica, Lda., Horta, 725, Author's Edition.

Anon. 1964a. Azores refugees taken to haven. *New York Times*, February 20, p. 2.

Anon. 1964b. Azores earthquake rescue starts. *The Times (London)*, February 20, p. 10.

Anon. 1964c. Earthquake rocks Azores Isle: Removal of 20,000 starts. *New York Times*, February 19, p. 1.

Anon. 1964d. Azores earthquake dies down. *The Times (London)*, February 21, p.1.

Anon. 1975. A quake in the Azores. *The New York Times*, November 25, p.7.

Anon. 1980a. At least 26 dead in the Azores earthquake. *The Times (London)*, January 2, p. 1.

Anon. 1980b. New earth tremors hit Azores. *The Times (London)*, January 4, p. 4.

Anon. 1980c. Earthquake kills 52 in the Azores. *New York Times*, January 2, p. A3.

Anon. 1980d. Tremors shake 2 islands in the Azores. *The Globe and Mail (Toronto)*, January 4, p. 3.

Anon. 1983. *Problemática da Recostrução - sismo de 1 de Janeiro de 1980*. Instituto Açoriano de Cultura, Angra do Heroísmo.

Anon. 2001. *Capelo 400 anos*. Do Registro da Memoria Colletera, Azores.

Anon. 2008. Interview with an anonymous farmer in Praia da Norte. April 18 2008 (personal communication).

Anon. 2009. *Reconstrução do sismo de 1998 foi um dos maiores projectos realizados nos Açores*. Governo Regional dos Açores. http://www.azores.gov.pt/Portal/pt/entidades/pgra/noticias/20090814+Reconstru%C3%A7%C3%A3o+do+sismo+de+1998+foi+um+dos+maiores+projectos+realizados+nos+A%C3%A7ores.htm

Anon. 2013a. Nicholas Neocles Ambraseys 1929–2012. *Journal of Earthquake Engineering* **17**, 301–303.

Anon. 2013b. Lajes Field History - Humanitarian efforts, 1980s and. 1990s. https://web.archive.org/web/20080507105531/http://www.lajes.af.mil/library/factsheets/factsheet.asp?id=4004

Anon. 2014. *Fieldwork in the Azores. Archaeology of Mediterranean Earthquakes in Europe 1000–1550.* https://armedea.wordpress.com/2014/12/10/fieldwork-in-the-azores/

Anon. 2015.*Vila Franca do Campo.* https://en.wikipedia.org/wiki/Vila_Franca_do_Campo

Anon. 2016a. *Saudade* expresses two emotions–sadness and longing. Live Azores, January 9. http://liveazores.com/saudade-expresses-two-emotions-sadness-and-longing/

Anon. 2016b. United States Census Bureau. Government of the United States of America. https://data.census.gov/cedsci/table?q=AncestryandhidePreview=trueandtid=ACSDT1Y2016.B04006andvintage=2016andt=Ancestry

Anon. 2016c. Sismo de 1998 mudou a ilha do Faial para sempre. *Observador* 8 August. https://observador.pt/2016/08/08/sismo-de-1998-mudou-a-ilha-do-faial-para-sempre/

Anon. 2020a. The Dabneys: A Boston family in the Azores. https://www.azores-adventures.com/2014/03/the-dabneys-a-boston-family-in-the-azores.html

Anon. 2020b. *Ficha Bio-Bibliográfica de José Silvestre Ribeiro (1807–1891).* Faculdade de Direito de UNL, Universidade Nova de Lisboa (https://www.fd.unl.pt/ConteudosAreasDetalhe_DT.asp?I=1andID=1082).

Anon. 2020c. *Comenius Project 'Gastronomy of the Soul': Holy Spirit Festivities.* Escola Profissional Praia da Vitória. (http://www.feppv.pt/wp-content/uploads/2013/05/holy_spirit.pdf).

Anon. 2020d. *Relembrar o terramoto 9 de Julho de 1998.* Instituto de Investigação en Vulcanologia e Avaliação de Rismo (IVAR), Ponta Delgada (http://www.cvarg.azores.gov.pt/noticias/Paginas/cms_173_Relembrar-o-terramoto-de-9-de-Julho-de-1998-.aspx).

Araújo, V. 2007. *O Resgatar de uma Memória: Sub-Prefeitos, Administradores Gerais, Governadores do Distrito Autónomo da Horta (1831–1975),* Author's edition. Nova Gráfica, Lda., Horta

Arêde, A., Costa, A, Moreira, D. and Neves, N. 2012, Seismic analysis and strengthening of Pico Island churches. *Bulletin of Earthquake Engineering* **10**, 181–209.

Avellar, J.C.S. 1902. *Ilha de S. Jorge (Açores). Apontamentos para a sua história.* Typ. Minerva Insulana, Horta.

Ávila, J.I. and Mendoça, L. 2008a. Azorean emigration: Causes, waves and destinations. A brief historical summary. *In:* Goulart, T. (ed) *Capelinhos, a volcano of synergies. Azorean emigration to America.* Portuguese Heritage Publications of California, San Jose, 17–24.

Ávila, J.I. and Mendoça, L. 2008b. Azorean emigration: Causes, waves and destinations. A brief historical summary, *In:* Goulart, T. (ed) *Capelinhos, a volcano of synergies. Azorean emigration to America.* Portuguese Heritage Publications of California, San Jose, 17–24.

Ávila, J.I. and Mendoça, L. 2008c. The impact of the Capelinhos emigration on the Azorean population and economy. *In:* Goulart, T. (ed) *Capelinhos, a volcano of synergies. Azorean emigration to America.* Portuguese Heritage Publications of California, San Jose, 67–71.

Azevedo, J., Serrano, S. and Oliveira, C.S. 2009. The next 1755- myth and reality; Priorities and actions to develop in case of an earthquake in the Lisbon

Metropolitan area. *In:* Mendes-Victor, L.A., Oliveira, C.S., Azevedo, J. and A. Ribeiro A. (eds) *The 1755 Lisbon Earthquake Revisited.* Springer, Berlin (Geotechnical, Geological and Earthquake Engineering vol. 7), 559–579.

Baklanoff, E.N. 1992. The political economy of Portugal's Estado Novo: A critique of the stagnation thesis. *Luso-Brazilian Review* **20** (1), 1–17.

Bassnett, S. 2006. Faith, doubt, aid and prayer: The Lisbon earthquake of 1755 revisited" *European Review* **14**, 321–328.

Bebo, C. 1992. Política financeira da reconstrução. Fontes de financiamento e situação actual. *In:* Oliveira, C., Lucas, A. and Guedes, J.H.C. (eds) *Monografia 10 anos após o sismo dos Açores de Janeiro de 1980.* Secetaria Regional de Habitação e Obras Públicas - delagação de ilha Terceira. Laboratório Nacional de Engenharia Civil, Lisboa **1**, 797–806.

Beier, R. and Kramer, J. 2018. A Portrait of the Azores: From natural forces to cultural identity. *In:* Kueppers, U. and Beier, C. (eds) Volcanoes of the Azores, Active Volcanoes of World, Springer-Verlag, Germany, 3–26.

Bento, C.M. 1993. *História dos Açores.* Empresa Gráfica Açoreana, Ponta Delgada (translated by Pilar Vaz Rego Pereira).

Birmingham, D. 1993. *A Concise History of Portugal.* Cambridge University Press, Cambridge.

Boid, E. 1835. *A Description of the Azores, or Western Islands.* Schulk and Co., London.

Brough, B. 1998. 10 die, 90 injured in Azores earthquake. *The Irish Times,* July, p. 10.

Burton, I. and Kates, R.W. 1964. The Perception of Natural Hazards in Resource Management. *Natural Resources Journal* **3**, 412–4.

Burton, I., Kates, R.W. and White, G.F. 1978. *The Environment as Hazard.* Oxford University Press, New York.

Cabral, L.M. 2015. Serviço Regional de Proteção Civil e Bombeiros dos Açores "Uma Tarefa de Todos para Todos." *Nação e Defesa* **141**, 29–43.

Callender, J.M. and Henshall, J.D. 1968. The land use of Faial in the Azores. *In:* Cook, A.N. (ed) *Four Island Studies.* The World Land Use Survey. Monograph 5, Geographical Publications, Bude, 367–395.

Campos, V., Machado, F. and Garcia, J.A.S. 1960. *Relatório da Missão Técnica do Ministerio das Obras Públicas para remediar as primeiras consequencias da erupção vulcânica da Ilha do Faial.* Arcqivo do Governo Civil do Distrito Autónomo da Horta, Horta, 19 pp. (Reproduced in: Anon 1962. *Le Volcanisme de l'Île de Faial et l' Eruption du Volcan de Capelinhos.* Direcção Geral de Minas e Serviços Geológicos de Portugal, Memória (Nova Série) 9, Lisboa.

Canto, E.P. 1879. Transporte de Colonos para o Brazil. *Archivo dos Açores* **1** (5), 371–372.

Canto, E.P. 1880. Erupção no Ilha de S. Jorge. Anno de 1580. *Archivo dos Açores* **2** 188–193.

Canto, E.P. 1881. Erupção na Ilha do Fayal. Anno de 1672. *Archivo dos Açores* **3** (16), 344–51.

Canto, E.P. 1883. A Jorge Goularte Pimentel, agradecendo os serviços pres- tados por occasião do vulcão e instrucções para o embar- que de 100 casaes de colonos para o Maranhão: 22 de dezembro de 1674. *Archivo dos Açores* **5** (27), 265–266.

Canto, E.P. 1884. Erupção na Ilha de S. Jorge. Anno de 1808. *Archivo dos Açores* **5** (29), 437–447.

Canto, E.P. 1888. Erupção no Capello, Ilha do Fayal. Anno de 1672. *Archivo dos Açores* **9** (53), 425–432.

Carneiro, A., Simões, A., Diogo, M.P. and Mota, T.S. 2013. Geology and religion in Portugal. *Notes and Records of the Royal Society of London* **67**, 331–354.

Carroll, R.P. 1990. Cognitive dissonance. *In:* Coggins, R.J. and Houlden, J.L (eds) *A Dictionary of Biblical Interpretation.* SCM, London, 123–4.

Castello Branco, A., Zbyszewski, G., Moitinho de Almedia, F. and da Veiga Ferreira, O. 1959. Rapport de la Première Mission Geológique. *Memórias Serviços Portugal* **4**, 9–27.

Chapin, F.W. 1989. *Tides of Migration: A Study of Migration Decision-Making and Social Progress in São Miguel, Azores.* AMS Press, New York.

Chester, D.K. 2002. Hazard and risk: Introduction and overview. *In:* Allison, R.J. (ed) *Applied Geomorphology: Theory and Practice.* John Wiley, Chichester, 251–263.

Chester, D.K. 2005. Volcanoes, society and culture. *In:* Marti, J. and Ernst, G.J. (eds) *Volcanoes and the Environment.* Cambridge University Press, Cambridge, 404–439.

Chester, D.K. and Duncan, A.M. 2007. Geomythology, theodicy and the continuing relevance of religious worldviews on responses to volcanic eruptions. *In:* Grattan, J. and Torrence, R. (eds) *Living under the Shadow.* Left Coast Press, Walnut Creek, CA, 203–224.

Chester, D.K., Duncan, A.M. and Dibben, C.R.J. 2008. The importance of religion in shaping volcanic risk perceptions in Italy, with special reference to Vesuvius and Etna. *Journal of Volcanology and Geothermal Research* **172**, 216–228.

Chester, D.K., Duncan, A.M. and Sangster, H. 2012a. Religion, faith communities and disaster. *In:* Gaillard, J-C., Wisner, B. and Kelman, I. (eds) *Routledge Handbook of Natural Hazards and Disaster Risk Reduction and Management.* Routledge, London, 109–120.

Chester, D.K., Duncan, A.M. and Sangster, H. 2012b. Human responses to eruptions of Etna (Sicily) during the late-Pre-Industrial Era and their implications for present-day disaster planning. *Journal of Volcanology and Geothermal Research* **225–226**, 65–80.

Chester, D., Duncan, A., Coutinho, R., Wallenstein, N. and Branca, S. 2017. Communicating information on eruptions and their impacts from the earliest times until the 1980s. *In:* Bird, D., Fearnley, C., Haynes, K., Jolly, G. and McGuire, B. (eds) *Observing the Volcano World; Volcano Crisis Communication.* Springer, Berlin, 419–443.

Chester, D.K., Duncan, A.M. and Speake, J. 2019a. Earthquakes, volcanoes and God: Comparative perspectives from Christianity and Islam. *GeoHumanities* **5** (2), 444–467.

Chester, D.K., Duncan, A.M., Coutinho, R. and Wallenstein, N. 2019b. The role of religion in shaping responses to earthquake and volcanic eruptions: A comparison between Southern Italy and the Azores, Portugal. *Philosophy, Theology and the Sciences* **6**, 33–45.

Cole, P.D., Guest, J.E. and Duncan, A.M. 1996. Capelinhos: The disappearing volcano. *Geology Today* 12 68–72.

Cole, P.D., Guest, J.E., Duncan, A.M. and Pacheco, J-M. 2001. Capelinhos 1957–1958, Faial, Azores: Deposits formed by an emergent surtseyan eruption. *Bulletin of Volcanology* **63**, 204–220.

Costa, R.M.M. 2008a. 1808- 'O Ano do Fogo'. Breve nota evocativa da erupção da Urzelina. *Boletim do Núcleo Cultural da Horta* **17**, 279–284.

Costa, R.M.M. 2008b. A brief introduction to the settlement of the Azores and a historical outline of the island of Faial. *In:* Goulart, T. (ed) *Capelinhos, a Volcano of Synergies. Azorean Emigration to America.* Portuguese Heritage Publications of California, San Jose, 3–16.

Costa, S.G. 2008. *Açores. Nove Ilhas, Uma História Azores. Nine Islands, One History.* Institute of Government Studies Press, University of California, Berkeley.

Coutinho, R., Chester, D.K., Wallenstein, N. and Duncan, A.M. 2010. Responses to, and the short and long-term impacts of the 1957/1958 Capelinhos volcanic eruption and associated earthquake activity on Faial, Azores. *Journal of Volcanology and Geothermal Research* **196**, 265–280.

Cunha, J.C. 1959. A erupção dos Capelinhos – sua influência na economia da Ilha do Faial. *Agros* **42** (6), 4.

Cunha, J.C. 1962. A erupão dos Capelinhos – Seus reflexos na economia de Ilha do Faial. *Memória dos Serviços de Portugal* **9**, 51–54.

Cutter, S.L., Barnes, L., Berry, M., Burton, C., Evans, E., Tate, E. and Webb, J.A. 2008. A place-based model for understanding community resilience to natural disasters. *Global Environmental Change* **18**, 598–606.

Dabney, J.B. 1808. New volcano: A letter from J.B. Dabney ESQ. American Consul, *In*: Richardson, J. (ed) *The American Reader: A Selection for Reading and Speaking Wholly from American Authors.* Lincoln and Edmants, Boston, MA, 45–48.

Davie, G. 1990. Believing without belonging: Is this the future of religion in Britain? *Social Compass* **37**, 455–469.

Degg, M. and Homan, J. 2005. Earthquakes vulnerability in the Middle East. *Geography* **90** (1), 54–66.

der Heide, E.A. 2004. Common misconceptions about disasters: Panic, the "disaster syndrome, "and looting. *In*: O'Leary, M. (ed) *The First 72 Hours: A Community Approach to Disaster Preparedness.* iUniverse Publishing, Lincoln, 340–380.

Dias, U.M. 1933. *Literatos dos Açores - Estudo histórico sobre os escritores açorianos.* Empresa Tipográfica, Vila Franca do Campo.

Dias, U.M. 1936. *História do Vale das Furnas.* Empresa Tipográfica, Vila Franca do Campo.

Dibben, C. and Chester, D.K. 1999. Human vulnerability in volcanic environments: The case of Furnas, Sao Miguel, Açores. *Journal of Volcanology and Geothermal Research* **92**, 133–150.

der Heide, E.A. 2004. Common misconceptions about disasters: Panic, the "Disaster Syndrome, "and looting. *In*: O' Leary, M. (ed) *The First 72 Hours: A Community Approach to Disaster Preparedness.* iUniverse Publishing, Lincoln, 340–380.

Dynes, R.R. 2003. *The Lisbon Earthquake in 1755: The First Modern Disaster.* Disaster Research Center, University of Delaware.

Dynes, R.R. 2005. The Lisbon earthquake of 1755: The first modern disaster. *In*: Braum, T.E.D. and Radner, J.B. (eds) *The Lisbon Earthquake of 1755: Representations and Reactions.* Voltaire Foundation, London, 34–49.

Etkin, D. 2016. *Disaster Theory: An Interdisciplinary Approach to Concepts and Causes.* Elsevier, Amsterdam.

Faria, D. 2014. Juízes indesejados? A contestação aos juízes de fora no Portugal Medieval (1354–1521). Unwanted judges? The protest against juízes de fora in medieval Portugal (1352–1541). *Cadernos do Arquivo Municipal* **2** (2), 19–37.

Feio, A., Bologna, R., Monteiro, D. and Félix, D. 2015. The earthquakes of Lisbon (1755) and Angra do Heroísmo (1980). *In*: Filion, P., Sands, G. and Skidmore, M. (eds) *Cities at Risk: Planning for and Recovering from Natural Disasters*, Routledge, Abingdon, 81–95.

Feraz, R. 2020. The Portuguese development plans in the postwar period: How much was spent and where? *Investigaciones de Historia Económica – Economic History Research* **16**, 45–55.

Fernandes, G. 2019. *The Pilgrim Nation: The Making of the Portuguese Diaspora in Post-War North America.* University of Toronto Press, Toronto, ON.

Ferreira, M.A., Proença, JM. and Oliveira, C.S. 2008. Vulnerability assessment in educational buildings - inference of earthquake risk: A methodology based on school damage in the July 9th 1998, Faial earthquake in the Azores. The 14th World Conference on Earthquake Engineering, October 12–17, 2008, Beijing China.

Ferreira, T.M., Maio, R. and Vicente, R. 2017. Analysis of the impact of large scale seismic retrofitting strategies through the application of a vulnerability-based approach on traditional masonry buildings. *Earthquake Engineering and Engineering Vibration* **16** (2), 329–348.

Forjaz, V.H. 2007a. Memórias de uma Geógrafa 40 anos depois. Professora Doutora Raquel Soeiro de Brito. In: Forjaz, V.H. (ed) *Vulcão dos Capelinhos 1957–2007.* OVGA Obervatório Vulcanológico e Geotérmico dos Açores, Ponta Delgada, Açores, 674–675.

Forjaz, V.H. 2007b. *Vulcão dos Capelinhos Memórias 1957–2007.* OVGA - Observatório Vulcanológico e Geotérmico dos Açores, Ponta Delgada.

Forjaz, V.H. 2007c. Em Memória dum Govenador 50 anos depois. Governador Dr. António de Frietas Pimentel. In: Forjaz, V.H. (ed) *Vulcão dos Capelinhos Memórias 1957–2007.* OVGA - Observatório Vulcanológico e Geotérmico dos Açores, Ponta Delgada, 670–672.

Forjaz, V.H. 2008. Historical eruptions in the Azores. In: Goulart, T. (ed) *Capelinhos: A Volcano of Synergies. Azorean Emigration to America.* Portuguese Heritage Publications of California, San Jose, 25–44.

Forlin, P. 2016. Under the volcano: Excavating in the Azores. *Medieval Archaeology* **56**, 3.

Forlin, P. 2020. Ritual of Resilience: The Interpretive Archaeology of Post-Seismic Recovery in Medieval Europe. In: Gerrard, C.M., Forlin, P. and Brown, P. (eds) *Waiting for the End of the World? New Perspectives on Natural Disaster in Medieval Europe*, Routledge, London, 21–42.

Fraião, O. 2005. Interview with Sr. Norberto de Oliveira Fraião cartographer employed by *Federação dos Municípios* (i.e. Federation of Local Councils), December 15, 2005 (personal communications).

França, J-A. 1977. *Lisboa Pombalina e Illuminismo.* Livraria Bertrand, Lisboa.

Garcia, M.V. 2005. Interview with Sr. Manuel Vargas Garcia – Technical Officer to the Junta Geral do Distrito Autónomo da Horta. December 15, 2005.

Garcia, F.N. 2007. Interview with Sr. Francisco Norberto Garcia. At the time of the eruption a farmer, but later he emigrated to the USA and worked for Delta and Pan American Airlines. Interviewed in 2007 in New York by Rui Coutinho.

García-Acosta, V. 2002. Historical disaster research. In: Hoffman, S.M. and Oliver-Smith, A. (ed) *Catastrophe and Culture: The Anatomy of Disaster.* School of American Research Press, Santa Fe, 49–66.

Gaspar, J.L., Queiroz, G., Ferreira, T., Medeiros, A.R., Goulart, C. and Medeiros, J. 2015. Earthquakes and volcanic eruptions in the Azores region: Geodynamic implications from major historical events and instrumental seismicity. In: Gaspar, J.L., Guest, J.E., Duncan, A.M., Barriga, F.J.A.S. and Chester, D.K. (eds) *Volcanic*

Geology of São Miguel Island (Azores Archipelago), Geological Society, London Memoir, 33–49.

Gerrard, C., Forlin, P., Froude, M., Petley, D., Gutiérrez, A., **44**Treasure, E., Milek, K. and Oliveira, N. 2021. The Archaeology of a Landslide: Unravelling the Azores Earthquake Disaster of 1522 and its Consequences *European Journal of Archaeology* 2021 (pre-publication).

Goulart, T. 2008. *Capelinhos, a volcano of Synergies. Azorean Emigration to America.* Portuguese Heritage Publications of California, San Jose.

Halikowski-Smith, S. 2010. The Mid-Atlantic Islands: A theatre of early modern ecocide. *International Review of Social History* **55** (suppl.), 51–77.

Hewitt, K. 1983. *Interpretations of Calamity.* Allen and Unwin, London.

Hinshaw, R.E. 2006. *Living with Nature's Extremes: The Life of Gilbert Fowler White.* Johnson Books, Boulder, CO.

Jolliffe, J. 1980. Relief operation begins after Azores earthquake. *The Guardian*, January 3, p. 6.

Junta-Geral (Horta). 1958a. *Inquérito Pecuário.* Intendência de Pecuária, Horta, Faial (cyclostyled).

Junta-Geral (Horta). 1958b. *Missão Técnica de Estudo das Consequências da Erupção Vulcânica da Ilha do Faial.* Ministério das Obras Públicas, Horta, Faial (cyclostyled).

Kelman, I. 2018. Lost for words amongst disaster risk science vocabulary. *International Journal of Disaster Risk Science* **9** (3), 281–291.

Kendrick, T.D. 1956. *The Lisbon Earthquake.* Methuen, London.

Lains, P. 2003. *Portugal's Growth Paradox 1870–1950.* Investigação – Trabalhos em Curso, Faculdade de Economia, Universidade do Porto.

Lobão, C.M.G. 2008. Capelinhos, fifty years later. The chronology of the eruption. *In*: Goulart, T. (ed) *Capelinhos, a Volcano of Synergies. Azorean Emigration to America.* Portuguese Heritage Publications of California, San Jose, 45–52.

Luz, J.B. 1996. O homem e a história em Gaspar Frutuoso. Homenagem ao Prof. Doutor Lúcio Craveiro da Silva. *Revista Portuguesa de Filosofia* **52**, 475–486.

Macdonald, N., Chester, D., Sangster, H., Todd, B. and Hooke, J. 2012. The significance of Gilbert F. White's 1945 paper, *Human Adjustment to Floods,* in the development of risk and hazard management. *Progress in Physical Geography* **36** (1), 125–133.

Macedo, A.L.S. 1871. *História das quatro ilhas que forman o Distrito do Horta.* Typographia de L.P. Silva Correa, Horta, 3 vols. (reissued in facsilile by Secretaria Regional da Educação Região Autónoma dos Açores, 1981.

Machado, F. 1959. A erupção do Faial em 1672. *Serviços Geológicos de Portugal,* Memória 4 (Nova Série), Lisboa.

Machado, F. and Forjaz, V.H. 1966. A crise síesmica de S. Jorge de Fevereiro 1964. *Boletim Sociedade Geológica de Portugal* **16** (1–2), 10–36.

Madeira, J.E. 1998. *Estudos de Neotectónica nas ilhas do Faial, Pico e S. Jorge: uma contribuição para o conhecimento Geodinâmico da junção tripla dos Açores.* Dissertação apresentada à Universidade de Lisboa para a obtenção do grau de doutor em geologia, na especialidade de Geodinâmica Interna. Faculdade de Ciências da Universidade de Lisboa, Departamento de Geologia.

Malheiro, A. 2006. Geological hazards in the Azores archipelago: Volcanic terrain instability and human vulnerability. *Journal of Volcanology and Geothermal Research* **156**, 158–171.

Martins, L.M. 1990. *Esbocete Histórico Sobre o Valle das Furnas.* Unpublished dissertation of the University of the Azores, Portugal.

Matias, L., Dias, N.A., Morais, I., Vales, D., Carrilho, F., Madeira, J., Gaspar, J.L., Senos, L. and Silveira, A.B. 2007. The 9th July 1998 Faial Island (Azores, North Atlantic) seismic sequence. *Journal of Seismology* **11**, 275–298.

Masson, F. and Banks, J. 1778. An account of the Island of St. Miguel. By Mr Francis Masson, in a letter to Mr William Aiton, Botanical Gardener to his Majesty. *Philosophical Transactions of the Royal Society of London* **68**, 601–610.

Mendonça, J.L.A. 1983. Assisténcia e culto na situação de emergéncia. *In*: Anon (ed) *Problemática da Reconstrução - sismo de 1 de Janeiro de 1980*. Instituto Açoriano de Cultura, Angra do Heroísmo **1**, 122–129.

Meyer, W.B. 2019. Urban Primacy before Mark Jefferson. *Geographical Review* **109** (1), 131–145.

Miller, G.W. 2011. *An Uncommon Man: The Life and Times of Senator Claiborne Pell*. University of Chicago, Press, Chicago, IL.

Ministry of the Interior. 1957. Letter sent to Governor Pimentel dated December 17 1957. *In*: Araújo, V. (ed) *O Resgatar de uma Memória: Sub-Prefeitos, Administradores Gerais, Governadores do Distrito Autónomo da Horta (1831–1975)*. Nova Gráfica Lda., Horta, 712, Author's Edition.

Neves, J.L.C. 1994. *The Portuguese Economy: A Picture in Figures: XIX and XX Centuries, with Long Term Series*. Universidade Católica Editora, Lisboa.

Neves, F. 2008. *Avaliação da vulnerabilidade sísmica do parque habitacional da ilha do Faial*. Civil Engineering. MSc thesis, University of Aveiro, Aveiro.

Neves, F., Costa, A., Vicente, R., Oliveira, C.S. and Varum, H. 2012. Seismic vulnerability assessment and characterisation of the buildings on Faial Island, Azores. *Bulletin of Earthquake Engineering* **10**, 27–44.

Nunes, J.C. 1999. *A actividade vulcânica na Ilha do Pico do plistocénico ao holocénico: mecanismo eruptivo e hazard vulcânico*. PhD thesis, Dept. Geociências, Universidade dos Açores.

Oliveira, A. de. 2008. Emotional ties and humanitarian support by the Azorean communities. 1500 U.S. Visas via a volcano. *In*: Goulart, T. (ed) *Capelinhos. A Volcano of Synergies*. Portuguese Heritage Publications of California, San Jose, 99–103.

Oliveira, C.S., Lucas, A.R. and Guedes, J.H. (eds) 1992. *Monografia 10 anos após o sismo dos Açores de 1 de Janeiro de 1980*. Secretaria Regional de Habitação e Obras Públicas - delagação da ilha Terceira. Laboratório Nacional de Engenharia Civil, Lisboa.

Oliveira, C.S., Ferreira, M.A. and Mota de Sá, F. 2012. The concept of a disruption index: application to the overall impact of July 9, 1998 Faial earthquake (Azores Islands). *Bulletin of Earthquake Engineering* **10**, 7–25.

Ordem dos Engenheiros. 1955. *Actas do Simpósio sobre a acção dos sismos*. Ordem dos Engenheiros Lisboa, no page numbers.

Paice, E. 2008. *Wrath of God*. Quercus, London.

Pell, C. 1980. *Earthquake in the Azores, a Tragedy and the American Response: A Report of the Committee on Foreign Relations, United States Senate*. US Government Printing Office, Washington, DC.

Pimentel, A. de Freitas. 1957a. Account of the eruption from Governor Pimentel to the Minister of the Interior October 27 1957. *In*: Araújo, V. (ed) 2007, *O Resgatar de uma Memória: Sub-Prefeitos, Administradores Gerais, Governadores do Distrito Autónomo da Horta (1831–1975)*. Nova Gráfica, Lda., Horta, 700–702, Author's Edition.

Pimentel, A, de Freitas. 1957b. Account of the eruption from Governor Pimentel to the Minister of the Interior October 28 1957. *In*: Araújo, V. (ed) 2007, *O Resgatar de* uma *Memória: Sub-Prefeitos, Administradores Gerais, Governadores do Distrito Autónomo da Horta (1831–1975)*. Nova Gráfica, Lda., Horta, 708–711, Author's Edition.

Pimentel, A, de Freitas. 1957c. Telegram from Governor Pimentel to the Minister of the Interior. *In*: Araújo, V. (ed) 2007, *O Resgatar de uma Memória: Sub-Prefeitos, Administradores Gerais, Governadores do Distrito Autónomo da Horta (1831–1975)*. Nova Gráfica, Lda., Horta, 711–712, Author's Edition.

Pimentel, A, de Freitas. 1957d. Letter from Governor Pimentel to the Minister of the Interior. *In*: Araújo, V. (ed) 2007, *O Resgatar de uma Memória: Sub-Prefeitos, Administradores Gerais, Governadores do Distrito Autónomo da Horta (1831–1975)*. Nova Gráfica, Lda., Horta, 715–716, Author's Edition.

Pinto, M.S. 2003. Gaspar Frutuoso, a Portuguese volcanologist of the 16th century. *Açoreana* **10**, 207–226.

Pires, A.M. 1995. *Açorianidade. Enciclopéfia Açoriana*. Cultura, Governo dos Açores.

Portuguese Government. 1964a. *Decreto-Lei 45 685*. Ministérios do Interior, das Finanças, das Obras Públicas, da Economia e da Saúde e Assistência, Sectretaria de Estado da Agricultura, Lisboa. https://dre.pt/application/dir/pdf1sdip/1964/04/10000/05810584.PDF

Portuguese Government. 1964b. Decreto 45 687. Ministérios do Interior, das Finanças, das Obras Públicas, da Economia e da Saúde e Assistência, Sectretaria de Estado da Agricultura, Lisboa. https://dre.pt/application/dir/pdf1sdip/1964/04/10000/05860586.PDF

Proença, J.M. 2004. Damage in schools in the 1998 Faial earthquake in the Azores Islands, Portugal. *In:* Anon (ed) *Keeping Schools Safe in Earthquakes*. OECD, Paris, Chapter 8, 131–139.

Ramos Villar, C.M. 2006. *The metaphorical 'tenth island' in Azorean literature. The theme of emigration in the Azorean imagination*. The Edward Mellon Press, Leviston, Queenston, Lampeter.

Ribeiro, J.S. 1844. *Collecção dos escriptos administrativos e litterarios do snr José Silvestre Ribeiro*. Imprensa do Governo, Angra do Heroísmo.

Robinson, R.A.H. 1977. The Religious Question and the Catholic Revival in Portugal, 1900–30. *Journal of Contemporary History* **12** (2), 345–362.

Rodrigues, R. 1991. *Notícia Biográfica do Dr. Gaspar Frutuoso*. Cultural Institute of Ponta Delgada, Ponta Delgada.

Rodrigues, L.N. 1996. *A Legião Portuguesa. A Milícia do Estado Novo, 1936–1944 (The Portuguese Legion: A New State Militia 1936–1944)*. Editorial Estampa, Lisboa.

Rosa, M.E. and Pereira, J.F. 2008. Capelo and Praia do Norte, before during and after the volcano 1957–1958. *In*: Goulart, T. (ed) *Capelinhos, a Volcano of Synergies. Azorean Emigration to America*. Portuguese Heritage Publications of California, San Jose, 53–66.

Sá, M. da Glória Pires de. 2008. The Portuguese in the United States Census of 2000. *In*. Goulart, T. (ed) *Capelinhos, a Volcano of Synergies. Azorean Emigration to America*. Portuguese Heritage Publications of California, San Jose, 87–97.

Scarth, A. and Tanguy, J-C. 2001. *Volcanoes of Europe*. Terra, Harpenden.

Scofield, J. 1958. A new volcano bursts from the Atlantic - off Fayal, in Portugal's verdant Azores, a jack-in-a-box eruption smothers villages, awes visitors and even catches whales. *National Geographic* **113** (6), 735–757.

Senos, M.L., Gaspar, J.L., Cruz, J., Ferreira, T., Nunes, J.C., Pacheco, J., Alves P., Queiroz, G., Dessai, P., Coutinho, R., Vales, D. and Carrilho, F. 1998. O terramoto do Faial de 9 de Julho de 1998. *Proceedings do 1° Simpósio de Meteorologia e Geofísica da APMG*, Lagos, Azores, 61–68.

Shapleight, A., Sterne, M., Tavares, J., Trejo, L. and Lucas, C. 1983. January 1, 1980. Azores earthquake reconstruction. A report after two and one half years. *In:* Anon (ed) *Problemática da Reconstrução - sismo de 1 de Janeiro de 1980*. Instituto Açoriano de Cultura, Angra do Heroísmo **2**, 210–270.

Silva, H.G. 2008. Portuguese-Azorean cultural identity: Common traits of Acorianidade/Azoreanness. *In:* Goulart, T. (ed) *Capelinhos, A Volcano of Synergies. Azorean Emigration to America*. Portuguese Heritage Publications, San Jose, 73–83.

Silveira, D.I.R.P. 2002. *Caracterização da sismicidade histórica da ilha de S. Miguel com base na reinterpretação de dados de macrossísica: contribuição para a avaliação do risco sísmo.* Departamento de Geociências Universidade dos Açores.

Simões, A., Carneiro, A. and Diogo, M.P. 2012. Riding the wave to reach the masses: Natural events in the early twentieth century Portuguese daily press. *Science and Education* **21**, 311–333.

Sousa Martins, R. 2015. Ruínas, patrimonialização e cultura cerâmicas na Praia da Vitória. Terceira, Açores. https://www.iac-azores.org/iac2018/projetos/IPIA/terceira/praiavitoria/ruinas.html.

SREA. 2007. Séries Estatísticas 1995–2005. Serviço Regional Estatística dos Açores, Açores.

SREA. 2011. *Censos 2011: Resultados Preliminares.* Serviço Regional de Estatística dos Açores, Açores.

SREA. 2020a *Demografia : Séries Annuais.* Serviço Regional de Estatística dos Açores, Açores. https://srea.azores.gov.pt/Conteudos/Relatorios/lista_relatorios.aspx?idc=6194andidsc=6825andlang_id=1

SREA. 2020b. *Demografia: Séries Longas.* SREA - Serviço Regional de Estatística dos Açores. https://srea.azores.gov.pt/Conteudos/Relatorios/lista_relatorios.aspx?idc=6194andidsc=6825andlang_id=1_

Teletin, A. and Manole, V. 2015. Expressing cultural Identity through saudade and dor: A Portuguese-Romanian comparative study. *In:* Baptista, M.M. (eds) *Identity: Concepts, Theories, History and Present Realities (a European Overview) Vol. 1* Grácio, Editor, Coimbra, 155–171.

Teves-Costa, P., Oliveira, C.S. and Senos, M.L. 2007. Effect of local site and building parameters on damage distribution in Angra do Heroísmo - Azores. *Soil Dynamics ad Earthquake Engineering* **27**, 986–999.

Teves-Costa, P. and Veludo, L. 2013. Soil characterization for seismic damage scenarios purposes: Application to Angra do Heroísmo (Azores). *Bulletin of Earthquake Engineering* **11**, 401–421.

Udías, A. 2009. Earthquakes as God's punishment in 17th and 18th century Spain. *In:* Kölbl-Edert, M. (ed) *Geology and Religion: A History of Harmony and Hostility.* The Geological Society, London, Special Publication **310**, 41–48.

United Nations. 2006. *Hyogo Framework for Action 2005–2015.* United Nations, Geneva.

United Nations. 2014. *Hyogo Framework for Action (HFA).* United Nations, Geneva.

United Nations. 2015. *Sendai Framework for Disaster Risk Reduction 2015–2030.* UN-ISDR, United Nations, Geneva.

Veludo, I., Teves-Costa, P. and Bard, P-Y. 2013. Damage seismic scenarios for Angra do Heroísmo. *Bulletin of Earthquake Engineering* **11**, 423–453.

Wallenstein, N., Chester, D., Coutinho, R., Duncan, A. and Dibben, C. 2015. Volcanic hazard vulnerability on São Miguel Island, Azores. *In:* Gaspar, J.L., Guest, J.E., Duncan, A.M., Barriga, F.J.A.S. and Chester, D.K. (eds) *Volcanic Geology of São Miguel Island (Azores Archipelago),* Geological Society, London Memoir **44,** 213–225.

Wallenstein, N., Duncan, A., Coutinho, R. and Chester, D. 2018. Origin of the term nuée ardentes and the 1580 and 1808 eruptions on São Jorge Island, Azores. *Journal of Volcanology and Geothermal Research* **358,** 165–170.

White, G.F. 1936. The Limit of Economic Justification for Flood Protection. *The Journal of Land and Public Utility Economics* **12,** 133–148.

White, G.F. 1945. *Human Adjustments to Floods. A Geographical Appeal to the Flood Problem in the United States.* Department of Geography, Research Paper 29, University of Chicago.

White, G.F. 1974. Natural hazards research: Concepts, methods, and policy implications. *In:* White, G.F. (eds) *Natural Hazards Local, National, Global.* Oxford University Press, New York, 3–16.

Williams, A.M. and Fonseca, M.G. 1999. The Azores between Europe and North America. *In:* King, R. and Connell, J. (eds) *Small Worlds, Global Lives.* Continuum, London, 55–76.

Wisner, B., Blaikie, P., Cannon, T. and. Davis, I. 2004. *At Risk: Natural Hazards, People's Vulnerability and Disasters.* Routledge, London.

Zaman, M.Q. 1999. Vulnerability, disaster and survival in Bangladesh: Three case studies. *In:* Oliver-Smith, A. and Hoffman, S. (eds) *The Angry Earth: Disaster in Anthropological Perspective.* Routledge, London, 192–212.

Zbyszewski, G. and Veiga Ferreira, O. 1959. Rapport de la deuxiéme mission géologique sur le volcanisme de L' Ille de Faial. *Serviços Geológicos de Portugal* 4 (new series), 29–55.

6 Conclusion

Developing policies of disaster risk reduction

The development of policies of disaster risk reduction (DRR) in the Azores is reflective of both changing domestic and international contexts. Domestically DDR policies are one manifestation of the profound political and cultural adjustments that occurred as the Portuguese State transitioned from the dictatorial regime of the *Estado Novo*, to representative democracy following the 1974 revolution (see Chapter 1). Key initial steps on this journey to effective policy development were the establishment in 1976 of both the Regional Government (*Governo Regional dos Açores*) and the Department of Geosciences of the University of the Azores (Departamento de Geociências da Universidade dos Açores). Further steps and stimuli were marked by the responses to and lessons learnt from the 1980 Terceira emergency; new academic initiatives undertaken from the start of 1990s and the emergence of aspects of *proactive* policy which were evident as the Azores coped with the exigencies of the 1998 Faial earthquake. Internationally a major change occurred in 1986 when Portugal became a member of the European Economic Community and from 1993 the European Union. This not only allowing the Azores to benefit from regional economic development funds (see Chapter 1), but also encouraged collaborative research programmes to be developed with other member states (see below).

Within these domestic contexts the principal academic initiatives have been as follows:

a   A re-focusing of research with increased support for the earth sciences generally and, more specifically, greater stress on themes related to risk reduction and the ways resilience may be enhanced. This began with participation of university-based geologists and geophysicists in the 1980 emergency and was given greater emphasis and range by the *European Laboratory Volcanoes* programme, which began in the early 1990s and concentrated on Furnas volcano (São Miguel).[1] Focus on applied earth science continued at a quickening pace over the next three decades. New research directions involved collaboration with a wide-range of scholars who have included, *inter alia*: civil engineers; mathematicians; physical geographers and medical researchers, both Portuguese and foreign (see

DOI: 10.4324/9780429028007-6

Duncan et al. 1999b; Gaspar et al. 2015a). Research has also called on the expertise of scholars with training in the social sciences having interests which include social, cultural, informational and perceptual aspects of hazard vulnerability (Wallenstein et al. 2015); planning and civil protection (see below). Research on the ways in which people perceive and respond to environmental risks has been particularly noteworthy and has included a wide-breadth of domestic and international scholarship (e.g. Dibben 1999; Dibben & Chester 1999; Estrela Rego & Arroz 2012; Estrela Rego et al. 2018; Anderson & Mileu 2020; Cabral 2020; Lotteri et al. 2021).

b   As argued in Chapters 3 and 5, it is buildings and not earthquakes *per se* that usually kill and injure people, and in eruptions over-loading of roofs by tephra-fall may transform traditional dwellings into dangerous structures. In certain areas of the Azores, ingress of gases into dwellings also poses a threat to public health through asphyxiation (see Chapter 4). Research on new resilient buildings and design changes that seek to make older building more resilient have been major research themes. This has involved, not only building inspection in the field such as occurred in the pioneer research of Pomonis et al. (1999) on the exposure of domestic architecture to tephra-loading in Furnas village (São Miguel), but also the use of databases of damaged buildings, particularly but not exclusively those compiled following the 1998 Faial earthquake. In both cases the aim is to promote design improvements including retrofitting (Ferreira et al. 2016; see also Chapters 3 and 5).

Research on hazard reduction has evolved since it was first introduced by Gilbert White and others in the 1940s. In the 1980s, scholarship underwent a paradigm shift as it moved from a physically deterministic conceptualisation to one more sensitive to society and culture (see Chapter 5). Today both the Portuguese State and the Regional Government are committed to boosting resilience and reducing vulnerability by embracing policies that accord more closely to those proposed by the United Nations and which it has commended to its member states.[2] In the Azores this has involved, not only the formation of an administrative structure, but also and more importantly the initiation of *proactive* policies in advance of future emergencies and an expressed desire to 'build-back better,' both of which were foreshadowed by the responses of the Regional Government to the 1998 earthquake (see Chapter 5).

Academic initiatives: applied earth science and civil engineering

The principal application of applied earth science in the Azores has been the long and short-term prediction of volcanic eruptions and earthquakes and, as a reflection of its importance, one chapter of the present volume (Chapter 4) is devoted exclusively to an in-depth discussion of this topic. A further applied theme which has engaged the interest, not just of earth scientists but of civil engineers, medical scientists and architects, is building safety and two inter-linked research questions have been addressed:

1 On the basis the success or failure of different building designs during past events what are the lessons that may be learnt with respect to new construction?

2 The Azores has a strong heritage of distinctive, valued yet highly vulner-able buildings. What lesson may be drawn from previous events so build-ings can be modified to make them more resilient? To use the United Nations' terminology how might Azoreans 'build-back better' (United Nations 2015).

In approaching these questions two methodologies have been adopted and two concerns addressed. Methodologies may be characterised as being either inductive or deductive: the former using surveys and databases of damage to draw general trends of inappropriate building practice which exacerbate vul-nerability; whereas the latter proposes different designs and tests these against catalogues of damage. In both cases policy recommendations are forthcom-ing. The two concerns are buildings at risk on the flanks of active volca-noes, which includes the dangers of ash loading, seismic vulnerability and the threats posed to homes subject to the ingress of harmful gases; and the more widespread exposure of settlements to the effects of tectonic earthquakes and earthquake-related phenomena.

Improving building safety on the flanks of active volcanoes

The first published research emerged under the auspices of the *European Lab-oratory Volcanoes* programme, when an in-depth study of Furnas volcano on São Miguel was carried out. Field work was undertaken by a team from the University of Cambridge (Pomonis et al. 1999) and, adopting a sub-Plinian '1630-type eruption' scenario and assuming that a similar area would be impacted (Chapter 3), the team surveyed all residential buildings in the settlements of Furnas and Ribeira Quente, the central 'core' area of Pov-oação and the eastern suburbs of Ponta Garça (Figure 3.3). Some 80% of buildings were found to be of traditional 'rubble-stone' construction (see Chapters 3 & 5), with many having been improved through the addition of reinforced concrete slabs and columns. More recent buildings were erected using load-bearing un-reinforced concrete blocks and reinforced concrete frames. Three aspects of a '1630-type eruption scenario' were considered: volcanic earthquakes; pyroclastic density currents (PDCs) and tephra fall. In all cases a large proportion of the building stock was found to be de-ficient. For instance, assuming an M 5.5 volcanic earthquake with a focal depth of 10 km, modelling indicated that 800 rubble-stone buildings and 20 un-reinforced concrete buildings would totally[3] or partially collapse.

With respect to PDCs, the maximum lateral pressure a typical Azorean house can withstand without collapsing was calculated at 5 kN/m^2, whereas pressures of up 20 kN/m^2 might be expected should a PDC sweep through Furnas village. Assuming a roof loading (i.e. thickness) of wet tephra in excess of 25 cm, some 4,500 houses could be affected and up to 20% of the population would be either injured or killed (Pomonis et al. 1999, p. 128).[4]

This ground-breaking paper set the standard for subsequent work by carefully suggesting policy options. These included improved construction methods for new homes to make them more seismically resistant, and the adoption of building codes and design principles from regions of the world where heavy snowfall necessitates strong roof construction. One notable feature highlighted by this study is its focus on design, construction methods and work standards. In the Furnas area the authors noted 'poor reinforcement ... and anchoring, irregular frames, slender columns connected to deep beams, short columns and soft storeys.'[5] These are some of the frequently occurring mistakes known to affect reinforced concrete building vulnerability in earthquakes zones around the world. Notwithstanding these issues, the authors still found that 'the overall strength of these (reinforced) buildings is superior to the older rubble-stone masonry' (Pomonis et al. 1999, p. 121).

Research by the Cambridge group continued into the following decade. Spence et al. (2005) compared findings on roof-loading from the Azores with those contained in a global database of 27 historical eruptions that produced significant quantities of tephra, with additional information being added from a number of volcanic regions including: Vesuvius (Italy); Tenerife (Canary Islands, Spain) and Guadeloupe (Caribbean). Detailed published studies of tephra-producing eruptions on Pinatubo (Philippines 1991; Spence et al. 1996) and Rabaul (Papua New Guinea 1994; Blong 2003) were also incorporated into the study. These data allow the authors to propose generic policy recommendations about tephra loading which may be applied across volcanically active regions of the world and, more specifically, in the Azores.

Table 6.1 summarises these recommendations, but Spence et al. (2005) go on to make a series of perspicacious comments that are of relevance to Civil Protection more generally. First even new buildings designed to high standards can never be wholly 'tephra-safe' under all future eruptions. On the basis of past activity, the most likely expected future scenario is selected for planning purposes, but there is always the possibility that the magnitude of an eruption would be greater and/or it could take an unexpected course. Second, just as river defences may encourage incautious development of flood plains (Kelman 2001), so building codes may encourage sub-optimal behaviour by falsely reassuring people that they can remain in their homes during an eruption and so potentially put themselves at greater risk. Finally, Spence et al. (2005) argue that prioritisation is vital. Funds for retrofitting are usually limited and utilitarian policies[6] may have to be applied in prioritising the retrofitting of buildings which will be used as shelters (e.g. churches and schools) and/or as control centres when managing an emergency (e.g. police headquarters, fire stations, government control rooms and hospitals).

Table 6.1 Mitigation of roof collapse from tephra loading: some international recommendations for building codes

Codes of Practice for new construction	The strength of roofs should be assessed using tephra maps showing the recurrence of an eruption with a 500-year interval. Using the nomenclature commonly adopted in earthquake codes (Kircher & Hamburger 2000), roofs should be designed for 10% probability of exceeding such an event once in 50 years.
	Increasing the roof angle (i.e. pitch) is not recommended because its effects are not well understood, and there may be material, technological and/or aesthetic concerns.
	Ventilation openings should be designed to prevent the ingress of small particulars as a result of adverse health effects on inhabitants and because of the additional damage they can cause to interiors.
Modifications to existing buildings	The straightforward measure is retrofitting to make roofs stronger, according to the design criteria for new buildings and the probability calculus suggested above.
	If buildings are immediately threatened, then research is required on the best ways to shore-up roofs given the particularities of design and the materials available.
	Damage due to the ingress of particulates requires openings to be sealed and glass to be covered with strong materials such as plywood or metal sheeting.
	Complete sealing is not likely be successful, but the infiltration rate of particulates may be successfully reduced.
	Success may require materials to be available on-site well in advance of an eruption.
Emergency management/	Sub-plinian or plinian events will involve the deposition of lithics and pumice clasts of differing calibres. These can cause injury and death, and people need to shelter in robustly constructed buildings.
	Cleaning tephra off roofs when it is still falling is challenging and unsafe, given the poor visibility and atrocious atmospheric conditions. This is not a recommended policy option.
	Community centres, such as church buildings and schools, are places where people naturally congregate and such edifices often have wide roof spans. Preventing people spontaneously entering such spaces is challenging. Strengthening may be a priory and/or additional safer locations for people to congregate should be identified by Civil Protection authorities well in advance of an eruption.
	Advising people to evacuate in advance of an eruption and from places not generally perceived to be threatened, is likely to be unsuccessful. Such policies require detailed pre-disaster planning and public acceptance.

Only those of relevance to the Azores are included. Based on Spence et al. (2005).

Following the start made by the Cambridge Group, further research on the hazard exposure of buildings located on the flanks of the active volcanoes was continued by members of *the Centro de Vulcanologia e Avaliação de Riscos Geológicos* (CVARG – Centre for Volcanology and Geological Risk

Assessment – see below) and has used a similar inductive methodology (Gomes et al. 2006). Buildings in 10 *freguesias* (i.e. parishes) located on the flanks of Sete Cidades volcano on São Miguel were identified and classified in terms of their exposure to earthquakes (Figure 6.1). Out of a total of over 7,000 buildings, 62% housed an estimated 11,429 inhabitants. Using the European Macroseismic Scale (EMS-98; Grünthal 1998; see Chapter 3),[7] the authors found that some 76% and 17% of houses (i.e. 93% of the total) fell, respectively, into Classes A and B: the least resilient categories. They estimate that, should the area be affected by an Intensity IX earthquake, then between 57 and 77% of dwellings would partially or totally collapse, 15%–25% would require some form of rehabilitation and up to 92% of the population could be affected. By constructing various scenarios CVARG researchers were able to show that even relatively modest seismic swarms, such as those that accompanied the 1811 submarine eruption, would severely impact the parishes of Candelária, Ginetes and Mosteiros (Figure 6.1). These settlements would have to cope with maximum seismic Intensities of IX and the remainder of the study area Intensities of VIII; with 28 to 63% of dwellings suffering partial or total collapse and 57%–81% of the population being adversely affected. Gomes et al. (2006) point out that for most future scenarios the outlook is bleak and that unless action is taken the authorities will have to contend with high levels of

Figure 6.1 Location of the 10 *freguesias* (parishes) that were included in the CVARG survey of housing on Sete Cidades volcano. Two parish names are abbreviated, full names: Pilar da Bretanha and Ajuda da Bretanha. Based on Gomes et al. 2006, fig. 6, page 46. Copyright permission Dr Ana Gomes This work is licensed under a Creative Commons License.

mortality and morbidity, mass evacuation and, in the longer-term, a possible brake being placed on the island's economic development for some years.

The severe threat posed to human health by the ingress of volcanic gases into homes has been discussed in Chapter 4 and some measures have been adopted and others proposed to improve resilience. Across the Azores volcanic-hydrothermal gases are released by low temperature fumaroles (95°C to 100°C), thermal springs, steaming ground, cold CO_2-rich springs and from areas of diffuse soil degassing (Ferreira et al. 2005). In areas of diffuse degassing CO_2, radon (^{222}Rn) and H_2S are gases that may be injurious to health and clearly the best solution is for new construction to avoid areas of high concentration. Maps have been produced by CVARG[8] to assist the Regional Government in future development planning. With respect to soil diffuse CO_2, Fátima Viveiros and her colleagues (Viveiros et al. 2015, p. 193) make the important point that ideally such hazardous locations should be avoided and that the most desirable solution would be to remove the population at risk, but recognise that this is not politically feasible because of an unwillingness of people to move from the area. There are mitigations measures that have either already been introduced or could be adopted in the future to make buildings safer, and these are summarised in Table 6.2.

Table 6.2 Building design improvement in areas of high gas emission

Gases	Suggested Building Design Improvements
Carbon Dioxide (CO_2)	Cellars and basements are to be avoided, systems of ventilation should be introduced (e.g. ventilated air spaces between the foundations and the ground floor) and impermeable membranes between the soil surface and the dwelling should be fitted. It is suggested by Viveiros et al. (2015) that these measures should be considered a minimum and made a legal requirement. Artificial ventilation could also be introduced in exposed existing buildings and this should involve improving natural under-floor ventilation. Forced draft and/or air conditioning should be considered (Durand & Scott 2005).
Radon (^{222}Rn)	Following measurement of ^{222}Rn in every potentially exposed building, remedial measure might include (as above): natural and/or artificial ventilation and use of impermeable membranes. Also in buildings within areas at high risk, permanent continuous monitoring of ^{222}Rn should be considered, because in the Furnas area measured values often exceed the reference level of 150 Bq/m^3, which is defined in the Azorean legal code as being hazardous (Silva et al. 2015).[1]

1 Under Portuguese Law the value is 400 Bq/m^3 and under Word Health Organisation (WHO) Guidelines 100 Bq/m^3.
Based on information in Silva et al. (2015) and Viveiros et al. (2015).

Improving the resilience of buildings to future earthquakes

In contrast to volcanic eruptions and large landslides, the products of which may sterilise land and preclude immediate re-settlement, there are few instances anywhere in the world where re-location of a city, town or village has been a policy response to earthquake losses.[9] There are two reasons why relocation has not been adopted in the Azores. First as the history of earthquake losses makes clear (Chapter 3) and *long-term prediction* (Chapter 4) confirms, with the exception of Corvo and Flores no location in the Azores is without risk of earthquake damage, though some areas are more exposed than others. Hence relocation may reduce risk but cannot eliminate it entirely. Second, although it is established practice to locate new development to avoid substrates that are known to amplify seismic waves (Chapter 3 and Teves-Costa et al. 2007) and to reduce vulnerability at the coast by being aware of the risk of tsunami inundation (Chapter 4 and Andrade et al. 2006), overall policy in the Azores is to rebuild settlements *in situ* following disasters. In the aftermath of the Ribeira Quente landslide disaster of October 1997 (see Chapter 1 and below), the Regional Government initially planned to move the population to a safer location. Some families agreed to move, but the majority preferred to stay and this desire was expressed politically in the recovery plan (Cunha 2003, p. 32). Although smaller settlements could theoretically be re-sited there are few if any sites where cities on the scale of Ponta Delgada, Angra do Heroísmo or Horta could successfully be re-located and enjoy similar locational advantages as they do today.

The majority of Azorean buildings are old having been constructed from traditional rubble-stone with or without reinforcement (see above), and complete replacement is ruled out on several grounds. Not only would the cost of replacing the historic building stock be prohibitive, but on both safety and aesthetic grounds there is also an established need to make existing buildings less vulnerable and more resilient. In addition, there is a clear imperative to maintain the attractiveness of the built environment and to keep the islands' heritage safe and attractive for tourists, who now make a significant contribution to economic wellbeing (Chapter 1). These considerations have combined to stimulate research on the techniques that can be used to make older buildings safer through retrofitting.

Retrofitting involves modifying buildings to make them more resilient to seismic shaking (Chuang & Zhuge 2015, see Chapter 4) and in the Azores two approaches have been followed. A minority of researchers have modelled the impact of future earthquake on 'typical' Azorean buildings and have used the results to propose retrofitting measures. In contrast, the majority of scholars have worked inductively through the interrogation of databases of damage, together with information collected in the field.

A good example of the modelling approach is Fagundes' (2015) research, where the behaviour of both individual structures and groups of joined buildings (i.e. rows) are considered. Selecting a townscape frequently found in

Angra do Heroísmo (Terceira), where a three-storey building is sandwiched between two-storey buildings, the effects of seismic activity are modelled. Both the separate buildings and the row showed considerable damage, with the latter being more badly impacted because of the differing seismic responses of the three structures, varying floor heights being particularly significant (Fagundes et al. 2017). On the basis of these results retrofitting measures are suggested, specifically: insertion of resistant vertical elements to reduce torsion (i.e. twisting) loads; replacement of flexible with rigid floors to enhance overall stiffness and the strengthening of load-bearing walls.

Inductive research using databases and field surveys of damage first began in the immediate aftermath of the 1998 earthquakes (Spencer et al. 2000)[10] and, in Horta (Faial) where damage was moderate (Maximum Intensity VI EMS), it was noted that: there were no collapses; some buildings needed extensive repairs with cracks of several centimetres in each storey due to rotation about the foundations; walls were deformed; interiors damaged and some parapets collapsed. One interesting finding was that buildings repaired after the 1926 earthquake using ties (i.e. metal rods with screw bolts) performed far better than those strengthened post-1973 using collar beams. Collar beams, or seismic bands, are continuous horizontal runners of reinforced concrete or wood going round buildings with strong connections at corners and at the 'T-junctions' between walls. After the 1973 earthquake these were only inserted in walls and did not provide a complete loop of reinforcement (Spencer et al. 2000, p. 5). Clearly if retrofitting is to be successful then it has to accord to best civil engineering practice.

Since 2000 many papers have been published using the 1998 database of damage (Costa & Arêde 2004, 2006; Zonno et al. 2010; Neves et al. 2012; Gomes et al. 2015; Ferreira et al. 2016, 2017a; Maio et al. 2017; Fontiela et al. 2020), and a flavour of research may be gleaned by examining one paper in detail. Based on a survey of over 300 buildings in Horta, Ferreira et al. (2016) make the point that retrofitting is expensive and argue that it should only take place if justified by a cost-benefit analysis. Retrofitting is generally not justified when either the required capital investment exceeds the value of the building, or when a structure is so badly damaged that it is classified as a pre-ruin. 'Costs associated with demolition, debris disposal and reconstruction determine the feasibility of each retrofit(ting) project. Moreover, legal issues arise when the safety of a building is dependent on adjacent housing units.' For example, when several owners share a building and/or housing unit and share a common wall or walls. 'Retrofitting a single house in row housing has low benefit when adjacent units are seismically deficient' (Ferreira et al. 2016, p. 295). Where justified the authors suggest a number of bespoke solutions which include wall to wall connection using tie rods; floors stiffening by diagonal bracing and new timber planking; concrete strapping of the roofs to the walls and 'masonry consolidation' (i.e. using special reinforcing plaster).

In conclusion, Ferreira et al. (2016) make an additional point. Retrofitting they argue ought not be the first choice policy because mitigation should ideally focus on land use zoning based on *long-term forecasting* (Chapter 4), together with the enforcement of building codes for new and substantially re-built structures (see below).

Academic initiatives: social science

Like many aspects of hazards research in the Azores, social science involvement dates from the time of the *European Laboratory Volcanoes Project* of the 1990s and its work on Furnas volcano (see above and Duncan et al. 1999a). This project spawned three social science firsts: a report on the issues involved in planning evacuations (Chester et al. 1995, see below); a paper on hazard vulnerability (Dibben & Chester 1999) and doctoral research related to social science aspects of hazard exposure (Dibben 1999).[11] Research has continued and some of the findings by social scientists have already been reviewed: in Chapter 5, where vulnerability was defined as comprising physical, de-mographic/economic, social/cultural and perceptual/ informational dimensions (Wallenstein et al. 2015); in Chapter 4, where it was argued that the Azores are hazard exposed even between discrete eruptions and earthquakes (Wallenstein et al. 2005, 2007) and again in Chapter 5, where the importance of population, economy, society, culture and religion in shaping responses to emergencies both in *pre-industrial* and *industrial* times, was discussed. In addition, Chapter 1 is based on a combination of long-standing historical and more recent social science scholarship. Other social science projects have been important and have been concentrated on two themes: the selection of routes suitable for evacuation, should an area be badly impacted by a volcanic eruption and analysis of the ways in which Azoreans perceive threats from extreme, yet rare, events that have implications for the policy development.

Selection of evacuation routes

Research was initiated during the Furnas-focused *European Laboratory Volcanoes Project* during the 1990s (Chester et al. 1995, 1999; Duncan et al. 1999b), was later extended to encompass Fogo (Wallenstein et al. 2005, 2007) and, finally, Sete Cidades (Wallenstein et al. 2015). The technique used has been a combination of cartographic analysis and detailed on-site surveys of the road network. So far only roads on the island of São Miguel have been studied, reflecting not only the relative demographic and economic importance of this island, but also the fact that it has three active and potentially dangerous volcanoes. Furnas and Sete Cidades villages are located within active calderas (Figure 6.2) and would have to be evacuated as soon as they were threatened by an eruption, with isolated villages also being particularly dangerous locations. For instance, the road linking the coastal village of Ribeira Quente (Figure 6.2) follows the valley of the Ribeira Amarela, which drains the crater lake and flows through Furnas village, reaching the coast at Ribeira Quente.

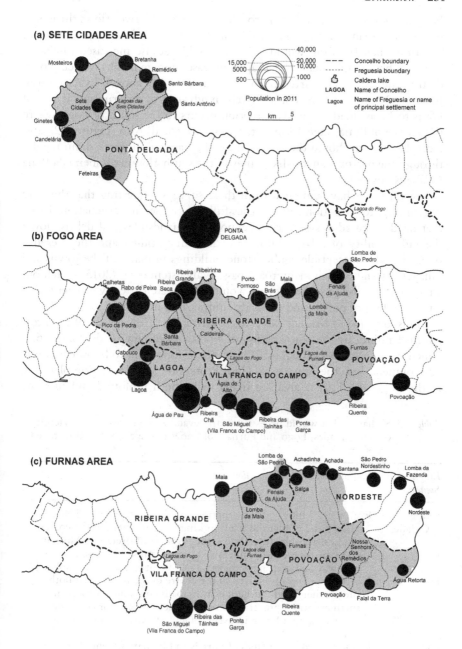

Figure 6.2 The areas most likely to be affected by eruptions of Sete Cidades, Fogo and Furnas volcanoes, and the population numbers of each *freguesia*. From Wallenstein ét al. 2015, Fig. 16.2, p. 216). Copyright permission Geological Society of London.

For part of its course and as it approaches the coast, the river flows through a narrow steep-sided valley and, should the Furnas crater lake drain during an eruption, then flooding would be exacerbated because increased discharge can only be accommodated by a rapid increase in stage height, which would destroy the road, as occurred in the 1997 disaster (see above). Evacuation of Ribeira Quente will be required well in advance of an eruption if loss of life is to be avoided. A similar situation occurs to the south of Fogo where the village of Praia, sited 1 km to the west of Água d' Alto (Figure 6.2), also faces destruction should the crater lake – the Lagoa do Fogo – drain rapidly through the narrow canyon-like valley of the south-flowing Ribeira da Praia (Wallenstein 1999).

Several issues have emerged from these surveys suggesting that the road system requires attention if it is successfully to be used for evacuation. Issues that need to be addressed include exposure to landslides and debris-flows; the vulnerability of a large number of masonry bridges and the presence of earthquake-susceptible rubble-stone buildings in many of the towns and villages through which many roads pass. Wallenstein et al. (2015) sketch out a number of possible scenarios that would necessitate the early evacuation of large numbers of people and these are summarised in Table 6.3.

Transport on São Miguel was transformed in the first decade of the present millennium by the construction of a network of new roads (Figure 6.3) which were fully operational by 2011.[12] These were built to boost economic development, but 'the impacts on Civil Protection and evacuation planning in the Fogo and Furnas area are likely to be both profound and in some

Table 6.3 Scenarios that would necessitate early evacuation of areas in the vicinity of Sete Cidades, Fogo and Furnas volcanoes in the event of a threatened eruption

Eruption Scenario	Possible problems in evacuation
Sete Cidades	Even a small intra-caldera eruption could devastate Sete Cidades village and render the roads linking the settlement to the coastal road un-useable. In addition, the *freguesias* of Bretanha, Mosteiros and Ginetes are also vulnerable to any disruption of the coastal highway and at its most extreme *c.*4,000 people could be isolated.
Fogo	It is because Fogo is located in the centre of the island, that both the northern and southern coastal roads could be cut by even a small eruption, pre-eruption seismic activity and/or landslides, so isolating a population in excess of 45,000 who occupy the *Fogo area*.
Furnas	In addition to the villages of Furnas and Ribeira Quente (see text), an eruption could impact *c.*9,600 people living in the *freguesias* of Salga, Achadinha, Achada, Santana, Nordestinho, Lonba da Fazenda, Nordeste, Água Retorta, Faial da Terra, Nossa Senhora dos Remédios and Povoação. Again, the road network is vulnerable to eruptive activity, earthquakes and landslides.

Based on information in Wallenstein et al. (2015).

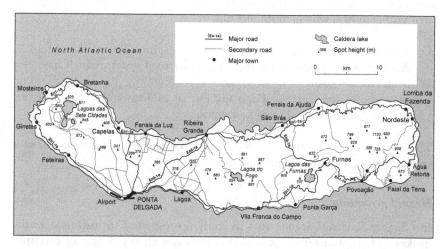

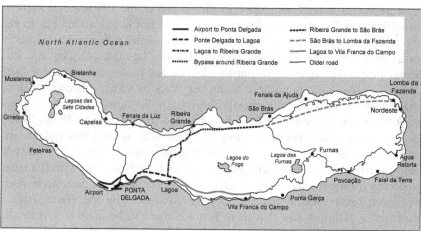

Figure 6.3 The road network of São Miguel: a. (above) before road improvements
carried out between 2007 and 2012 and b. (below) roads constructed under the SCUTS Programme (see text). From Wallenstein et al. 2015, Fig.
16.4, page 218. Copyright permission Geological Society of London.

respects uncertain,' the new roads having no impact upon the Sete Cidades
area (Wallenstein et al. 2015, p. 217), Figure 6.3. The main positive impacts are that: several towns (i.e. Água de Pau, Água d'Alto, Vila Franca do
Campo) and the city of Ribeira Grande are now bypassed; and roads are
built to a higher standard, are faster and have been constructed further inland
and at higher altitudes bridging many rivers in a more satisfactory manner
than was the case hitherto. It seems likely that severe flooding, land-sliding,
lahars and even small-scale pyroclastic density currents (PDCs) could be contained without damaging these substantial bridges. As we have argued before
(Wallenstein et al. 2015) and notwithstanding many positives, there may be

instances of what David Alexander (1997, p. 292) has termed 'newly generated vulnerability' which could include:

a   Weather at high altitude (e.g. stronger winds, higher rainfall and fog).
b   Access points to new roads from the older road system which have not been risk assessed.
c   New routes are nearer to Fogo and Furnas calderas and could be at risk of heavy ash loading during an eruption.

These caveats point to the need for a continual updating of evacuation surveys as new vulnerabilities emerge.

Hazard perception

In Chapter 5, it was argued that in studying responses to historical earthquakes and eruptions it is not possible to assess with any degree of certainty how people perceived the risks they faced. Before the 1990s social surveys, based on questionnaires and/or interviews had not been carried out in the Azores and recourse had to be made to archives and published accounts in order to gauge public opinion. Such sources may be misleading because observers, especially those from outside the Azores, often misunderstood local culture and/or were prejudiced about the ways in which Azoreans seek to make sense of disasters, the strongly anti-Catholic views of the British naval officer Captain Edward Boid being a case in point (Boid 1835). Today generalisations based on untested opinions are still encountered and, for instance, there is the oft repeated claim that there are differences in attitude towards risk between mainland Portuguese and the Azoreans, with the latter knowing from experience that 'earthquakes cause extensive damage to non-engineered structures; know that they also have to... (be concerned with)... emergency planning and that there is much that can be done to reduce the impact of an earthquake' (Azevedo et al. 2009, p. 563). As the discussion which follows will show, such statements are highly misleading.

The first person to study risk perception was Christopher Dibben, in both his doctoral thesis (Dibben 1999) and in a publication which emerged from the *European Laboratory Volcanoes Project* (Dibben & Chester 1999). Focusing on Furnas village (São Miguel), in 1994 Dibben questioned 50 subjects on 5 themes: length of residence and reasons for moving to the village; attitudes towards Furnas; perceptions of volcanic threats; disaster preparedness and their views on mitigation measures. Dibben's findings are summarised in Table 6.4. The principal lesson that may be drawn from his research is that there was a wide gulf between the views of residents towards risk and those that were based on research into seismic and volcanic processes (see Chapter 2). Interviews reveal disbelief about the possibility of future activity, which is exacerbated by a lack of trust in those in authority and a desire to live life without considering what might be a bleak future, these traits being deeply embedded within the consciousness of the population. On the basis of experience elsewhere (Quarantelli 1980; Donald & Canter 1990), Dibben argues that there is a severe danger of people ignoring early warnings and so delaying timely evacuation. In places that are threatened by eruptions, being forewarned is to be forearmed and experience from other disaster-prone areas implies that families need at least 72 hours notice in order to decide how to respond especially if plans include evacuation (Russell et al. 1995). In contrast and more positively, Christopher Dibben found respect for university-based scientists with their detailed knowledge of volcanology and seismology.

Table 6.4 A summary of the principal findings of study of attitudes to eruptions and earthquakes in Furnas Village (São Miguel)

Perception	Details
Future eruptions	
Preparation	None of the residents interviewed had made any preparations, mental or physical, for a future eruption. Some people were shocked by the question. It is not surprising that few preparations had been made by the people of Furnas because hardly any respondents expressed any concern about a future volcanic eruption (see below). The majority replied that given a warning they would run away.
Source of warnings and advice	Most interviewees did not think they would receive a warning and from whom they should seek advice. Television was viewed as a reliable source of information as were scientists from the University of the Azores, who appeared regularly to discuss seismic and volcanic activity.
Awareness of volcanic activity	All respondents were aware that the area was volcanically active and no one discounted the possibility of a future eruption, but all held beliefs which minimised their concerns. These included: a. Some 34% of respondents who replied that fumaroles in the village meant that eruptions where unlikely and earthquakes would be smaller. b. Hot springs were commonly believed to be a relief valve. c. Some interviewees coped by holding a belief that eruptions were very rare events, and 32% stated that damaging eruptions were unlikely because nothing had occurred during their lifetimes.
Control over future events	Some 28% of respondents believed they had little control over future events and many were fatalistic. Sometimes fatalism was framed within a religious worldview and God was perceived as controlling the future. As a result of this perceived lack of control over their futures, many interviewees deliberately avoided thinking about a future eruption.

(Continued)

Perception	Details
Furnas is a relatively safe place despite its active volcanism	It is because of the supposed mitigation provided by fumaroles and springs, that many argued that Furnas was a safer location than many other places in the Azores. This attitude had behavioural implications and these perceptions were shared by some people living outside the village. For example, when two seismic crises occurred in Ribeira Quente in 1952, some people with family connections to Furnas were relocated. This may be a potentially very dangerous attitude as seismic activity frequently precedes eruptions and incautious in-migration to Furnas caldera could be encouraged.
Earthquakes	
Reactions to past event	Between 1900 and 2000 three major earthquake affected São Miguel and many minor tremors were experienced. Respondents recorded a number of behaviours ranging from 'staying in bed' to 'curling up in a corner.' No respondent had tried to leave either their homes or Furnas village during episodes of felt seismic shaking. There were two reasons behind this attitude: a belief that the ground 'would open up' and consume them if they left their homes and, more prosaically, civil defence guidelines which advised people to stay indoors and shelter by door frames.

Interviews were carried out in middle 1990s. Based on: Dibben (1999); Dibben and Chester (1999).

Since the 1990s, there has been a major effort in the Azores to promote disaster awareness on the part of the Regional Government, its agencies and the university. Education programmes devised by the Civil Protection Service (*Serviço Regional de Proteção Civil e Bombeiros dos Açores* – SRPCBA – see below), have been particularly noteworthy (Anon 2020d).[13] Three groups have undertaken surveys to monitor the degree to which attitudes have changed.

The first team, working under the auspices of IVAR (*Instituto de Investigação em Vulcanologia e Avaliação de Riscos* – see below), has questioned respondents across the Azores (Estrela Rego & Arroz 2012; Estrela Rego et al. 2018). In the first study (Estrela Rego & Arroz 2012), semi-structured interviews were carried out in five of the nine islands: São Miguel; Santa Maria; Terceira; Faial and Flores. The principal findings are as follows:

a   Earthquakes are regarded as the more serious threat and in some cases volcanic activity is not mentioned.
b   Regardless of age and gender, there are strongly expressed feelings of fear and anxiety.

c There is an awareness that seismic and volcanic risks differ both across the Azores and within individual islands.

d Whilst accepting that *short-term* prediction[14] of earthquakes is not possible, local people often place great faith in highly dubious precursory weather patterns and animal behaviours which are believed to presage seismic activity.

e Some respondents place reliance on strengthened buildings and individual preparedness, but the majority do nothing at all to protect themselves.

f There is an expectation that the authorities will be responsible for crisis communications and disaster management.

An overall finding is that people show a strong attachment to 'place,' which involves a complex amalgam of long-standing personal and family ties, quality of life and the scenic beauty of the area, all of which militate against any desire to move from or, indeed, evacuate their neighbourhood on even a temporary basis.

A second report by the IVAR group (Estrela Rego et al. 2018), was confined to interviews conducted on São Miguel and participants were asked to construct a *Household Emergency Plan* (HEP) which were then evaluated. Results show that: there is a general recognition of seismic, but not volcanic risks; knowledge of threats is deficient, although people are generally aware of the authorities' work on civil protection, and that there is a lack of engagement with preparedness. The report concludes by painting a bleak picture, the sense of which is encapsulated in two conclusions: (a) 'most (people) recognise earthquakes as damaging and severe, yet still report feeling safe at home, pointing to an unrealistic optimistic bias and suggesting that participants may be resilient to educational programs (sic) and to the adoption of seismic adjustments' and (b) 'overall participants are not prepared to face an earthquake or volcanic eruption. Therefore, there is a need to foster awareness and preparedness for volcanic hazard and analyse the efficiency of... current awareness and preparedness campaigns (i.e. *programmes*) for earthquakes' (Estrela Rego et al. 2018, pp. 202 &503, our *italics*).

The IVAR group continued their investigations by studying preparedness within families with children on São Miguel (Pacheco et al. 2020, p. 9), through interviewing 125 family representatives and then further questioning a sub-set of 105 who had done little to prepare for a future emergency. Their findings are stark with 'families (being) poorly prepared to face a major emergency (and that) only a small number of families included children in the process of developing preparedness measures, thus their impact (on) household preparedness seems to be limited.' This was despite that the fact the *Serviço Regional de Proteção Civil e Bombeiros dos Açores* (SRPCBA) specifically targeting their training and awareness programmes to children of school age with the hope that knowledge would diffuse through families.

This conclusion of very slow progress in raising public awareness is confirmed by the findings of the second research team (Lotteri 2020; Lotteri et al. 2021). Research involved a longitudinal study and, by means of the same questions used by Dibben (1999), charts changing perceptions of risk in Furnas village over a quarter of a century. The authors conclude that, even though advances have been made in increasing awareness and in the dissemination of information about the ways in which people might protect themselves, only slow progress has been made in changing attitudes. Increased trust in the Civil Protection Service is noted and today it is more likely than in 1994 that the people of Furnas will follow instructions and evacuate should this prove to be necessary. It remains the case, however, that boosting public awareness of danger on a volcano that has not erupted since 1630 and where within living memory earthquakes have been small-scale, requires a long-term commitment to public education on the part of the authorities if policies of risk reduction are to be successful (Lotteri et al. 2021).[15] This finding is given additional weight by a recently published study of the perception of a sample of elderly residents on São Miguel (Anderson & Mileu 2020) which reveals an underestimation of volcanic risks, with nearly a third of respondents believing that such events are of historic interest only and that there will be no further eruptions. Interviewees also showed a lack of confidence in the authorities.

Policy development

Just as an increase in the pace of academic research coincided with wider political change in the Azores, so policy has developed to reduce vulnerability and enhance resilience. Overall development has been evolutionary, but there have been episodes of more rapid change with important policy enactments following the earthquakes of 1980 and 1998, and the landslides which devastated Ribeira Quente in 1997. New policies have also been introduced as a result of legislation which has included the Earthquake Building Code of 1983 and its later modification to comply with European-wide standards.

The evolution of administrative structures

At the time of the 1980 Terceira earthquake, the University of the Azores had been in existence for four years and the presence of an active Earth Sciences Department (*Departamento de Geociências*) allowed more detailed seismic monitoring to take place than had been possible in earlier emergencies (see Chapter 4). The 1980 Terceira earthquake was only recorded by the three seismographic stations in the Azores which were maintained by the Portuguese *Instituto de Meteorologia* (Institute of Meteorology).[16] In the aftermath of this disaster a temporary network was deployed, a few months later six stations were operating in Terceira and eight on São Miguel (Fontiela et al. 2018) and today there is a well-established grid of stations across the islands which is able to detect earthquakes and seismic swarms.

Changes in policy and administrative systems followed in the wake of the 1997 Ribeira Quente (Figure 6.2) emergency. As discussed in Chapter 1, this landslide killed 29 people, injured 7 and left 69 homeless and was dealt with by the authorities in a purely *reactive* manner because no preparedness measures were in place (Cunha 2003). In view of these administrative deficiencies, a report was commissioned from an expert group which included representatives from the university, the *Laboratório Regional de Engenharia Civil* (Regional Laboratory of Civil Engineering) – LREC in Ponta Delgada, University College London and the Nordic Volcanological Institute (Gaspar et al. 1997). Some recommendations were specific to Ribeira Quente and included controlling construction particularly in hazard-prone zones; building a new fishing harbour and improving communications – including the provision of a heliport – to prevent isolation should the village be impacted by a future disaster. More fundamental were actions to make the system of emergency planning more *proactive*, which, together with the added stimulus provided by the 1998 earthquake, led to the development of the administrative structures which are in place today.

The interface between the university and the Regional Government is crucial and finds expression in the *Centro de Informação de Vigilância Sismovulcânica dos Açores* (CIVISA – Centre for Information and Seismovolcanic Surveillance of the Azores). In addition to CIVISA, the university and more specifically the *Departamento de Geociências*, established a research body, the *Centro de Vulcanologia e Avaliação de Riscos Geológicos* (CVARG – Centre for Volcanology and Geological Risk Assessment), which from early 2017 was upgraded to become the *Instituto de Investigação em Vulcanologia e Avaliação de Riscos,* or IVAR (Institute for Volcanology and Risk Assessment Research).[17] A control/conference room within the university enables scientific information to be shared with the Civil Protection Service.

Other changes were instituted in the wake of the 1997 report (Gaspar et al. 1997). These included, not just general support for evacuation planning plus pure and applied earth science research, but further changes in administration. One major innovation was the approval for each *concelho* (i.e. county) to develop a civil protection plan (*Plano Munucipal de Emergência*), the first for Santa Cruz (Graciosa) being completed in May 1998 and the last, for Calheta (São Jorge), in December 2016. The Povoação Plan, which includes the village of Ribeira Quente, was approved in 2002 and a document for the whole region (*Plano Regional de Emergência dos Açores*) was published in March 2007 (Anon 2020a). Taking the Povoação plan as an example, this was jointly authored by CIVISA and the Civil Protection Service and comprises three substantial documents dealing with: hazard exposure; the organisation of responses and areas where interventions will be required (Anon 2020b). Finally, an integrated Fire and Civil Protection Service was formed: the *Serviço Regional de Proteção Civil Bombeiros dos Açores* (SRPCBA). As wellbeing responsible for firefighting and the provision of ambulances, SRPCBA is charged with providing information on and raising awareness of the likely impacts of environmental extremes through education programmes and supporting the population when disaster strikes (Cunha 2003).

Decision-making at the time of an emergency operates as a cascade. 'Minor and localized problems are dealt with within individual *freguesias* and larger-scale problems are successively the responsibility of: the *concelho*; the Azores Autonomous Region and a minister within the national government' in Lisbon. A repetition of a disaster on the scale of 1957/8 or 1980 'would be under the control of the General responsible for Civil Protection who is located on Terceira... and who would be able to exercise powers very similar to those so successfully employed by Governor Pimentel' on Faial in the 1950s or *Gabinete de Apoio e Reconstrução* (Office of Assistance and Reconstruction – GAR) in 1980 (Coutinho et al. 2010, p. 278).

Seismic building codes

Portugal made an early start in designing buildings to be more resilient to earthquakes and, following the 1755 Lisbon disaster, the Marquês de Pombal's principal architect General Manuel da Maia was responsible for enacting what is often claimed to be the world's first earthquake code (França 1977; Chester & Chester 2010).[18] Unfortunately, even in Lisbon lessons were soon forgotten, the code was never applied in the Azores and until well into the twentieth century most buildings were constructed of highly vulnerable rubble-stone with 'soft-storeys,' a lack of torsional stiffness and poorly connected walls and floors (see above). In 1955 a symposium was held to observe the 200th anniversary of the Lisbon disaster (Ordem dos Engenheiros 1955) and, because of the widespread acceptance that Portuguese buildings were highly vulnerable, it was agreed that a new code was urgently required. This was introduced in 1958 (RSCCS 1958), additional regulations for bridges and buildings followed in 1961 (RSEP 1961) and a more rigorous code was enacted in 1983 (RSA 1983), which has operated since 1985. On the basis of these codes and the dates of their introduction, Tang et al. (2012) have classified buildings in Portugal and the Atlantic Islands into three age-based vulnerability classes: 1961 and earlier; 1962–1985 and constructed after 1985; pre-1961 construction being the most dangerous.[19] Later the 1983 code was modified so that standards conformed to those that currently apply across the European Union and which have been published as Eurocode 8 (CEN 2005; Campos Costa et al. 2008). Under this code the standards to which buildings have to conform vary according to the 'seismic zone' into which a particular location falls. In the Azores most of the islands fall into the fifth (i.e. highest) category – peak ground acceleration 2.5 m/s^2, with the exception of Santa Maria which is placed in the fourth – peak ground acceleration 2.0 m/s^2 and Flores and Corvo which are in the second – peak ground acceleration 1.1 m/s^2.

Moving forward: current planning issues

Today the Regional Government is in a far better position to handle disasters than was the case following the 1997 and 1998 emergencies, but there are still issues that have been highlighted by research and which have not been addressed adequately by the authorities. First, in view of the large number of buildings constructed before 1961, there is much discussion amongst civil engineers about the most appropriate future policies to adopt to encourage retrofitting. As discussed above for individual buildings retrofitting may not always be cost-effective and doubts have been expressed over larger-scale interventions. Ferreira et al. (2017b, p. 345) estimate that a total of 330 million Euros was spent following the 1998 earthquake on reconstruction and retrofitting around 3,000 buildings on Faial, which represented *c*.17% of Azorean GDP at the time.[20] They argue that if retrofitting had been carried out before the earthquake, then costs would have been negligible in comparison and make a strong case for more comprehensive building codes to encourage the introduction of pre-disaster safety measures. The importance of raising public awareness of risk by means of education has already been highlighted, but Ferreira et al. (2017b, p. 331) argue that 'society is still not acknowledging the absolute necessity of retrofitting and strengthening buildings' and that additional education is required, not only of the inhabitants, but also of planners, architects, builders, investors and – not least – government, who have often rehabilitated buildings 'without considering the principles of seismic design.'

Second, there is a need better to integrate disaster, development, urban and coastal planning. This is a major theme within the United Nations' *Sendai Framework* document (Anon 2015) and, according to Alexander and Davis (2012), is one of the five themes that frequently emerge at international gatherings and which governments find it very difficult to address because of conflicting domestic priorities.[21] In the Azores a fine start has been made in incorporating disaster preparedness into regional and municipal plans, but as Ferreira et al. (2017b, p. 331) note with reference to seismic safety, although their observations apply across the gamut of environmental threats, amending legislation 'for landuse zoning of development area, building codes and empowerment for implementation is needed on an urgent basis.... Moreover, seismic safety certification of both new and existing buildings has been set as one of the main goals' for the future.

The observation that planning policies, though commendable, are far from ideal may be illustrated by examining in detail the *concelho* (i.e. county) of Vila Franca do Campo which is located on the south coast of São Miguel (Figure 6.2). Martins et al. (2012) examined this *concelho* with respect to planning policies, urban expansion and environmental threats from the middle of the 1990s to 2005, with forecasts up to 2016 and assuming established trends continued. Development was controlled by two pieces of legislation. The first was the Municipal Master Plan (*Plano Diretor Municipal* – PDM) which included measures to prevent building in fault zones (Articles 8 & 24). The second enactment was the South Coast Management Plan (*Plano de Ordenamento da Orla Costeira* – POCC), which defined a zone of protection within 500 m of the coastline and attempted to marry urban growth and rehabilitation, on the one hand, with protection against the many environmental risks associated with coastal areas, on the other. Martins et al. (2012) found that policies have not been wholly successful, with urban expansion occurring in the more risky of two seismic zones (see Chapter 3). As far as tsunami exposure was concerned, in the late 1990s only 16 ha of land was at risk, representing the urban 'cores' of the *freguesias* (i.e. parishes) of São Miguel and São Pedro, whereas by 2016 this was projected to rise to 420 ha. Plans in Vila Franca do Campo and elsewhere in the Azores are under constant review as research identifies both further aspects of risk and dangerous building practices.[22]

Conclusion

Perhaps the most important lesson to emerge from this chapter and one that is implicit throughout the book may be simply stated but is profound in its implications, transcending questions of policy development and involving an ethical dilemma which lies at the heart of political decision-making. As discussed above, long-term rehabilitation following the 1998 earthquake was expensive at an estimated cost of 17% of Azorean GDP (Ferreira et al. 2017b). What is more the 1998 event was an earthquake of modest size (M 5.8; Gaspar et al. 2015b), which only impacted a relatively small number of people. Some 2,500 people were made homeless representing just over 1% of the population of the Azores at the time. Politicians were motivated to be generous by an implicit social contract which places an obligation on democratic governments and their leaders to mitigate the effects of disasters almost without reference to cost. This is encapsulated in a press report published two decades later in which the Mayor of Faial and Regional Director of Housing, not only take great pride in the reconstruction and rehabilitation of buildings but also in the commitment of government to the people and their welfare (Anon 2020c).[23] Within living memory all disasters have been of limited spatial impact and of modest size, raising questions over whether this from of response would be sustainable if the 1998 emergency had been on the scale of, say, a '1630 type eruption.' Two questions arise:

a Could compensation be as generous, even if the Autonomous Region enlisted the financial support of the National Government and the European Union?
b Would the majority who were unaffected and their political leaders be prepared to sanction spending a high proportion of GDP on a small minority of inhabitants?

Answering these questions is difficult, but probably involves being even more *proactive* by working to reduce risk through more cost-effective forward planning specifically involving: the encouragement – and if necessary the funding – of retrofitting; planned evacuation drills and locating new development in less exposed areas. As the experience of the Ribeira Quente emergency has shown, people are reluctant to re-locate even over short distances when it is clearly in their own best long-term interests to do so. Education is a second priority made more pressing by the wildly inaccurate perceptions of risk that have been revealed through the research of social scientists. Finally, a lack of faith in the ability of the authorities successfully to manage a future disaster implies that confidence building and restoring trust should be further priorities for the Regional Government and its agencies. All these avenues are being actively pursued by the Civil Protection Service as is clear from their publications and web-based information (SRPCBA 2020).

Globally emergency planning is strongly influenced by ethical theory and, whereas pre-disaster procedures are normally strongly utilitarian, being based on the assumption of the greatest good of the greatest number, when disaster strikes there is an understandable tendency for planners and politicians to adopt alternative ethical positions. Doing the right thing out of sense of duty (i.e. deontology) and assuming that a social contract exists between government and the governed tilts policy in the opposite direction in the aftermath of a disaster. Both ethical approaches have well-rehearsed problems, not least that minorities suffer under utilitarianism and the majority under both deontology and through the application of social contract theory (Etkin 2016). A third theory of ethics, known as the ethics of virtue, stresses 'right being' over 'correct action,' and emphasises traits within individuals and decision-makers which include honesty, caring, empathy, compassion and trust (Etkin & Timmerman 2016). This may also be a useful way forward for emergency planning in the Azores in two respects. First when applied to disaster planning, advocates of virtue ethics argue that a successful response is strongly correlated with the trust people place in the decisions of politicians and planners. Second, the ethics of virtue explains why family, kinship and friendship networks are important and often successful agents of mitigation in *pre-industrial* societies. An additional policy strand for the future could be to build trust in institutions and encourage family, extended family and community support networks that, were not only important in the past, but could also have the potential to be so in the future.

Notes

1 At the start of the 1990s Furnas volcano was identified as one of the European Laboratory Volcanoes which were selected for detailed research during the United Nations' *International Decade of Natural Disaster Reduction* (IDNDR). The project was funded by the European Community's Environment Programme and was also under the auspices of the European Science Foundation. The project involved academics from Portugal, France, Iceland, Spain and the United Kingdom (Duncan et al. 1999a).

2 Of particular relevance have been the *Hyogo Framework for Action* (2005–2015) and the *Sendai Framework for Disaster Risk Reduction* (2015–2030), both of which were developed and published under the auspices of *International Strategy for Disaster Reduction* which began in 2001 (United Nations 2006, 2015). In 2015, a statement of support and commitment was read at the Sendai meeting in Japan

by the Professor Anabela Rodrigues the Portuguese *Ministério da Administração Interna* or *MAI* (Minister for Internal Administration) (Rodrigues 2015). The major policy goals of the Sendai Framework are to: 1. understand disaster risk; 2. strengthen disaster risk governance better to manage disaster risk; 3. invest in disaster risk reduction to boost resilience and 4. enhance disaster preparedness and 'build back better', in recovery, rehabilitation and reconstruction (United Nations 2015).

3 Total collapse describes a building that has lost at least half of its plan area and interior volume (Pomonis et al. 1999, p. 119)

4 The authors calculate that in the more proximal villages tephra loading will exceed 25 cm by a wide margin and calculate figures of 300 cm for Furnas and Ribeira Quente, 100–300 cm for Ponta Garça and 100 cm for Povoação.

5 A soft-storey refers to one level, usually the ground floor, of a building that is significantly more flexible (i.e. with weak lateral load resistance) than the storeys located above it. Examples of soft-storeys include: large garage doors; openings made for shop windows and, in residential dwellings, large picture windows (Guevara-Perez 2012).

6 Utilitarianism is an ethical theory which is often encountered in planning. It holds that some individual harm may have to be accepted, in this case not retro-fitting all homes, in order to be able to benefit the many by making emergency buildings more 'tephra proof' (Etkin 2016, pp. 280–282).

7 EMS-98 classifies buildings into 6 categories, A to F, with respect to building vulnerability. A is least resistant and F the most resistant. An in-depth discussion of EMS-98 is presented in Chapter 3

8 Since 2017 CVARG has been named IVAR (*Instituto de Investigação o em Vulcano-logia e Avaliação de Riscos* (See below and Chapter 3).

9 One of the few historical exceptions to this generalisation was the relocation of ten settlements, including the city of Noto, following 1693 earthquake in Sicily (Branca et al. 2015). Usually because of a process known as 'geographical inertia', even in a ruined state an existing settlement may have many economic, social and cultural advantages over any alternative location. These include: the fact that not all buildings are destroyed; that the ruined settlement is still a focus of com-munications; that it may be a natural harbour, a centre of administration and that people maintain strong cultural and social ties rooted in a particular location. On São Miguel the re-siting of Vila Franca do Campo to the west of its original site following the 1522 disaster was not due to the earthquake *per se*, but because of landslide debris (see Chapter 3).

10 The first author of this paper is actually Professor Robin Spence.

11 Although this project was funded by the University of Luton (now University of Bedfordshire), it was associated with the Furnas *Laboratory Volcano Project*.

12 These new roads are known by the acronym SCUTS (*estradas sem custos para utilizador* - roads without charge to the user). This was a private finance initiative costing 325 million Euros (ca. 433 million Euro 2020) and the debt is to be ser-viced from general taxation over a 30 year period.

13 In this there is a parallel with other countries where 'disaster education' has been highlighted as being of major importance in future planning (e.g. Bird & Gísladóttir 2018).

14 The terminology is that adopted in Chapter 4.

15 Making the population aware of the possible future impact of environmental extremes is a major undertaking. In addition to outreach from the university and teaching in schools, the Civil Protection Service has numerous on-going pro-grammes. In 2019 there were at total of 211 courses, 63 more than in 2017. These included courses for fire fighters, health professionals and many private bodies.

The service has also undertaken drills based on possible future emergency situations (Anon 2020d).

16 In 1997 the two networks came together into a consortium known as the *Sistema de Vigilância Sismológica dos Açores* (SIVISTA), but in 2007 once again spit. Since 2012 the *Instituto de Meteorologia* has been known as the *Instituto Português do Mar e da Atmosfera* (IPMA).

17 IVAR comprises the Azores Volcanological and Seismological Observatory and is a member of the World Organisation of Volcano Observatories (WOVO). IVAR is a full operational unit of the university. It is also the operational arm of CIVISA - a private non-profit organisation that pursues scientific and technological research and which was established in 2008 as a joint venture by the University of the Azores and the Regional Government. From 1976 a leading early figure in the establishment of earth sciences in the University of the Azores was Professor Victor Hugo Forjaz (born in 1940).

18 The Marquês de Pombal (1699–1782) was a Portuguese statesman and dictator who was in power before, during and after the 1755 Lisbon earthquake.

19 Lotteri (2020, p. 134) reports that insurance companies are unwilling to insure buildings constructed before 1985 for seismic risks, so placing a severe constraint on the property market for older properties.

20 The comprehensive programme of retrofitting carried out following the 1998 earthquakes was undertaken under a special Regional Legislative Decree of September 25, 1998 (Maio et al. 2019).

21 The others which do not apply in the Azores are: the public right to information; the effects of explosive population growth in some countries; corruption and discrimination against women (Alexander & Davis 2012).

22 This is clear if websites for individual municipalities are consulted.

23 In fact this press report is strongly party political, with the Regional Director of Housing remarking that reconstruction was one of the biggest undertakings that socialist government of the Azores has ever faced' and lauding its success (Anon 2020c, no page numbers, our translation).

References

Alexander, D. 1997. The study of natural disasters, 1977 to 1997: Some reflections on a changing field of knowledge. *Disasters* **21** (4), 284–304.

Alexander, D. and Davis, I. 2012. Editorial: Disaster risk reduction: An alternative viewpoint. *International Journal of Disaster Risk Reduction* **2**, 1–5.

Anderson, M. and Mileu, N. 2020. Ponta Delgada, a volcanic city. Main natural risks and reasons that dictate the coexistence of populations. *In*: Fernandes, F., Malheiro, A. and Chaminé, H.I. (eds) *Advances in Natural Hazards and Hydrological Risks: Meeting the Challenge*. Advances in Science, Technology and Innovation, Springer Nature, Switzerland, 97–100.

Andrade, C., Borges, P. and Freitas, M.C. 2006. Historical tsunami in the Azores archipelago (Portugal). *Journal of Volcanology and Geothermal Research* **156**, 172–185.

Anon. 2015. *Sendai Framework for Disaster Risk Reduction 2015–2030*. United Nations, Geneva.

Anon. 2020a. *Planos de Emergência na Região Autónoma dos Açores*. Governo dos Açores, Ponta Delgada. http://www.azores.gov.pt/Portal/pt/entidades/srs-srpcba/textoTabela/Planos+de+Emerg%C3%AAncia+na+Regi%C3%A3o+Aut%C3%B3noma+dos+A%C3%A7ores.htm

Anon. 2020b. *Plano Municipal de Emergência de Proteção Civil da Povoação.* Povoação Municipal. https://www.cm-povoacao.pt/index.php/component/edocman/plano-municipal-de-emergencia/plano-municipal-de-emergencia-de-protecao-civil-da-povoacao-nov-2019

Anon. 2020c. Sismo de 1998 mudou a ilha do Faial para sempre. *Observador* (Portugal). https://observador.pt/2016/08/08/sismo-de-1998-mudou-a-ilha-do-faial-para-sempre/

Anon. 2020d. Training provided by the Azores Civil Protection has increased more than 40% since 2017. Governo dos Açores, Ponta Delgada. http://www.azores.gov.pt/Portal/en/entidades/srs-srpcba/noticias/Training_provided_by_the_Azores_Civil_Protection_has_increased_more_than_40_since_2017.htm

Azevedo, J., Serrano, S. and Oliveira, C.S., 2009. The next 1755 – myth and reality; Priorities and actions to develop in case of an earthquake in the Lisbon Metropolitan area. *In*: Mendes-Victor, L.A., Oliveira, C.S., Azevedo, J. and Ribeiro, A. (eds) *The 1755 Lisbon Earthquake Revisited.* Springer, Berlin (Geotechnical, Geological and Earthquake Engineering vol. 7), 559–579.

Bird, D. and Gísladóttir, G. 2018. Responding to volcanic eruptions in Iceland: From the small to the catastrophic. *Palgrave Communications* **4**, 151. https://doi.org/10.1057/s41599-018-0205-6

Blong, R.J. 2003. Building damage in Rabaul, Papua New Guinea. *Bulletin of Volcanology* **65**, 43–54.

Boid, E. 1835. *A Description of the Azores, or Western Islands.* Schulk and Co, London.

Branca, S., Azzaro, R., De Beni, E., Chester, D. and Duncan, A. 2015. Impacts of the 1669 and the 1693 earthquakes of the Etna Region (Eastern Sicily, Italy: an example of recovery and response of a small area to extreme events. *Journal of Volcanology and Geothermal Research* **303**, 25–40.

Cabral, B. 2020. Perceção de Riscos Naturais e Ambientais Numa Comunidade Insular: O Caso de Santa Maria (Açores). Dissertação de Mestrado em Vulcanologia e Riscos Geológicos, Faculdade de Ciências e Tecnologia, Universidade dos Açores.

Campos Costa, A., Sousa, M.L. and Carvalho, A. 2008. Seismic Zonation for Portuguese National Annex of Eurocode 8. The 14th World Conference on Earthquake Engineering, October 12–17, 2008, Beijing, China. https://www.iitk.ac.in/nicee/wcee/article/14_07-0167.PDF

CEN. 2005. *Eurocode 8: Design of Structures for Earthquake Resistance - Part 1: General Rules, Seismic Actions and Rules for Buildings.* European Committee for Standardisation, Brussels, Belgium.

Chester, D.K. and Chester, O.K. 2010. The impact of eighteenth century earthquakes on the Algarve Region. Southern Portugal. *Geographical Journal* **176** (4), 350–370.

Chester, D.K., Dibben, C. and Coutinho, R. 1995. *Report on the Evacuation of the Furnas District, São Miguel, Azores.* CEC Environment/ESP Laboratory Volcano Furnas Azores. Image Centre University College, University of London, Open File Report.

Chester, D.K., Dibben, C., Coutinho, R., Duncan, A.M., Cole, P.D., Guest, J.E. and Baxter, P.J. 1999. Human adjustments and social vulnerability to volcanic hazards: The case of Furnas Volcano, São Miguel, Açores. *In*: Firth, C.R. and Mc Guire, W.J. (eds) *Volcanoes in the Quaternary.* Geological Society, London, Special Publication **161**, 189–207.

Chuang, S.W. and Zhuge, Y. 2015. Seismic retrofitting of unreinforced masonry buildings - a literature review. *Australian Journal of Structural Engineering* **6** (1), 25–36.

Costa, A. and Arêde, A. 2004 Strengthening of structures damaged by the Azores earthquake of 1998. *In*: Lourenço, P.B., Barros, J.Q. and Oliveira, D.V. (eds) *International Workshop 'Masonry Walls and Earthquakes'*, 6° Congresso Nacional de Sismologia e Engenharia Sísmica Livro de Aetas, Universidade do Minho, April 14–16 2004, Guimarães, Portugal, no page numbers.

Costa, A. and Arede, A. 2006. Strengthening of structure damaged by the Azores earthquake of 1998. *Construction and Building Materials* **20**, 252–368.

Coutinho, R., Chester, D.K., Wallenstein, N. and Duncan, A.M. 2010 Responses to, and the short and long-term impacts of the 1957/1958 Capelinhos volcanic eruption and associated earthquake activity on Faial, Azores. *Journal of Volcanology and Geothermal Research* **196**, 265–280.

Cunha, A. 2003. The October 1997 landslides in São Miguel Island, Azores, Portugal. *In*: Hervás, J. (ed) *Lessons Learnt from Landslide Disasters in Europe*. European Commission Joint Research Center, Ispra, Italy, 27–32.

Dibben, C.J.L. 1999. *Looking Beyond Eruptions for an Explanation of Volcanic Disasters: Vulnerability in Volcanic Environments*. Ph.D. thesis, University of Luton.

Dibben, C. and Chester, D.K. 1999. Human vulnerability in volcanic environments: The case of Furnas, Sao Miguel, Açores. *Journal of Volcanology and Geothermal Research* **92**, 133–150.

Donald, I. and Canter, D.V., 1990. Behavioural aspects of the Kings Cross disaster. *In*: Canter, D. (ed) *Fires and Human Behaviour*. David Fulton, London, 15–30.

Duncan, A.M., Gaspar, J.L., Guest, J.E. and Wilson, L. 1999a. Introduction: Special Issue: Furnas Volcano, São Miguel Azores. *Journal of Volcanology and Geothermal Research* **92** (1–2), vii–ix.

Duncan, A.M., Gaspar, J.L., Guest, J.E. and Wilson, L. 1999b. Special issue: Furnas volcano, São Miguel Azores. *Journal of Volcanology and Geothermal Research* **92** (1–2), 1–214.

Durand, M. and Scott, B.J. 2005. Geothermal ground gas emissions and indoor air pollution in Rotorua, New Zealand. *Science of the Total Environment* **345**, 69–80.

Estrela Rego, I. and Arroz, A.M. 2012. Places of fear and attachment. How Azoreans perceive seismic and volcanic risk. *Global Journal of Community Psychology Practice* **3** (4), 1–9.

Estrela Rego, I., Pereira, S.M., Morro, J. and Pacheco, M.P. 2018. Perceptions of seismic and volcanic risk and preparedness at São Miguel (Azores, Portugal). *International Journal of Disaster Risk Reduction* **31**, 498–503.

Etkin, D. 2016. *Disaster Theory: An Interdisciplinary Approach to Concepts and Causes*. Elsevier, Amsterdam.

Etkin, D. and Timmerman, P. 2016. Virtue ethics. *In*: Etkin, D. (ed) *Disaster Theory: An Interdisciplinary Approach to Concepts and Causes*. Elsevier, Amsterdam, 290–292.

Fagundes, C. 2015. Seismic assessment of a typical building from Azores (*sic*). Técnico, Lisboa.

Fagundes, C., Bento, R. and Cattari, S. 2017. On the seismic response of buildings in aggregate: Analysis of a typical masonry building from the Azores. *Structures* **10**, 184–196.

Ferreira, T., Gaspar, J.L., Viveiros, F., Marcos, M., Faria, C. and Sousa, F. 2005. Monitoring of fumarole discharge and CO_2 soil degassing in the Azores: Contribution to volcanic surveillance and public health risk assessment. *Annals of Geophysics* **48**, 787–796.

Ferreira, T.M., Maio, R., Vicente, R. and Costa, A. 2016. Earthquake risk mitigation: The impact of seismic retrofitting strategies on urban resilience. *International Journal of Strategic Property Management* **20** (3), 291–304.

Ferreira, T.M., Maio, R and Vicente, R. 2017a. Seismic vulnerability assessment of the old city centre of Horta, Azores: Calibration and application of a seismic vulnerability index method. *Bulletin of Earthquake Engineering* **15**, 2879–2899.

Ferreira, T.M., Maio, R. and Vicente, R. 2017b. Analysis of the impact of large scale seismic retrofitting strategies through the application of a vulnerability-based approach on traditional masonry buildings. *Earthquake Engineering and Engineering Vibration* **16** (2), 329–348.

Fontiela, J., Oliveira, C.S. and Rosset, P. 2018. Characterisation of seismicity of the Azores archipelago: An overview of historical events and a detailed analysis for the period 2000–2012. *In*: Kueppers and Beier, C. (eds) *Volcanoes of the Azores, Active Volcanoes of World*, Springer-Verlag, Germany, 127–153.

Fontiela, J., Rosset, P., Wyss, M., Bezzeghoud, M., Borges, J. and Cota Rodgriques, F. 2020. Human losses and damage expected in future earthquakes on Faial Island - Azores. *Pure and Applied Geophysics* **177**, 1831–1844.

França, J-A. 1977. *Lisboa Pombalina eo Iluminismo*. Livraria Bertrand, Lisboa.

Gaspar, J.L., Wallenstein, N., Coutinho, R., Ferreira, T., Queiroz, G., Pacheco, J, Guest, J.E., Tryggvason, E. and Malheiro, A. 1997. *Considerações sobre a ocorrência dos movimentos de massa registados na madrugada de 31 de Outubro de 1997 na ilha de S. Miguel, Açores*. Relatório Técnico-Científico 17/DGUA/97, Centro de Vulcanologia, Ponta Delgada.

Gaspar, J.L., Guest, J.E., Duncan, A.M., Barriga, F.J.A.S. and Chester, D.K. (eds) 2015a. *Volcanic Geology of São Miguel Island (Azores Archipelago)*, Geological Society, London Memoir **44**.

Gaspar, J.L., Queiroz, G., Ferreira, T., Medeiros, A.R., Goulart, C. and Medeiros, J. 2015b. Earthquakes and volcanic eruptions in the Azores region: Geodynamic implications from major historical events and instrumental seismicity. *In*: Gaspar, J.L., Guest, J.E., Duncan, A.M., Barriga, F.J.A.S. and Chester, D.K. (eds) *Volcanic Geology of São Miguel Island (Azores Archipelago)*, Geological Society, London Memoir **44**, 33–49.

Gomes, A., Gaspar, J.L. and Queiroz, G. 2006. Seismic vulnerability of dwellings at Sete Cidades Volcano (S. Miguel Island, Azores). *Natural Hazards and Earth System Sciences* **6**, 41–48.

Gomes, F., Correia, M.R., Carlos, G.D. and Lima, A. 2015. The high and intense seismic activity in the Azores. *In*: Correia, M.R., Lourenço, P.B. and Varum, H. (eds) *Seismic Retrofitting: Learning from Vernacular Architecture*. Taylor and Francis, London, 197–200.

Grünthal, G. 1998 *European macroseismic scale 1998*. Conseil de L'Europe Cahiers du Centre Européen de Géodynamique et de Séismologie vol. 15, Luxembourg.

Guevara-Perez, L.T. 2012. "Soft Story" and "Weak Story" in Earthquake Resistant Design: A Multidisciplinary Approach. 15th World Conference on Earthquake Engineering, Lisboa 24–28.

Kelman, I. 2001. The autumn 2000 floods in England and flood management. *Weather* **56**, 346–348 and 353–360.

Kircher, C.A. and Hamburger, R.O. 2000. Preface: Theme issue on seismic design provisions and guidelines. *Earthquake Spectra* **16** (1), 7–13.

Lotteri, A. 2020. *São Miguel (Azores): Changing Characteristics of Seismic and Volcanic Vulnerability and Resilience.* PhD thesis, Liverpool Hope University.

Lotteri, A., Speake, J., Kennedy, V., Wallenstein, N., Coutinho, R., Chester, D., Duncan, A., Dibben, C. and Ferreira, F. 2021. Changing hazard awareness over two decades: The case of Furnas, São Miguel (Azores). In: Marotta, E. (ed) *Volcanic Island: From Hazard Assessment to Risk Mitigation.* 20th Anniversary: Cities on Volcanoes. Geological Society of London, Special Publication (in press).

Maio, R., Estêvão, J.M.C., Ferreira, T.M. and Vicente, R. 2017. The seismic performance of stone masonry buildings in Faial island and the relevance of implementing effective seismic strengthening policies. *Engineering Structures* **141**, 41–58.

Maio, R., Ferreira, T.M., Vicente, R. and Costa, A. 2019. Is the use of traditional seismic strengthening strategies economically attractive in the renovation of urban cultural heritage assets in Portugal? *Bulletin of Earthquake Engineering* **17**, 2307–2330.

Martins, V.N., Cabral, P. and Sousa e Silva, D. 2012. Urban modelling for seismic areas: the case study of Vila Franca do Campo (Azores Archipelago, Portugal). *Natural Hazards and Earth System Sciences* **12**, 2731–2741.

Neves, F., Costa, A., Vicente, R., Sousa Oliveira, C. and Varum, H. 2012. Seismic vulnerability assessment and characterisation of the buildings on Faial Island, Azores. *Bulletin of Earthquake Engineering* **21**, 27–44.

Ordem dos Engenheiros. 1955. *Actas do Simpósio sobre a acção dos sismos.* Ordem dos Engenheiros, Lisboa, no page numbers.

Pacheco, M.P., Pereira, S.M. and Estrela Rego, I. 2020. Seismic preparedness of families with children: Measures and dynamics. *International Journal of Psychology.* doi: 10.1002/ijop.12694 (pre-publication).

Pomonis, A., Spence, R. and Baxter, P. 1999. Risk assessment of residential buildings for an eruption of Furnas Volcano, São Miguel, the Azores. *Journal of Volcanology and Geothermal Research* **92** (1–2), 107–131.

Quarantelli, E.L., 1980. *Evacuation Behavior and Problems: Findings and Implications from the Research Literature.* Disaster Research Center, The Ohio State University.

Rodrigues, A. 2015. *Portugal: Statement made at the Third UN World Conference on Disaster Risk Reduction (WCDRR).* Minister of Internal Administration support for the Sendai Framework United Nation, Portugal. https://www.preventionweb. net/english/policies/v.php?id=44046andcid=0

RSA. 1983. *Regulamento de Segurança e Açores para Estruturas de Edifícios e Pontes.* Decreto-Lei 235/83, Lisboa.

RSCCS. 1958. *Regulamento de Segurança das Construções Contra os Sismo.* Decreto-Lei 41658, Lisboa.

RSEP. 1961. *Regulamento de Solicitações e, Edificico e Pontes.* Decreto-Lei 44041, Lisboa.

Russell, L.A., Goltz, J.D. and Bourque, L.B. 1995. Preparedness and hazard mitigation actions before and after two earthquakes. *Environment and Behavior* **27** (6), 744–770.

Silva, C., Viveiros, F., Ferreira, T., Gaspar, J.L. and Allard, P. 2015. Diffuse soil emanations of radon and hazard implications at Furnas Volcano, São Miguel (Azores) *In:* Gaspar, J.L., Guest, J.E., Duncan, A.M., Barriga, F.J.A.S. and Chester, D.K. (eds) *Volcanic Geology of São Miguel Island (Azores Archipelago),* Geological Society, London Memoir **44**, 197–211.

Spence, R.J.S., Pomonis, A., Baxter, P.J., Coburn, A.W., White, M. and Dayrit, M. 1996. Field Epidemiology Training Program Team 1996. Building Damage

Caused by the Mount Pinatubo Eruption of 15 June 1991. *In:* Newhall, C.G. and Punongbayan, R.S. (eds) *Fire and Mud: Eruptions and Lahars of Mount Pinatubo, Philippines.* University of Washington Press, London, 1055–1061.

Spencer, R.J.S., Oliveira, C.S., D'Ayala, D.F., Papa, F. and Zuccaro, G. 2000. *The Performance of Strengthened Masonry Buildings In Recent European Earthquakes.* Abstract: 12th World Conference on Earthquake Engineering, Auckland, New Zealand, January 2000 (The first author is actually Professor Robin Spence).

Spence, R.J.S., Kelman, I., Baxter, P.J., Zuccaro, G. and Petrazzuoli, S. 2005. Residential building and occupant vulnerability to tephra fall. *Natural Hazards and Earth System Sciences* **5**, 477–494.

'SRPCBA. 2020. Planeamento de Emergência. Serviço Regional de Proteção Civil Bombeiros dos Açores. https://www.prociv.azores.gov.pt/operacoes/plan-emergencia/

Tang, Y., Yin, K., Hill, V, Katiyar, A., Nasseri, A. and Lai, T. 2012. Seismic Risk Assessment of Lisbon Metropolitan Area under a Recurrence of the 1755 Earthquakes with Tsunami Inundation. The 15th World Conference on Earthquake Engineering, Lisboa. https://www.iitk.ac.in/nicee/wcee/article/WCEE2012_4460.pdf

Teves-Costa, P., Oliveira, C.S. and Senos, M.L. 2007. Effects of local site and building parameters on damage distribution in Angra do Heroísmo - Azores. *Soil Dynamics and Earthquake Engineering* **27**, 986–999.

United Nations. 2006. *Hyogo Framework for Action 2005–2015.* Geneva United Nations.

United Nations. 2015. *Sendai Framework for Disaster Risk Reduction 2015–2030.* UNISDR, United Nations, Geneva.

Viveiros, F., Gaspar, J.L., Ferreira, T., Silva, C., Marcos, M. and Hipólito, A. 2015. Mapping of soil CO_2 diffuse degassing in São Miguel Island and its public health implications. *In:* Gaspar, J.L., Guest, J.E., Duncan, A.M., Barriga, F.J.A.S. and Chester, D.K. (eds) *Volcanic Geology of São Miguel Island (Azores Archipelago),* Geological Society, London Memoir **44**, 185–195.

Wallenstein, N. 1999. *Estudo da histótia recente e do comporamento eruptivo do Vulcão do Fogo (S.Miguel, Açores). Avaliação preliminary do hazard.* Tese de doutoramento no ramo de Geologia, especialidade de Vulcanologia. Departamento de Geociências, Universidade dos Açores.

Wallenstein, N., Chester, D.K. and Duncan, A.M. 2005. Methodological implications of volcanic hazard evaluation and risk assessment: Fogo Volcano, São Miguel, Azores. *Zeitschrift für Geomorphologie Supplementband* **140**, 129–149.

Wallenstein, N., Duncan, A., Chester, D. and Marques, R. 2007. Fogo Volcano (São Miguel, Azores): a hazardous edifice Le volcan Fogo, un édifice générateur d'aléas indirects. *Géomorphologie: relief, processus, environnement* **3**, 259–270.

Wallenstein, N., Chester, D., Coutinho, R., Duncan, A. and Dibben, C. 2015. Volcanic hazard vulnerability on São Miguel Island, Azores. *In:* Gaspar, J.L., Guest, J.E., Duncan, A.M., Barriga, F.J.A.S. and Chester, D.K. (eds) *Volcanic Geology of São Miguel Island (Azores Archipelago),* Geological Society, London Memoir **44**, 213–225.

Zonno, G., Oliveira, C.S., Ferreira, M.A., Musacchio, G., Meroni, F., Mota-de-Sá, F. and Neves, F. 2010. Assessing seismic damage through stochastic simulation of ground shaking: The case of the 1998 Faial Earthquake (Azores Islands). *Surveys Geophysics* **31**, 361–381.

Index

Note: **Bold** page numbers refer to tables, *Italic* page numbers refer to figures and page number followed by "n" refer to end notes.

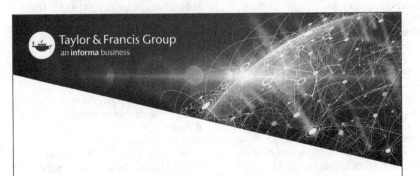

Printed in the United States
by Baker & Taylor Publisher Services